Christianto Wibowo, Ka Ming Ng
Conceptual Design of Crystallization Processes

Also of interest

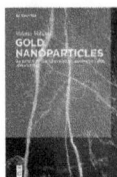

Gold Nanoparticles.
An Introduction to Synthesis, Properties and Applications
Valerio Voliani, 2020
ISBN 978-1-5015-1901-7, e-ISBN 978-1-5015-1145-5,
EPUB-ISBN 978-1-5015-1157-8

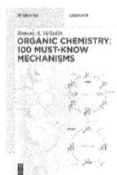

Organic Chemistry: 100 Must-Know Mechanisms
Roman Valiulin, 2020
ISBN 978-3-11-060830-4, e-ISBN 978-3-11-060837-3,
EPUB-ISBN 978-3-11-060851-9

Solubility in Pharmaceutical Chemistry
Christoph Saal, Anita Nair (Ed.), 2020
ISBN 978-3-11-054513-5, e-ISBN 978-3-11-055983-5,
EPUB-ISBN 978-3-11-055888-3

NMR Multiplet Interpretation.
An Infographic Walk-Through
Roman Valiulin, 2019
ISBN 978-3-11-060835-9, e-ISBN 978-3-11-060840-3,
EPUB-ISBN 978-3-11-060846-5

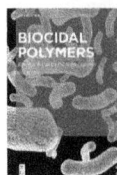

Biocidal Polymers
Narendra Pal Singh Chauhan (Ed.), 2019
ISBN 978-3-11-063855-4, e-ISBN 978-3-11-063913-1,
EPUB-ISBN 978-3-11-063863-9

Christianto Wibowo, Ka Ming Ng

Conceptual Design of Crystallization Processes

—

DE GRUYTER

Authors
Dr. Christianto Wibowo
ClearWaterBay Technology Inc.
671 Brea Canyon Road, Suite 5
Walnut, CA 91789
USA
E-Mail: cwibowo@cwbtech.com

Prof. Ka Ming Ng
Dept. of Chemical and Biological Engineering
Hong Kong University of Science and Technology
Clear Water Bay
Hong Kong
China
E-Mail: kekmng@ust.hk

ISBN 978-1-5015-1987-1
e-ISBN (PDF) 978-1-5015-1990-1
e-ISBN (EPUB) 978-1-5015-1352-7

Library of Congress Control Number: 2020948686

Bibliographic information published by the Deutsche Nationalbibliothek
The Deutsche Nationalbibliothek lists this publication in the Deutsche Nationalbibliografie;
detailed bibliographic data are available on the Internet at http://dnb.dnb.de.

© 2021 Walter de Gruyter GmbH, Berlin/Boston
Cover image: christianto wibowo
Typesetting: Integra Software Services Pvt. Ltd.
Printing and binding: CPI books GmbH, Leck

www.degruyter.com

To the memory of Jim Douglas

About the Authors

Christianto Wibowo is group manager and principal engineer at ClearWaterBay Technology, Inc., Walnut, California. He obtained a BS degree from Bandung Institute of Technology, Indonesia, and a PhD from the University of Massachusetts Amherst, both in chemical engineering. He was a postdoctoral fellow at the Hong Kong University of Science and Technology before joining ClearWaterBay Technology in 2002, where he has been managing consulting projects on a wide range of topics including process synthesis, process modeling, process improvement, and energy management. He was involved in the development of novel processes such as the patented processes for the production of methyl isobutyl ketone and neopentyl glycol. Dr. Wibowo is a senior member of the American Institute of Chemical Engineers (AIChE), where he received the Process Development Division Student Paper Award in 2002. He is an active member of the Process Development Division.

Dr. Wibowo has been developing a framework for the conceptual design of crystallization processes, starting from his PhD research work on a unified approach for crystallization process synthesis. Through his continuing collaborations with Prof. Ka Ming Ng's research group at the Hong Kong University of Science and Technology, the framework has been expanded in various directions to allow its use in different systems and applications. Concurrently, he was involved in consulting projects on the development of crystallization and solid–liquid separation processes, in which the framework was fine-tuned to solve real-life problems. The learning has been disseminated in public as well as in-house short courses around the world. He is the instructor of an online version of the course, which is a part of AIChE's eLearning program (http://www.aiche.org/academy/courses/ela101/crystallization-process-development).

Ka Ming Ng is currently CTO of CN Innovations and Professor Emeritus of Chemical and Biological Engineering at the Hong Kong University of Science and Technology. From 1980 to 2000, he served as professor of chemical engineering at the University of Massachusetts (UMass) Amherst. He joined the Hong Kong department in 2000 and served as Head from 2002 to 2005. He was CEO of the Nano and Advanced Materials Institute Ltd., a government funded R&D center from 2006 to 2013. In this positon, he worked with investors and industrialists in Hong Kong and Mainland China to develop products and the corresponding manufacturing processes focusing on the application of nano- and advanced materials. He served as corporate science and technology advisor for Mitsubishi Chemical, Japan, from 2001 to 2013. He held visiting positions at DuPont, MIT, and the National University of Singapore. His research interests center on product conceptualization, process design, and business development. He is an editor/editorial board member of various journals. He founded a number of companies in the United States and in Hong Kong. He has served as a board member for private and listed companies. Professor Ng is the recipient of the General Electric Outstanding Teaching Award. He is a fellow of the American Institute of Chemical Engineers where he received the Excellence in Process Development Research Award in 2002. In Hong Kong, he served as an advisor for teams winning top prizes in the annual One Million Dollar Entrepreneurship Competition.

While at DuPont Central R&D in 1995–1996, Professor Ng worked on reactor design and design of commodity and agricultural chemical plants. In particular, he articulated a multiscale perspective for the design of reaction systems. From 1998 to 2001, he was a team member at Mitsubishi Chemical, Japan, for the development of a novel 50,000 ton monomer plant that won the Japan Petroleum Institute Award for Technological Progress in 2004. In this project, his team conceptualized, scaled-up, and commercialized a novel 10-m tall trickle-bed reactor for acetoxylation reaction at high pressure and temperature. In academic research of process systems

https://doi.org/10.1515/9781501519901-202

engineering, he has been focusing on the design of solid processing plants, the conceptual design of crystallization systems, and consumer products.

Professor Ng started his research on product design of personal care products in the late 1990s at UMass. This was expanded to other products, classified into molecular, functional and formulated products, and devices. In 2006, along with R. Gani and K. Dam-Johansen, he edited *Chemical Product Design: Toward a Perspective Through Case Studies*, published by Elsevier. Based on his experience in entrepreneurship and interactions with product innovators, he formulated a systematic multidisciplinary, multiscale, hierarchical approach along with some of the necessary tools and methods to product design. These ideas are described in a text *Product and Process Design Principles: Synthesis, Analysis, and Evaluation, 4th Ed.* (with W. D. Seider, D. R. Lewin, J. D. Seader, S. Widagdo, and R. Gani) published by Wiley in 2017.

Preface

Crystallization can be viewed from two perspectives. The most common perspective, by far, views crystallization as an isolated chemical processing step or unit operation from which the crystals of a single compound of desired purity, crystal size, shape, chirality, and/or polymorphic form are recovered from a solution. A voluminous body of literature exists focusing on this point of view. Another perspective that views crystallization as a multicomponent separation process has not received adequate attention. The compound to be crystallized rarely exists by itself. It is important to understand how to design a process to recover the crystals of a desired component or components from a mixture of compounds in conjunction with other unit operations such as filtration, distillation, and so on.

Our research on crystallization process design started around 1990 at the University of Massachusetts Amherst (UMass) under the guidance and encouragement of Professor Jim Douglas, who pioneered conceptual process design. A crystallization separation process is conceptualized by representing it as a sequence of process points on a solid–liquid equilibrium (SLE) phase diagram. The disparate traditional crystallization techniques such as fractional, extractive, reactive, and antisolvent crystallization were unified under this approach. The mathematical framework for representing high-dimensional SLE phase diagrams for both electrolyte and nonelectrolyte systems were developed, thereby allowing the design of a multicomponent system with complex phase behavior. This effort was continued at the Hong Kong University of Science and Technology (HKUST) starting in 2000, focusing on experimental aspects of multicomponent crystallization. The recovery of isoflavone, ibuprofen, various inorganics from a salt lake, stereoisomers from natural herbs, and proteins were considered. Concurrently, the methods and experimental techniques developed were successfully applied to a large number of industrial design projects from companies in the United States, Japan, China, and other countries around the world through consultancy.

The aim of this book is to provide a comprehensive exposition of the fundamental and practical aspects of crystallization as a separation process. Much of the book is derived from the short courses at the American Institute of Chemical Engineers meetings, other international conferences, and various industrial sites, as well as the lectures delivered at UMass and HKUST. This book is intended for researchers and engineers interested in the design of crystallization processes. As crystallization is an integral part of chemical processes, it can also be used as a reference text for university courses on process design. Exercise problems are provided to enhance the pedagogical aspect of this monograph.

We would like to express our deep appreciation for the contribution of former students and associates at UMass and HKUST: Dr. David Berry, Mr. Marcos Cesar, Dr. Alex Chan, Mr. Stephen Chan, Dr. Wen-Chi Chang, Dr. Alice Cheng, Dr. Susan Dye, Dr. Kelvin Fung, Dr. Benny Harjo, Dr. Priscilla Hill, Dr. Vaibhav Kelkar, Dr. Martin

https://doi.org/10.1515/9781501519901-203

Kwok, Dr. Emily Lai, Mr. Arthur Lam, Ms. Carrie Lam, Dr. Candy Lin, Dr. Michelle Lin, Mr. Owen Luk, Dr. Thomas Pressly, Dr. Shankar Rajagopal, Dr. Ketan Samant, Dr. Joseph Schroer, Dr. Sze Kee Tam, and many others, to the contents in this book. We also thank Dr. Lionel O'Young and other colleagues at ClearWaterBay Technology, as well as our industrial partners, for valuable discussions during the course of various projects in which we were involved.

Contents

About the Authors —— VII

Preface —— IX

List of Symbols —— XV

1   Introduction —— 1
1.1   Crystallization in the Chemical Processing Industries —— 1
1.2   Crystallization as a Multiscale Design Problem —— 3
1.3   An Integrative Approach to Crystallization Process Design —— 4
1.4   Organization of the Book —— 6

2   Basics of Solid–Liquid Equilibrium Phase Behavior —— 9
2.1   Solid–Liquid Equilibrium —— 9
2.1.1   Melting point and solubility —— 9
2.1.2   Phase diagram and Gibbs phase rule —— 10
2.2   Molecular Systems —— 11
2.2.1   Single component systems —— 11
2.2.2   Binary systems —— 12
2.2.3   Ternary systems —— 19
2.2.4   Quaternary systems —— 24
2.2.5   Multicomponent systems —— 28
2.2.6   Reactive systems —— 35
2.3   Electrolyte Systems —— 39
2.3.1   Acid–base systems —— 42
2.3.2   Conjugate salt systems —— 44
2.3.3   Multicomponent salt system —— 48
2.4   Summary —— 51

3   Thermodynamic-Based Conceptual Design of Crystallization
         Processes —— 57
3.1   Movements in Composition Space —— 57
3.1.1   Lever rule —— 58
3.1.2   Representation of basic operations on SLE phase diagram —— 60
3.2   Maximum Recovery of a Pure Solid —— 62
3.3   Crystallization of Desirable Product —— 73
3.4   Complete Dissolution —— 79
3.5   Summary —— 83

4   **Synthesis of Crystallization-Based Separation Processes —— 89**
4.1   Crystallization as a Separation Process —— 89
4.2   Bypassing the Thermodynamic Boundary —— 89
4.2.1   Ternary system —— 89
4.2.2   Quaternary system —— 97
4.2.3   Multicomponent system —— 100
4.2.4   Conjugate salt system —— 102
4.3   Effect of Solvent —— 109
4.3.1   Solvent selection —— 109
4.3.2   Solvent switching —— 111
4.4   Hybrid Separation Process —— 113
4.5   Summary —— 120

5   **Crystallization Processes Involving More Complex Phase Behaviors —— 126**
5.1   Adductive Crystallization —— 126
5.1.1   SLE phase behavior involving adduct —— 126
5.1.2   Process synthesis —— 127
5.2   Chiral Resolution by Crystallization —— 131
5.2.1   SLE phase behavior of chiral systems —— 132
5.2.2   Process synthesis —— 134
5.2.3   Hybrid process for chiral resolution —— 138
5.3   Solid Solution Crystallization —— 141
5.3.1   SLE phase behavior of solid solution systems —— 141
5.3.2   Process synthesis —— 142
5.4   Amino Acid and Protein Crystallization —— 149
5.4.1   SLE phase behavior of amino acid systems —— 149
5.4.2   Process synthesis —— 151
5.5   Antisolvent Crystallization —— 156
5.5.1   Solid–liquid–liquid equilibrium phase behavior —— 156
5.5.2   Process synthesis —— 159
5.6   Supercritical Fluid Crystallization —— 162
5.6.1   Solid–fluid equilibrium phase behavior —— 162
5.6.2   Process synthesis —— 166
5.7   Summary —— 170

6   **Impact of Kinetics and Mass Transfer on Crystallization —— 176**
6.1   Kinetic Effects in Crystallization Process —— 176
6.1.1   Supersaturation —— 176
6.1.2   Nucleation and growth models —— 177
6.2   Actual Process Path —— 179
6.3   Preferential Crystallization —— 181

6.4   Kinetically Controlled Reactive Crystallization —— **184**
6.5   Polymorphic Crystallization —— **190**
6.5.1   SLE phase behavior of polymorphic systems —— **191**
6.5.2   Process synthesis —— **193**
6.6   Summary —— **202**

**7   Management of Particle Size Distribution and Impurities
           in Crystallization Processes —— 207**
7.1   Crystallizer Model —— **207**
7.1.1   Population balance —— **207**
7.1.2   Solutions of population balance equations —— **210**
7.2   Particle Size Distribution Management —— **211**
7.2.1   PSD manipulation in batch crystallization —— **211**
7.2.2   PSD targeting in continuous crystallization —— **220**
7.2.3   Scale-up consideration in PSD management —— **225**
7.3   Impurity Management —— **227**
7.3.1   Control of inclusion impurities —— **228**
7.3.2   Removal of inclusion impurities by melt crystallization —— **233**
7.3.3   Crystallization downstream processing system —— **238**
7.3.4   Simultaneous washing and melt crystallization —— **244**
7.4   Summary —— **247**

**8   Determination of Solid–Liquid Equilibrium Phase Behavior
           and Crystallization Kinetics —— 253**
8.1   Strategy for SLE Phase Behavior Determination —— **253**
8.2   Modeling of SLE Phase Behavior —— **254**
8.2.1   Solubility equation and solubility product equation —— **255**
8.2.2   Activity coefficient models —— **260**
8.2.3   Equation of state models —— **260**
8.3   Prediction of Solid–Liquid Equilibrium Phase Behavior —— **264**
8.3.1   Group contribution methods —— **264**
8.3.2   Quantum chemistry methods —— **270**
8.4   Experimental Determination of SLE Phase Behavior —— **270**
8.4.1   Thermal method for obtaining SLE data —— **271**
8.4.2   Synthetic method for obtaining SLE data —— **276**
8.4.3   Analytical method for obtaining SLE data —— **279**
8.4.4   Fitting of model parameters —— **280**
8.4.5   Boundary verification —— **287**
8.5   Experimental Determination of Crystallization Kinetics —— **292**
8.6   Summary —— **294**

9   **Concluding Remarks** —— **300**
9.1   A Hierarchical, Multiscale, Integrative Approach —— **300**
9.2   Outlook —— **303**

Appendix: Solutions to Selected Exercise Problems —— 307

References —— **331**

Index —— **347**

List of Symbols

a	Activity
a_{ij}	Binary interaction parameter in NRTL activity coefficient model
A	Pre-exponential factor (eq. (6.2)), $no/m^3.s$ or $no/kg.s$
A	Debye–Hückel constant (eqs. (8.50), (8.51), (8.53)), $mol^{-1/2}kg^{1/2}$
A_f	Filter area, m^2
A_S	Seed surface area, m^2
b	Power in nucleation rate power law equation (eq. (6.4))
b_{ij}	Binary interaction parameter in NRTL activity coefficient model
B	Nucleation rate, $no/m^3.s$ or $no/kg.s$
B	Baffle width (eq. (7.20)), m
c	Concentration, kg/m^3 or kg/kg
c	Mass of deposited solids per unit volume of filtrate (eqs. (7.32), (7.33), (7.36), (7.47)), kg/m^3
c^*	Solubility, kg/m^3 or kg/kg
c_1, c_2	Parameters (eq. (7.38))
C	Number of components
C	Distance between impeller and tank bottom (eq. (7.20)), m
D	Impeller diameter (eqs. (7.20)–(7.26)), m
D	Diffusivity of impurity in solvent (eq. (7.28))
D_L	Axial diffusivity, m^2/s
D_n	Dispersion number
D_S	Dielectric constant or relative permittivity
f	Fugacity
F	Mass, kg, or mass flow rate, kg/h
F	Number of degrees of freedom (eq. (2.1))
F	Faraday constant (eq. (8.51)), 96,485 C/mol
g	Gravitational acceleration, $9.81\ m/s^2$
g	Power in growth rate power law equation (eq. (6.6))
G	Crystal growth rate, m/s
G	Specific Gibbs free energy (eqs. (8.16)–(8.18)), J/mol
G_{ic}	Critical growth rate, m/s
H	Specific enthalpy, J/mol
$H(...)$	Heaviside function (eqs. (7.8), (7.9), (7.31))
i	Index variable
I	Ionic strength, mol/kg
j	Index variable
k	Boltzmann constant (eq. (7.30)), 1.3805×10^{-23} J/K
k	Cake permeability (eqs. (7.38), (7.44)–(7.47)), m^2
k_a	Surface factor
k_b	Nucleation rate constant
k_g	Growth rate constant
k_r	Reaction rate constant
k_v	Volume factor
K	Reaction equilibrium constant (eq. (2.15))
K	Impurity distribution coefficient between solid and liquid phases (eq. (7.27))
K_{eff}	Effective value of impurity distribution coefficient
K_{imp}	Constant in Langmuir adsorption isotherm
K_{SP}	Solubility product constant

https://doi.org/10.1515/9781501519901-205

L	Crystal size, m
L_d	Dominant crystal size, m
L_D	Length of belt filter section dedicated to deliquoring, m
L_F	Length of belt filter section dedicated to filtration, m
L_P	Target size of product crystals, m
L_S	Seed particle size, m
L_W	Length of belt filter section dedicated to washing, m
m	Molality, mol/kg
m_P	Mass of product crystals at maximum theoretical yield, kg
m_S	Mass of seed, kg
m_{solv}	Mass of solvent, kg
M_i	ith moment of particle size distribution
M_T	Magma density, kg/m^3
n	Population density, no/m^4 or no/m.kg
n	Number of moles (eqs. (8.2), (8.3), (8.6), (8.7))
n_2	Parameter (eq. (7.30))
N	Impeller rotational speed, rpm
N	Number of particles per unit volume or mass of slurry (eq. (7.4)), no/m^3 or no/kg
N_A	Avogadro number, 6.022×10^{23} mol^{-1}
N_{Ca}	Capillary number
N_{Da}	Damköhler number
N_{Gr}	Growth number
N_{Nu}	Nucleation number
N_P	Power number
N_P	Total number of two-dimensional projections that can be generated (eq. (2.13))
$N_{P,min}$	Minimum number of two-dimensional projections required to completely represent the phase behavior (eq. (2.14))
N_{Re}	Reynolds number
p	Pressure, Pa
p_c	Critical pressure, Pa
P	Number of phases (eq. (2.1))
q	Projection vector
q_f	Filtrate flow rate, m^3/s
Q	Volumetric flow rate, m^3/s
r_0	Ionic radius, m
R	Universal gas constant, 8.314 J/mol.K
R	Number of reactions (eqs. (2.16)–(2.20))
R	Ratio of wash liquid to product in wash column (eq. (7.42))
$R(. . .)$	Ionic coordinates
R_m	Filter medium resistance, m^{-1}
S	Supersaturation ratio
S	Saturation (eq. (7.38))
S	Specific entropy (eqs. (8.1)–(8.3)), J/mol.K
S_∞	Irreducible saturation (eqs. (7.38)–(7.39))
t_D	Deliquoring time, s
t_f	Filtration time, s
T	Temperature, °C or K
T	Coordinate of solventless projection (eqs. (5.4), (5.5))
T	Tank diameter (eq. (7.20)), m

T_b	Boiling point, °C or K
T_c	Critical point, °C or K
T_m	Melting point, °C or K
u	Filtrate linear velocity, m/s
U	Specific internal energy, J/mol
v	Specific volume or molar volume, m³/mol
v	Linear speed of moving belt in belt filter (eqs. (7.33), (7.36), (7.41)), m/s
V	Volume, m³
V_f	Cumulative filtrate volume, m³
w	Belt filter width, m
w	Weighting factor (eq. (8.77))
W	Wash ratio (mass of wash liquid to mass of residual liquid in wet cake)
W	Mass-based differential particle size distribution (eqs. (7.15), (7.16))
W_0	Activation energy for crystal growth (eq. (7.30)), J
x	Composition, mole fraction or mass fraction
x	Cake thickness (eqs. (7.35), (7.37), (7.38), (7.40), (7.44)–(7.47)), m
X	Transformed mole fraction
y	Solid composition, mole fraction or mass fraction
y	Cartesian coordinate (eqs. (2.2), (2.5), (2.10)–(2.12))
Y	Canonical coordinate
Y_i	Maximum recovery of component i
z	Ionic charge
Z	Liquid height (eq. (7.20)), m
Z	Compressibility factor (eqs. (8.63), (8.65)–(8.66))

Greek symbols

α	Specific cake resistance, m/kg
α	Effectiveness factor (eq. (6.8))
α_{ij}	Randomness parameter in NRTL activity coefficient model
γ, γ^*	Activity coefficient
γ	Linear edge free energy, J
δ	Length of diffusion boundary layer (eq. (7.28)), m
δ	Solubility parameter (eqs. (8.38), (8.40)), Pa$^{1/2}$
Δc	Supersaturation, kg/kg
ΔG^{EX}	Excess Gibbs free energy, J/mol
ΔH_f	Enthalpy (heat) of fusion, J/mol
ΔH_v	Enthalpy of vaporization, J/mol
Δp	Pressure drop, Pa
ΔT	Supercooling, °C or K
ΔT_{max}	Metastable zone width, °C or K
ΔU_v	Internal energy of vaporization, J/mol
ε	Cake porosity
ε	Screw dislocation activity (eq. (6.3))
ε_0	Cake porosity before compaction
ε_0	Vacuum permittivity (eq. (8.51)), 8.854×10^{-12} F/m
ε_f	Cake porosity after compaction
μ	Liquid viscosity, Pa.s

μ, μ^*	Chemical potential (eqs. (8.3), (8.7), (8.10)–(8.12), (8.31), (8.33)), J/mol
v	Stoichiometric coefficient
ρ	Liquid density, kg/m^3
$\rho_L{}^*$	Saturated liquid molar density, kg/m^3
ρ_S	Solid true density, kg/m^3
σ	Surface energy per unit area (eq. (6.2)), J/m^2
σ	Surface tension, N/m
σ	Standard deviation (eq. (7.43)), m
$\sigma_1{}'$	Parameter (eq. (6.3))
τ	Residence time, s

1 Introduction

1.1 Crystallization in the Chemical Processing Industries

Crystallization is an important separation technique in the chemical processing industries for various reasons. First, it is able to produce a high-purity final product in solid form. Most separation techniques such as extraction and absorption simply transfer the chemical product from one phase to another. Second, crystallization can be scaled up for large-scale production while techniques such as electrophoresis and chromatography are only suitable for the separation of chemicals in small quantity. Third, it can separate organics, biochemicals, inorganics, and polymers that are sensitive to heat or have high boiling point. This is important, as chemical companies, in search of a higher profit margin, shift their focus from commodity to specialty chemicals that often cannot be separated by distillation. Fourth, like distillation, crystallization allows the complete separation of a multi-component mixture of chemicals even if the system exhibits complex phase behavior. In distillation, this is achieved by bypassing the azeotropes using extractive distillation, azeotropic distillation, and so on. In crystallization, this is accomplished by bypassing the eutectics using fractional crystallization, extractive crystallization, and so on.

Table 1.1 lists a number of examples that illustrate the wide range of applications of crystallization in the chemical processing industries. The earliest applications of crystallization mostly deal with the recovery of inorganic salts from solution by evaporating water. The harvesting of salt from seawater has been practiced since antiquity [1]. Similarly, the separation of potash as a fertilizer from sodium chloride has been a major industrial operation to this day. Lithium carbonate, the major component for lithium batteries, is extracted from salt mines via crystallization from brines. Recently, there is a significant increase in the co-precipitation of compound salts such as $Ni_{0.6}Co_{0.2}Mn_{0.2}(OH)_2$ to meet the rising demand for cathode materials used in electric vehicle batteries. Crystallization is also widely used in the production of organic commodity chemicals such as p-xylene, bisphenol-A, terephthalic acid, and adipic acid, to separate the desired product from impurities and byproducts. Many specialty chemical and pharmaceutical products, such as diamondoids, azo dyes, monosodium glutamate, vitamin C, aspirin, and ibuprofen, are isolated in solid form by crystallization. In fact, most small molecule drugs (>90%) are delivered in crystalline form, and about 90% of newly developed active pharmaceutical ingredients are sparingly soluble in water [2].

In the design and development of crystallization processes, there are three requirements. First, the crystals have to meet product specifications such as size, shape, polymorphic form, and so on. If the particle size is too small, filtration may be difficult or take too long, and the presence of fines in the dried product can lead

https://doi.org/10.1515/9781501519901-001

Table 1.1: Application examples of crystallization in the chemical processing industries.

Market sector	Application	Reference
Inorganics	Recovery of potassium chloride from Lake Searles	Mumford [3]
	Isolation of potash (KCl) from sylvinite ore	Rajagopal et al. [4]
	Production of lithium carbonate from salt mine brines	Wilkomirsky [5]
	Production of precursors for NMC battery cathode material	Liang et al. [6]
Organics	Separation of *p*-xylene from *m*-xylene using extractive crystallization	Haines [7]
	Production of bisphenol-A using adduct crystallization	Moyers [8], Dermer [9]
	Reactive crystallization to produce technical grade terephthalic acid by liquid-phase air oxidation of *p*-xylene	Bernis et al. [10]
	Crystallization of adipic acid from a reaction mixture containing adipic, glutaric, and succinic acids with nitric acid and water as solvent	Oppenheim and Dickerson [11]
Specialty chemicals	Isolation of diamondoids from petroleum	Dahl et al. [12]
	Crystallization of azo dyes by salting-out from solution	Guccione [13]
	Crystallization of monosodium glutamate	Kawakita [14]
Pharmaceuticals	Purification of vitamin C (ascorbic acid)	Kuellmer [15]
	Production of aspirin	Thomas [16]
	Resolution of ibuprofen	Tung et al. [17]

to dusting problem. Needle-shaped particles are often difficult to handle because they tend to break. Also, getting the right polymorphic form of a drug is extremely important, as different polymorphs can have different bioavailability. Second, the desired product has to be recovered in high purity and at low cost. Thus, it is important to design the crystallizer as well as its operating conditions to produce low impurity crystals at high per-pass yield. High purity crystals can be obtained by minimizing the amount of inclusion impurities within the crystals, which is influenced by particle size, growth rate, and agglomeration rate. Impurities adsorbed on the surface of the crystals can be washed away in the product recovery process, downstream of the crystallizer. High per-pass yield can be effected by setting the operating conditions in such a way that only the desired compound precipitates out. That is, the concentration of the desired chemical is set at its lowest value but above its saturation solubility, while the concentration of each of the remaining components is set at or above its saturation solubility. Third, due to the intense competition in the chemical processing industries, the design of the crystallization processes has to be of high quality and completed fast. These requirements can be met by examining crystallization from all scales, as explained here.

1.2 Crystallization as a Multiscale Design Problem

Figure 1.1 shows the length and timescales of various events/entities related to crystallization [18, 19]. The crystallizer size is of the order of meters. Fluid flow, crystal nucleation and growth, and chemical reaction occur at smaller length scales within the crystallizer. The size range of a single crystal spans about ten orders of magnitude. Semiconductor nanocrystals such as CdSe [20] and nanocrystal drugs [21] are just nanometers in size, while the potassium dihydrogen phosphate crystals created at Lawrence Livermore National Laboratory in California can reach a size of about a meter [22]. The time duration for an event in the crystallizer ranges from hours (such as residence time) to seconds (such as mixing time). The overlap signifies the fact that the interplay of mixing, transport, reaction kinetics, and nucleation and growth kinetics has to be taken into account in designing a crystallization process properly. Molecular considerations are critical for getting the desired particle shape and polymorphic form. The crystallizer is located within a chemical plant, which has a size of the order of hundreds of meters. The overlap between the plant and crystallizer signifies the need to properly design the crystallizer, in relation to its surrounding equipment. In fact, the crystallizer and its downstream filtration and washing system need to be designed in an integrated manner. Further up the length scale is the global enterprise. Decisions at the corporate level can influence whether a plant will be built and how a plant is designed.

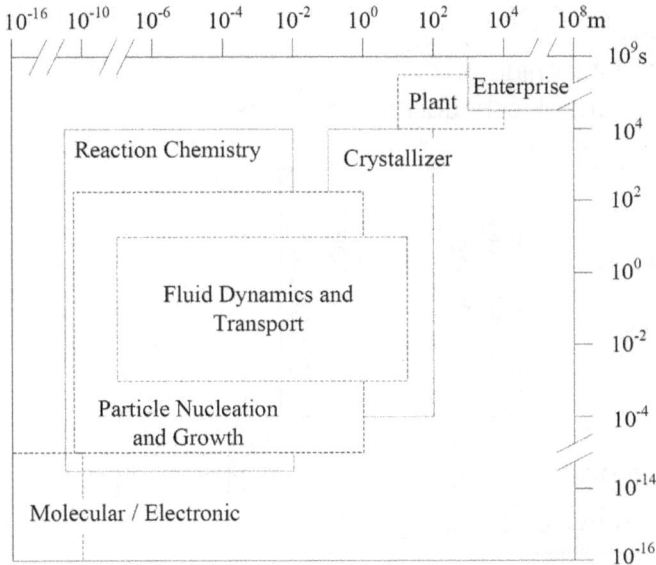

Figure 1.1: The time and length scales of the tasks involved in the design of a crystal product (reprinted from [19] with permission from Elsevier).

This multiscale perspective provides a bird's-eye view of what tasks need to be performed in the conceptual design of crystallization processes and the possibility of executing the tasks concurrently. The same approach has been applied to the design of reaction systems [23].

1.3 An Integrative Approach to Crystallization Process Design

Designing a crystallization process can be likened to building a house. As depicted in Figure 1.2, one begins with a sketch of the concept house, followed by building a strong foundation, before proceeding upwards. This foundation is the solid–liquid equilibrium (SLE) behavior of the system, which dictates the thermodynamic feasibility of the process. Obviously, SLE alone does not provide a complete picture. Selection of an appropriate cooling or evaporation profile, seeding policy, and crystallizer configuration – the equivalent of building the pillars and walls of the house – is next. Downstream processing units such as filters, washers, and dryers, which often cost more than the crystallizer and can be a source of serious operational problems, must be properly designed in conjunction with the crystallizer. Finally, process systems engineering techniques such as optimization, control, scheduling, and hazard and operability (HAZOP) analysis are applied to integrate the various aspects and achieve the best overall process performance. The enterprise and molecular tasks in Figure 1.1

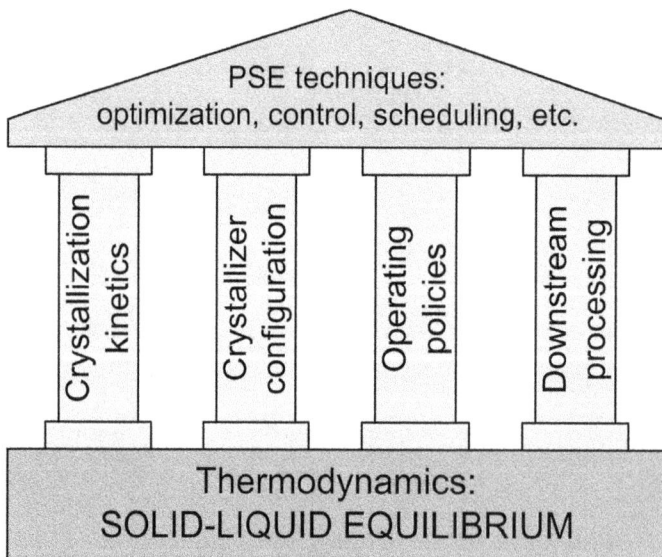

Figure 1.2: Crystallization process development can be likened to the construction of a house.

traditionally handled by management and product scientists, respectively, are not included as part of the design activities here.

To minimize the required time, effort, and money, the multiscale, integrative approach is realized by breaking down the design tasks into different subtasks involving synthesis, modeling, and experimental activities [24]. Figure 1.3 is a workflow diagram showing how the subtasks are performed either sequentially or concurrently. Often, a crystallization project begins with a target compound (or compounds) to be recovered by crystallization from a mixture of compounds. The first task is to collect the physicochemical properties of all the components. It is important to consider whether the target compound should be crystallized out. This question may seem to be unnecessary but is essential. A process designer is in a much better position than the chemistry team in ensuring an efficient separation process. For example, as discussed in Example 5.1, instead of crystallizing bisphenol A (BPA), the product, the BPA–phenol adduct is crystallized, followed by decomposition to obtain BPA in the commercial process. The second task is to determine the solubility of the relevant species and, preferably, the SLE phase diagram. These data would help determine how to come up with the best feed composition for crystallization of the desired compound and how the crystallization should be performed – for example, by cooling or by solvent evaporation. This is particularly important if more than one compound is recovered by crystallization. The third task is to perform a small-scale crystallization

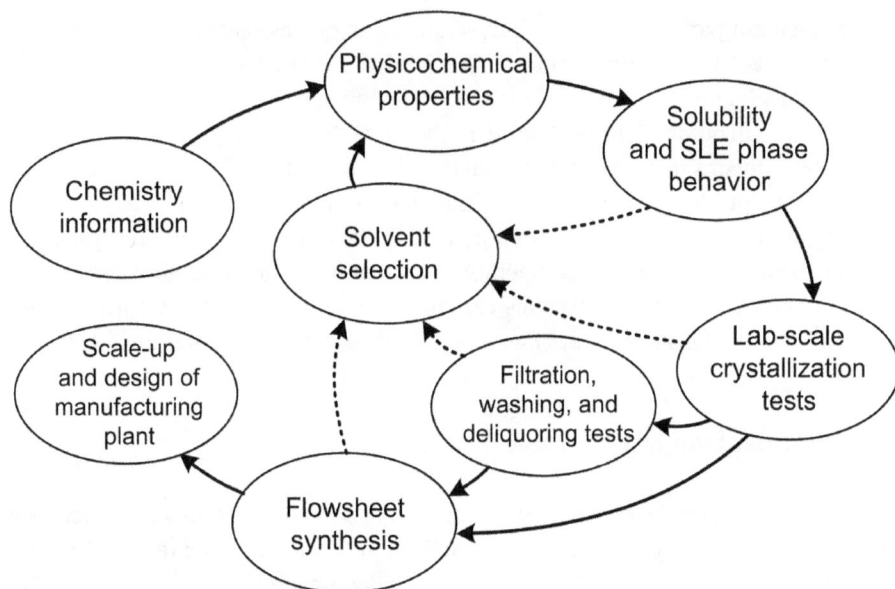

Figure 1.3: Workflow for the conceptual design of a crystallization process.

experiment to verify whether crystals of sufficiently good quality can be obtained. A good solvent for solubility may not yield good quality crystals. The fourth task is to perform filtration, washing, and dewatering tests. The crystallization downstream system can be more expensive than the crystallizer itself. The fifth task is to generate the entire flowsheet. This is important because the solvents need to be recycled, the impurities purged, and so on. In fact, flowsheet synthesis starts from the very beginning of the project, although it is formalized as the fifth task. At any point of the conceptual design process, if the design does not seem to work well, a new solvent is selected and tasks 1 to 5 are repeated. If the conceptual design is deemed acceptable, the project can proceed to process development, in which pilot plant testing and engineering design to fix the materials of construction, plant layout, and so on are considered [25].

The integrative approach is in line with the Quality by Design (QbD) concept in which quality is built into a product with a thorough understanding of the product and manufacturing process, along with knowledge of the risks involved in manufacturing the product and how best to mitigate those risks [26]. In crystallization process design, incorporating the fundamental understanding of all the key issues such as SLE behavior, kinetics, and mass transfer would lead to a process that produces the desired product with consistent quality.

This multiscale, integrative design method for crystallization plants originates from the hierarchical conceptual design approach by Douglas [27]. Starting with the input-output structure, the crystallization plant is conceptualized by initially focusing on the main parts of the plant – crystallizer and downstream solid–liquid separation process, then adding details to each of the main parts level by level, as illustrated in Figure 1.4. The top-level decisions involve the temperature and pressure of crystallization. The next level is the crystallizer configuration, which includes mechanisms to control the crystal size distribution, such as fines dissolution and classified product removal, as well as downstream operations such as filtration, washing, and drying. This is followed by mixing, detailed simulation of crystal nucleation, growth, breakage, agglomeration, and so on, to obtain a detailed design [28]. Monitoring and control systems can also be added to ensure that the desired product specifications are always met and a stable operation is maintained [29].

1.4 Organization of the Book

This book focuses on the conceptual design of crystallization processes, particularly the foundation of these processes – the SLE phase behavior as represented by the phase diagrams. Readers will learn what phase diagrams represent, how to properly model the SLE behavior, and how to collect the necessary data to obtain model parameters. Most importantly, the use of the phase diagrams to synthesize crystallization processes is discussed in detail. Based on the knowledge of SLE behavior, one

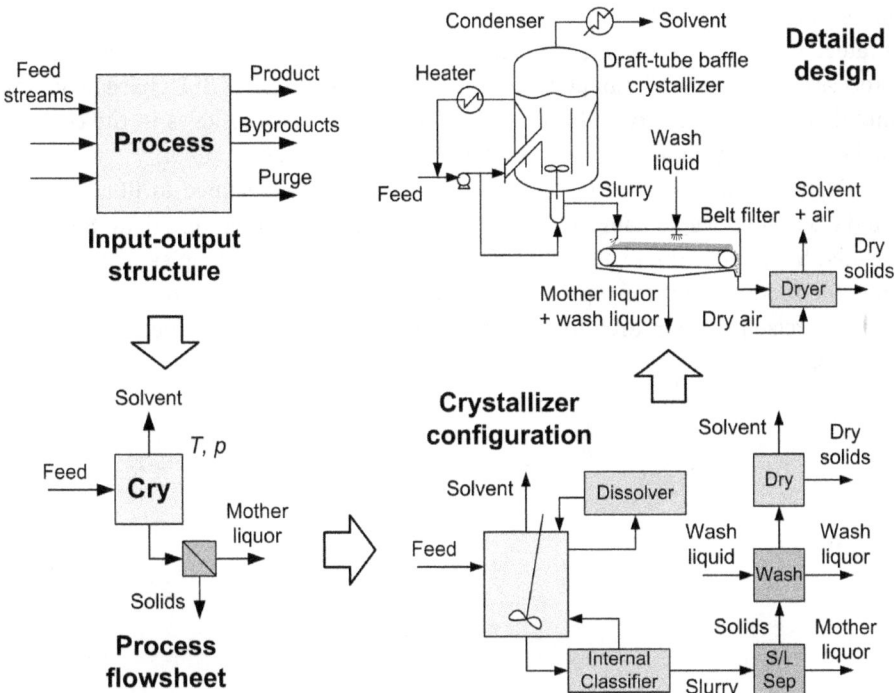

Figure 1.4: Hierarchical design of a crystallization process.

can generate thermodynamically feasible process alternatives, as well as evaluate and compare them to select the best option. Once the process has been conceptualized, kinetic issues are taken into account to ensure that the crystallizer would work properly to give a product with the desired attributes. The downstream unit operations such as filtration, washing, dewatering, and drying are also designed. The final step, which is the detailed design of the crystallizer accounting for issues such as fluid flow, mixing, breakage and agglomeration, and control have been covered elsewhere [30, 31] and will not be discussed in this book.

In Chapter 2, the basics of SLE phase behavior and phase diagrams are discussed to provide a solid foundation of the book. This is followed by a discussion of how the thermodynamic information is used for the conceptual design of crystallization processes in Chapter 3.

In Chapter 4, these concepts are applied to the synthesis of separation processes involving crystallization. Often, crystallization is used in combination with other techniques such as distillation and extraction to come up with an effective separation process. In Chapter 5, the synthesis of crystallization processes for phase diagrams with complex behavior – the presence of compound, adduct, solid solution, amino acid, and supercritical fluid – is discussed. The impact of kinetics and mass transfer effects on process design is considered in Chapter 6. In Chapter 7, strategies for particle size distribution and impurity management in crystallization processes are discussed. Chapter 8 focuses on the generation of SLE phase diagrams and determination of crystallization kinetics. Finally, the key ideas in the conceptual design of crystallization processes are summarized in Chapter 9.

Examples are used throughout the book. Some are designed to illustrate the fundamental concepts using actual chemical systems and literature data. Some indicated by (I) after the example number are real industrial examples, which show how the fundamental concepts are used creatively to solve real problems. In addition, exercise problems are provided, some with solutions, to reinforce the learning process.

2 Basics of Solid–Liquid Equilibrium Phase Behavior

2.1 Solid–Liquid Equilibrium

Solid–liquid equilibrium (SLE) provides the thermodynamic basis for the design and synthesis of crystallization processes. Therefore, before delving into the subject of conceptual design of crystallization processes, it is imperative to understand the basics of SLE phase behavior and its representation using phase diagrams.

2.1.1 Melting point and solubility

For a single-component system, SLE simply involves equilibrium at the pure component melting point (or freezing point). For example, water and ice are in equilibrium at 0 °C and 1 atm. The SLE of a binary system usually involves the solubility of one component (*solute*) in the other (*solvent*). For many systems, solubility varies significantly with temperature, such as that of sugar in water. The dependence of solubility on temperature is often plotted as the solubility curve, as illustrated in Figure 2.1a for sucrose in water. It is well known that solubility behavior may not be always that simple, with the potential presence of hydrates, solvates, or different polymorphic forms. There can be multiple solubility curves corresponding to the different forms of the solute.

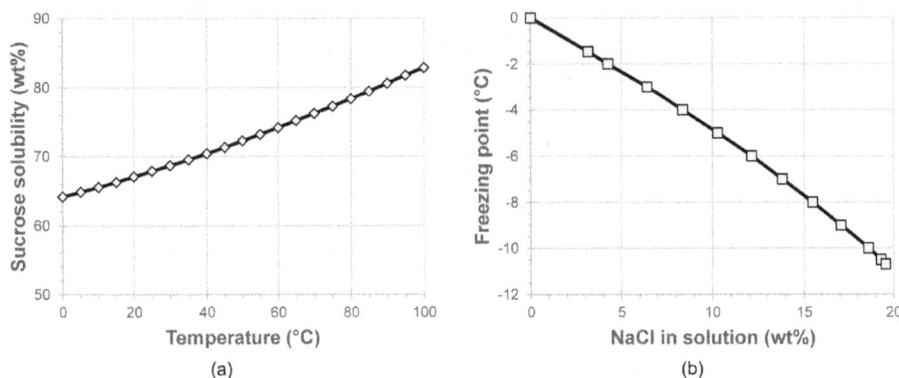

Figure 2.1: Examples of SLE relationships of binary systems: (a) sucrose solubility in water (data from [32]), and (b) freezing point depression in NaCl solution in water (data from [33]).

The freezing point of the solution is affected by solute concentration, a phenomenon commonly referred to as *freezing point depression*. A simple example is the use of salt to melt snow in the winter by lowering the freezing point of water to below the

https://doi.org/10.1515/9781501519901-002

ambient temperature. As shown in Figure 2.1b, the freezing point decreases to below −10 °C when about 20 wt% NaCl is present in the solution. Note that if the melting points of the two components in the binary system are close to each other, it is irrelevant which one should be referred to as the solute and which one should be called the solvent. Generally speaking, the SLE behavior concerns the temperature versus composition relationship for both components of the mixture.

Most industrial applications involve more than just two components. In such multicomponent systems, the presence of other species in the solution can greatly affect the solubility of a particular component. A case in point is the so-called *common ion effect* in an electrolyte mixture. For example, the solubility of NaCl in water would decrease significantly in the presence of NaOH because of the increased concentration of sodium ion, as shown in Figure 2.2a. Another common issue in industrial crystallization is the solubility of a solute in mixed solvents, which depends on the solvent composition. Sometimes the dependence can be strongly nonlinear, as in the case of betulin, a natural chemical with medicinal applications, which has a much larger solubility in mixtures of methanol and chloroform compared to the solubility in pure methanol or pure chloroform (Figure 2.2b).

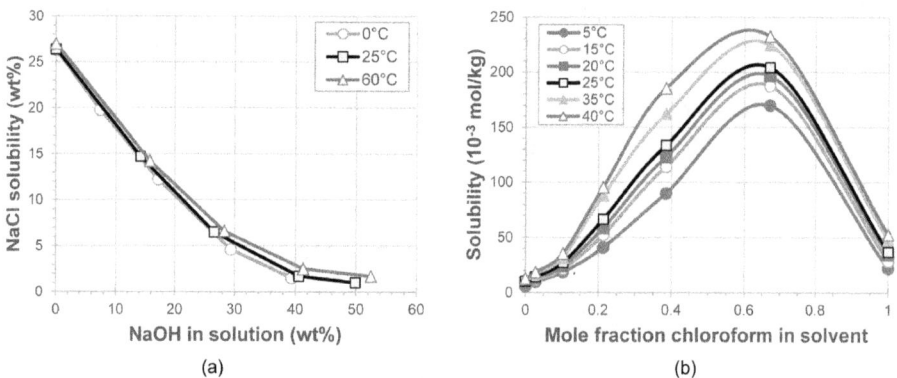

Figure 2.2: SLE behavior of multicomponent systems: (a) solubility of NaCl in water as a function of NaOH concentration (data from [34]), and (b) solubility of betulin in methanol/chloroform mixed solvent (data from [35]).

2.1.2 Phase diagram and Gibbs phase rule

The complex thermodynamic relationships among composition, temperature, and pressure in a multicomponent system can be visually represented in a coherent manner by using a phase diagram, which highlights the regions of composition, temperature, and pressure under which the system would exist as a single phase or a mixture of multiple phases. These regions are defined by various boundaries,

which can be present in the form of a curve, a surface, or even a hypersurface, depending on the dimension of the phase diagram. The dimension of a phase diagram
is determined by the degree of freedom, which is the minimum number of independent intensive parameters (such as temperature, pressure, and concentration of a
component) that must be specified to fully define the condition or state of the system. According to the *phase rule*, developed by J. W. Gibbs in 1876, for a closed system in the absence of any external effects and special conditions,

$$F = C - P + 2 \tag{2.1}$$

where F is the degree of freedom, C is the number of components in the system, and
P is the number of phases present. The number 2 refers to the temperature and pressure of the system. Since the minimum number of phases present in a system is 1,
the phase behavior of a C-component system can be represented using a phase diagram in a $(C+1)$-dimensional space, where the axes represent temperature, pressure,
and $(C - 1)$ composition variables (mole or mass fractions). Since SLE behavior is
normally insensitive to pressure, it is generally sufficient to consider an *isobaric*
(constant pressure) phase diagram with a dimension of C for an SLE phase diagram.

There are different types of SLE phase behavior based on the chemical and
physical nature of the components involved, such as whether or not they are electrolytes, their ability to form solid solutions, and the existence of compounds and
polymorphic forms. Correspondingly, the phase diagrams can take different geometrical shapes and include various features, from simple to highly complicated.
This chapter focuses on the basics of SLE phase diagrams, which can be classified
according to the type of system and number of components involved. Descriptions
of phase diagrams with different emphases and levels of detail are scattered in
various textbooks and monographs, some dating back to the mid-twentieth century
[36, 37, 38, 39, 40]. Here, the key features of relevance to the conceptual design of a
crystallization process are highlighted.

2.2 Molecular Systems

Many industrially important mixtures consist only of molecules that do not dissociate to form ions. The term *molecular systems* is used throughout this book to refer
to this type of chemical systems. For discussion purposes, molecular systems are
classified according to the number of components involved.

2.2.1 Single component systems

The simplest system to consider is a pure component, which typically has phase
behavior shown in Figure 2.3. The vertical and horizontal axes are pressure and

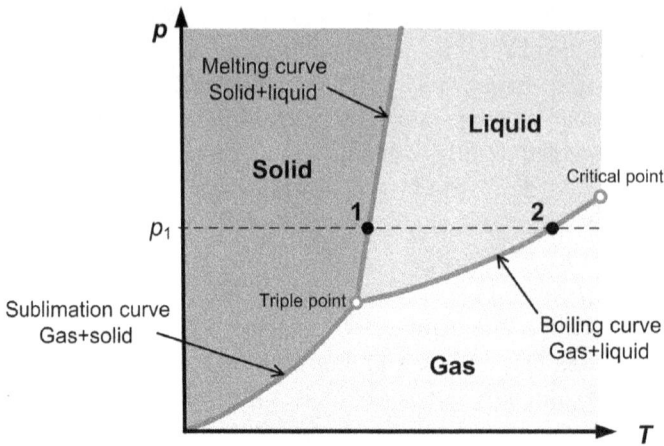

Figure 2.3: Typical pure component phase diagram.

temperature, respectively. Depending on the pressure and temperature, the pure component can exist as solid, liquid, or gas. Along a curve that forms the boundary between two regions, two phases are in equilibrium. For example, solid and liquid phases coexist in equilibrium along the melting curve, which describes the dependence of melting point on pressure. The boundary between liquid and gas region (boiling curve) ends at the *critical point*, beyond which liquid and gas are no longer distinguishable. The point of intersection of all three regions is referred to as the *triple point*, at which all three phases coexist and are in equilibrium with each other. At a fixed pressure (p_1) above the triple point, possible behaviors of the system are restricted to the dashed line, which intersects the melting curve at point 1 and the boiling curve at point 2. Depending on the temperature, the system can be in solid, liquid, or gas phase. SLE occurs at point 1, which is the melting point at the given pressure.

2.2.2 Binary systems

There are two common types of SLE behavior of a binary system. The first is *simple eutectic*, for which the solid phase is always pure regardless of the liquid composition. The second is *solid solution* behavior, which is similar to the typical vapor–liquid equilibrium behavior, where both phases are never 100% pure. The phase diagram of a binary system exhibiting simple eutectic behavior at a constant pressure (Figure 2.4a) is the most common type; examples of systems exhibiting this behavior are given in Table 2.1. In Figure 2.4a, the vertical axis indicates temperature, while the horizontal axis shows the composition in terms of weight or mole percent; this is why this isobaric diagram is often called the *T–x diagram*. At the top portion of the diagram is the

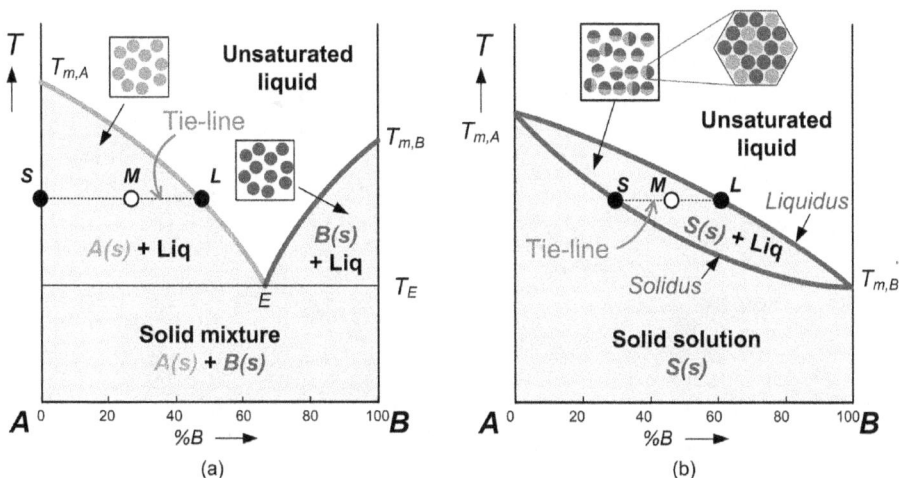

Figure 2.4: Binary SLE phase diagram: (a) simple eutectic system and (b) solid solution.

unsaturated liquid region, where there is only a single liquid phase. Below the melting point of pure A ($T_{m,A}$), there is a two-phase region where a solid phase consisting of pure A is in equilibrium with a liquid phase. Any mixture in this region, such as the one with an overall composition and temperature indicated by point M in the diagram, phase splits into pure solid A (point S) and a saturated liquid (point L), which lies on the *saturation curve* (also referred to as the *solubility curve*) of A. Line SL is called a *tie-line*. The region below the melting point of pure B ($T_{m,B}$) signifies the region where solids of pure B can be found in equilibrium with a liquid phase. The saturation curve of B, which forms a boundary of this region, meets the saturation curve of A at the *eutectic point* (E). At this point, both A and B are saturated and the liquid phase is in equilibrium with a solid mixture of the same composition. Below the eutectic temperature T_E, the system can only exist as a mixture of two solid phases. Clearly, the location of the eutectic point is an important characteristic of the SLE behavior, as it demarcates two regions where a different solid can be recovered in each region.

In contrast, if the pair of components forms a solid solution (Figure 2.4b), there is no pure solid region. There is an envelope within which a solid solution can be found in equilibrium with a liquid phase. The upper curve, adjacent to the unsaturated liquid region, is called the *liquidus* curve, and the lower curve is referred to as the *solidus* curve. Below the envelope, the system consists of just one solid phase. Microscopically, each of the solid solution crystals consists of molecules of A and B interspersed in the same crystal lattice.

The phase behavior of a binary system can be more complex. The phase diagram of a compound-forming system (Figure 2.5a) resembles two simple eutectic phase diagrams put together side by side. In addition to the A and B solubility

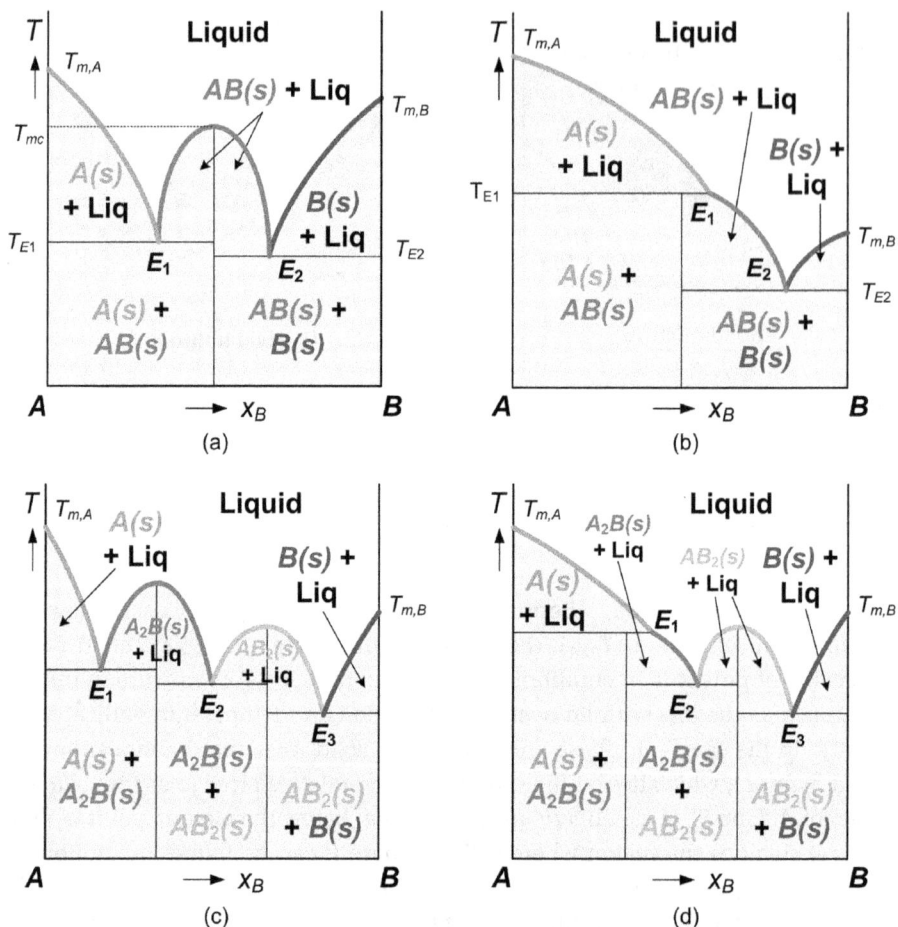

Figure 2.5: Complex binary SLE phase diagrams for systems without solid solution: (a) compound formation, (b) peritectic, (c) two congruently melting compounds, and (d) two compounds with different melting behaviors.

curves, there is also a solubility curve for compound AB, with a peak corresponding to the *congruent melting point* of the compound (T_{mc}). At this temperature, the compound coexists with a liquid of the same composition. An example is the 1:1 compound between tetrazole, a synthetic heterocyclic compound used as a gas generator in automobile airbags, and imidazole, a corrosion inhibitor on transition metals such as copper, which melts congruently at about 70 °C [41]. A compound is also formed in a *peritectic* system (Figure 2.5b). If compound AB is heated beyond T_{E1}, it would decompose into solid A and a liquid phase. Since it decomposes above its *incongruent melting point* (T_{E1}), it cannot coexist with a liquid of the same composition. This phase behavior is typically observed when the melting points of the components

forming the compound are very different. An example of a system exhibiting this behavior is phenol and bisphenol A (BPA), a precursor for polycarbonate plastic and
epoxy resins, which forms a 1:1 compound that melts incongruently around 100 °C
[42, 43]. Above this temperature, the compound would decompose into solid BPA
and a liquid phase.

It is also possible that more than one compound can be formed. Different scenarios
are possible, including all compounds melting congruently (Figure 2.5c) or some compounds melting congruently while others showing peritectic behavior (Figure 2.5d). For
example, phenol and 2-methyl-2-propanol can form two compounds with a molar
ratio of phenol to 2-methyl-2-propanol of 1:2 and 2:1, respectively [44]. Many
salts can form hydrates, which are compounds with water at different proportions. More examples of systems exhibiting these complex behaviors are given in
Table 2.1.

Table 2.1: Examples of binary systems without solid solution.

Behavior	Component A	Component B	Reference
Simple eutectic (Figure 2.4a)	p-Nitrochlorobenzene	o-Nitrochlorobenzene	Tare and Chivate [45]
	p-Dibromobenzene	p-Dichlorobenzene	Singh et al. [46]
	Picric acid	Trinitrotoluene	Moore [39]
	Triazole	Imidazole	Hilgeman et al. [41]
Compound formation (Figure 2.5a)	o-Cresol	Phenol	Jadhav et al. [44]
	Chloroform	Pyridine	Scholastica Kennard and McCusker [47]
	Phenol	Aniline	Moore [39]
	Tetrazole	Imidazole	Hilgeman et al. [41]
Peritectic (Figure 2.5b)	Bisphenol A	Phenol	Moyers [42], Wyrzykowska-Stankiewicz and Palczewska-Tulińska [43]
	4,4′-Diaminobiphenyl	2-Nitrophenol	Rai and George [48]
	Al_2O_3	SiO_2	Moore [39]
Two congruently melting compounds (Figure 2.5c)	Phenol	t-Butyl alcohol	Jadhav et al. [44]
	3-Nitrophenol	1,3-Diaminobenzene	Sangster [49]
	Water	Nitric acid	Lincoln [50]
Two compounds with mixed melting behavior (Figure 2.5d)	1,2-Diaminobenzene	Phenol	Sangster [49]
	1,2-Diaminobenzene	3-Nitrophenol	Sangster [49]

The solid solution behavior can also be more involved with the presence of a minimum
melting eutectic (similar to the azeotropic behavior in VLE), limited solubility (involving two different solid solutions that are not completely miscible), or combinations

thereof, as shown in Figure 2.6. Regions of single phase or multiple phases in equilibrium are marked on the figures along with the pure component melting points ($T_{m,A}$ and $T_{m,B}$).

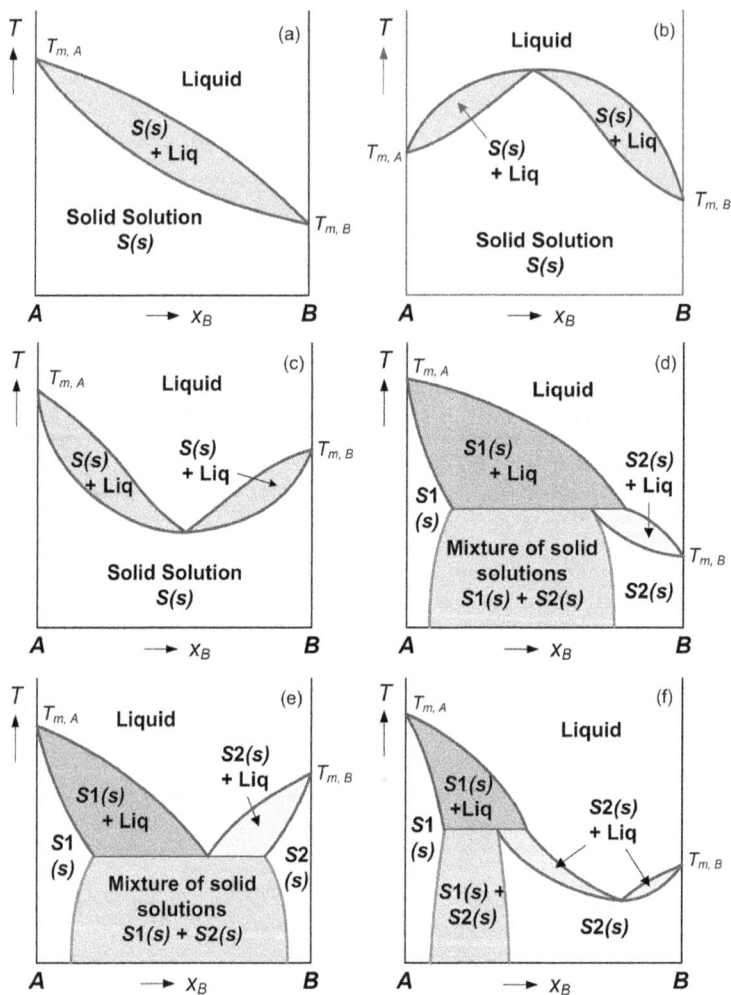

Figure 2.6: Binary SLE phase diagrams for systems with solid solution: (a) Roozeboom type I, (b) Roozeboom type II, (c) Roozeboom type III, (d) Roozeboom type IV, (e) Roozeboom type V, and (f) complex behavior.

There are five most common types of solid solution phase behavior known as Roozeboom types I to V [51]. Types I to III feature a solid solution with no discontinuity in the solid phase solubility, with type I (Figure 2.6a) showing a monotonous liquidus curve, while types II (Figure 2.6b) and III (Figure 2.6c) exhibit a maximum and minimum melting point, respectively. Types IV and V are characterized by gaps in solid solubility with the presence of a peritectic point in type IV (Figure 2.6d) and a eutectic point in type V (Figure 2.6e). The phase diagram for these two types features a region in which two solid solutions of different compositions (S1 and S2) are in equilibrium with each other. Different solid regions are separated by *solvus* curves. For example, in Figure 2.6d, a solvus curve separates the one-phase region S1(s) from the two-phase region S1(s)+S2(s) and another separates the two-phase region from the one-phase region S2(s). In addition, more complex behaviors can exist. For example, the system shown in Figure 2.6f has a peritectic point as well as a minimum melting point, making it essentially a combination of types III and IV. Examples of systems exhibiting each type of behavior are given in Table 2.2.

Table 2.2: Examples of binary systems exhibiting solid solution behavior.

Behavior	Component A	Component B	Reference
Type I (Figure 2.6a)	p-Dibromobenzene	p-Chlorobromobenzene	Campbell and Prodan [52]
	Dibenzofuran	Fluorene	Sediawan et al. [53]
	Copper	Nickel	Moore [39]
Type II (Figure 2.6b)	d-Carvoxime	l-Carvoxime	Oonk et al. [54]
Type III (Figure 2.6c)	Bibenzyl	Diphenylacetylene	Nojima and Akehi [55]
Type IV (Figure 2.6d)	Anthracene	Phenanthrene	Joncich and Bailey [56]
	Bibenzyl	Stilbene	Nojima [57]
Type V (Figure 2.6e)	Acridine	Anthracene	Kitaigorodsky [58]
	Coumarine	Naphthalene	Kitaigorodsky [59]
	Dibenzothiophene	Fluorene	Sediawan et al. [53]
	Naphthalene	Thianaphthene	Mastrangelo and Dornte [60]
	Diphenyl	Dipyridyl	Kitaigorodsky [59]
Complex (Figure 2.6f)	Dibenzothiophene	Dibenzofuran	Sediawan et al. [53]

Matsuoka [61] investigated many binary systems involving simple organic chemicals and reported that the simple eutectic behavior is the most common, comprising 54.3% of all systems studied. Compound formation and peritectic behavior together make up another 30%, while solid solution behavior is relatively uncommon. The distribution of the various behaviors is illustrated in Figure 2.7. However, the situation is different for pharmaceutical and biological molecules, many of which are chiral compounds. According to Jacques et al. [62], the compound formation type is by far the most common in nature, comprising 90% of all systems. Simple eutectic systems make up 5–7%, while solid solution behavior is rare.

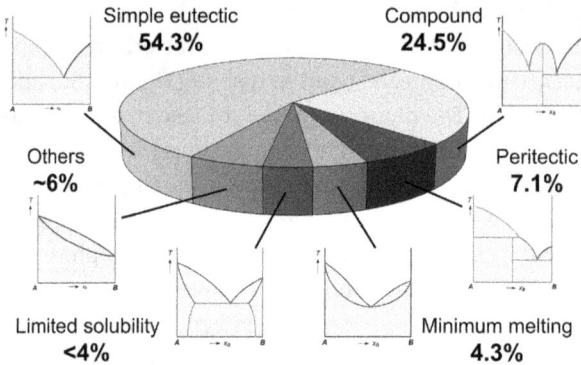

Figure 2.7: Distribution of various types of binary SLE phase behavior (data from [61]).

Example 2.1 – Binary simple eutectic system

Nitrochlorobenzene (NCB) isomers are industrially important intermediates in the chemical industry. The para isomer (p-NCB, melting point 83 °C) is an intermediate in the production of rubber
antioxidants, while the ortho isomer (o-NCB, melting point 33 °C) is a precursor to many dyes and
pesticides. The experimental data of Tare and Chivate [45] for this system are plotted on a binary
SLE phase diagram shown in Figure 2.8. The data clearly indicate the existence of two saturation
curves, extending from the pure component melting points and intersecting at the eutectic point,
which is located at approximately 68 mol% o-NCB and has a temperature of 14 °C. Drawing a horizontal line through the eutectic point divides the area below the saturation curves into three
regions: p-NCB solid + liquid, o-NCB solid + liquid, and solid mixture.

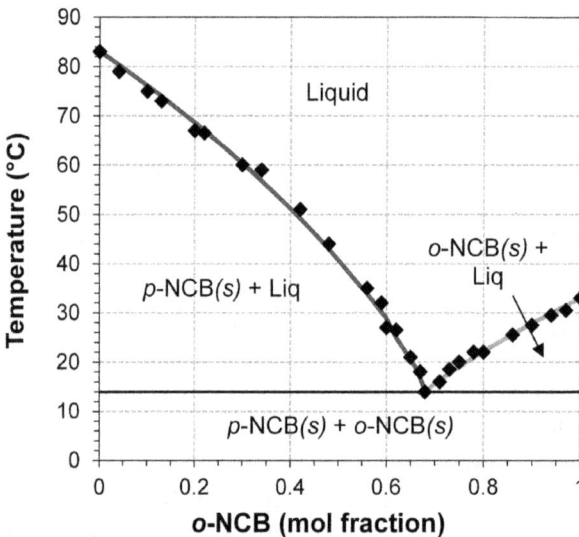

Figure 2.8: Binary SLE phase diagram of p-nitrochlorobenzene/o-nitrochlorobenzene system
(data from [45]).

Example 2.2 – Hydrate formation from aqueous salt solution
Magnesium nitrate ($Mg(NO_3)_2$), which is primarily used as a dehydrating agent (desiccant) in the preparation of nitric acid, can form several hydrates from an aqueous solution. This leads to a complex binary SLE phase diagram of $Mg(NO_3)_2/H_2O$ system, as shown in Figure 2.9. The existence of three hydrates as well as anhydrous magnesium nitrate was confirmed by Ewing et al. [63] and the solubility curves based on their experimental data are displayed on the phase diagram. The vertical lines on the diagram at 47.8, 57.8, and 80.5 wt% $Mg(NO_3)_2$ represent the nonahydrate, hexahydrate, and dihydrate, respectively. The eutectic and peritectic points define the horizontal lines that form the boundaries of different regions. The nonahydrate precipitates simultaneously with ice at a eutectic point located at −31.9 °C and 32 wt% $Mg(NO_3)_2$, which defines the boundary between the regions of ice + liquid, nonahydrate + liquid, and solid mixture of ice and nonahydrate. The boundary between the region of the nonahydrate, which melts incongruently, and that of the hexahydrate is defined by a peritectic point at −14.7 °C and 37 wt% $Mg(NO_3)_2$. The other two transition points are eutectic points: between hexahydrate and dihydrate at 52.7 °C and 67 wt% $Mg(NO_3)_2$, and between dihydrate and anhydrous magnesium nitrate at 130.5 °C and 82 wt% $Mg(NO_3)_2$. The presence of these two eutectic points corresponds to the fact that both hexahydrate and dihydrate melt congruently. Besides the ice+nonahydrate region, the phase diagram also features three other solid mixture regions.

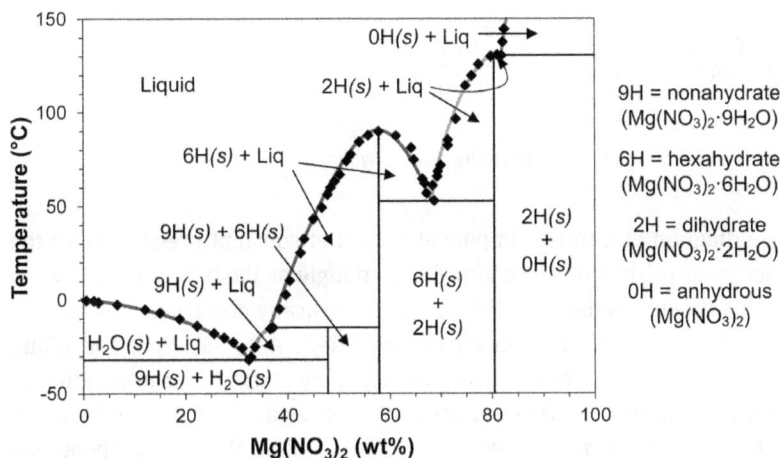

Figure 2.9: Binary SLE phase diagram of magnesium nitrate–water system (data from [63]).

2.2.3 Ternary systems

Just as the composition of a binary system is represented on a straight line, the composition of a ternary system is plotted on a composition triangle shown in Figure 2.10. The composition, either in mass or mole fraction, should be read along the sides of the triangle. The best way to read this diagram is to start from the vertex, at which the mass or mole fraction is equal to 1 (pure component). The composition of that particular component should be read along the parallel lines opposite the vertex. For example, point M in Figure 2.10 represents a mixture of 30% A (read along the left edge), 30% B (read along the bottom edge), and 40% S (read along the right edge).

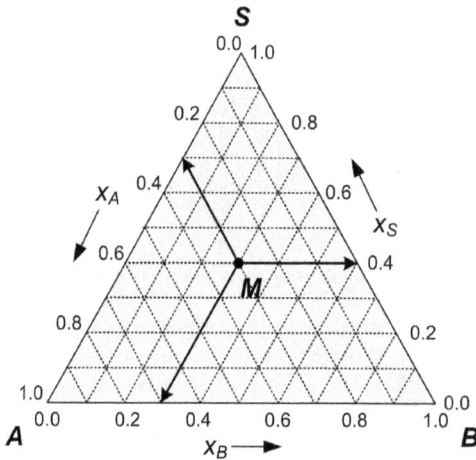

Figure 2.10: Composition triangle for representing ternary systems.

The SLE phase diagram of a three-component system at constant pressure takes the shape of a triangular prism with the composition triangle as the base, as depicted in Figure 2.11a. This three-dimensional (3D) diagram is basically constructed by putting together the $T–x$ diagrams of three binary systems ($A–B$, $A–S$, and $B–S$) and filling out the interior with the *saturation surfaces* (or *solubility surfaces*) corresponding to each of the three components. Two surfaces intersect at a *eutectic trough* and all three surfaces meet at the *ternary eutectic* point, at which all three components are saturated.

Since it is not easy to view a 3D figure, it is often desirable to reduce the dimensionality via projections and cuts. One way of doing this is by taking the so-called *polythermal projection* (Figure 2.11b), which is a projection in vertical direction parallel to the temperature axis. Basically, it is the picture that a viewer would see when they place their eyes right above the prism and look straight down. The saturation surface of A would appear as region A-AB-ABS-AS, which is called the *compartment* of A. This region signifies the range of composition from which pure A can be crystallized out. Similarly, the saturation surfaces of B and S appear as region B-BS-ABS-AB and S-BS-ABS-AS, respectively. Every point on this projection has

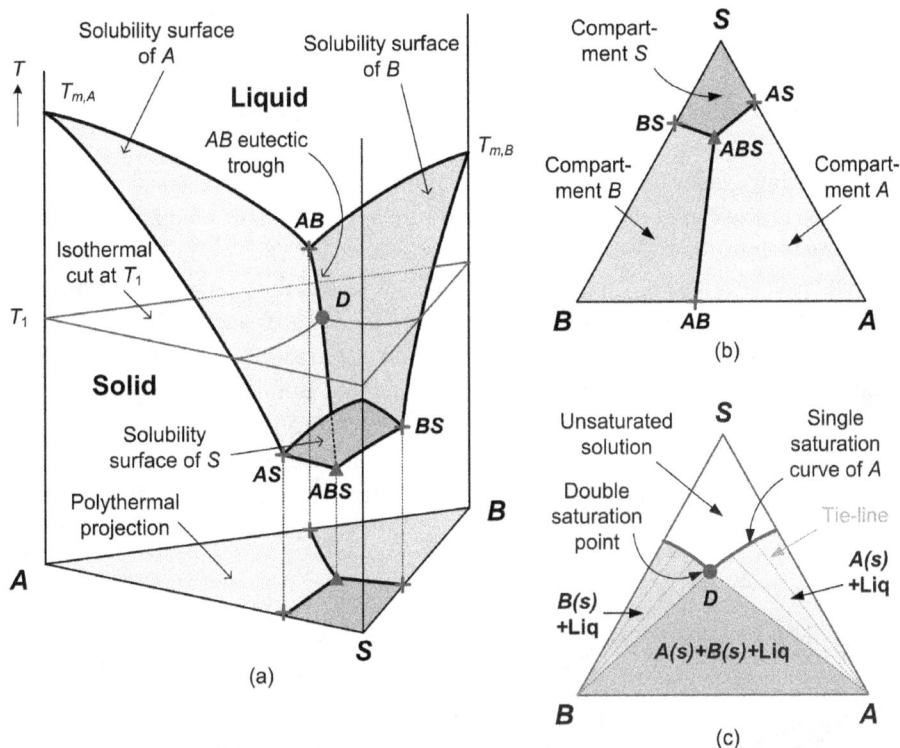

Figure 2.11: Ternary SLE phase diagram with simple eutectic behavior: (a) *T–x* diagram, (b) polythermal projection, (c) isothermal cut at T_1.

a different temperature, corresponding to the saturation surfaces in the 3D diagram, thus the name polythermal projection. Furthermore, since only the saturation surfaces are displayed, this projection does not provide any information on the solid mixture regions below the eutectic points.

Another way to reduce the dimensionality is an *isothermal cut*, which is obtained by slicing the prism at a constant temperature. A series of such cuts taken at different temperatures would provide a good mental picture of the 3D figure. Figure 2.11c shows an isothermal cut at temperature T_1. Since this temperature is above the melting point of component S, no region associated with solid S can be found. Instead, a region of unsaturated liquid is present at the S-rich portion of the diagram. The wedge-shaped region to the right, $A(s)$+Liq, is the region where solids of pure A exist in equilibrium with a liquid phase. If the composition of a mixture lies within the other wedge-shaped region on the left-hand side, solids of pure B are obtained at this temperature. Inside the lower triangle, $A(s)$+$B(s)$+Liq, both solids A and B crystallize out of the solution and the solid mixture is in equilibrium with a liquid phase having a composition indicated by point D, which is referred to as a *double saturation point*. Point D is actually a part of the eutectic trough in the 3D figure. It can be clearly seen that the location of the double saturation point – and hence the region boundaries in the isothermal cut – would change with temperature.

Cuts along planes that are perpendicular to the base can also be taken to visualize the various mixed solid regions below the solubility curves. Two such *isoplethal cuts*, each of which basically forms a pseudobinary system, are shown in Figure 2.12a. The first one is taken along a plane parallel to the back face of the prism, which signifies a constant composition of S. This plane sits on line pq on the triangular base and intersects the solubility surfaces of A and B at curves aD_1 and bD_1, respectively. Line cd represents the intersection between this plane and an isothermal cut at the ternary eutectic temperature T_{ABS}. The second cut is taken along a plane that passes through the axis of pure A sitting on line Am on the triangular base, which signifies a constant ratio between components B and S. The plane intersects the solubility surfaces of A and B at curves hD_2 and eD_2, respectively.

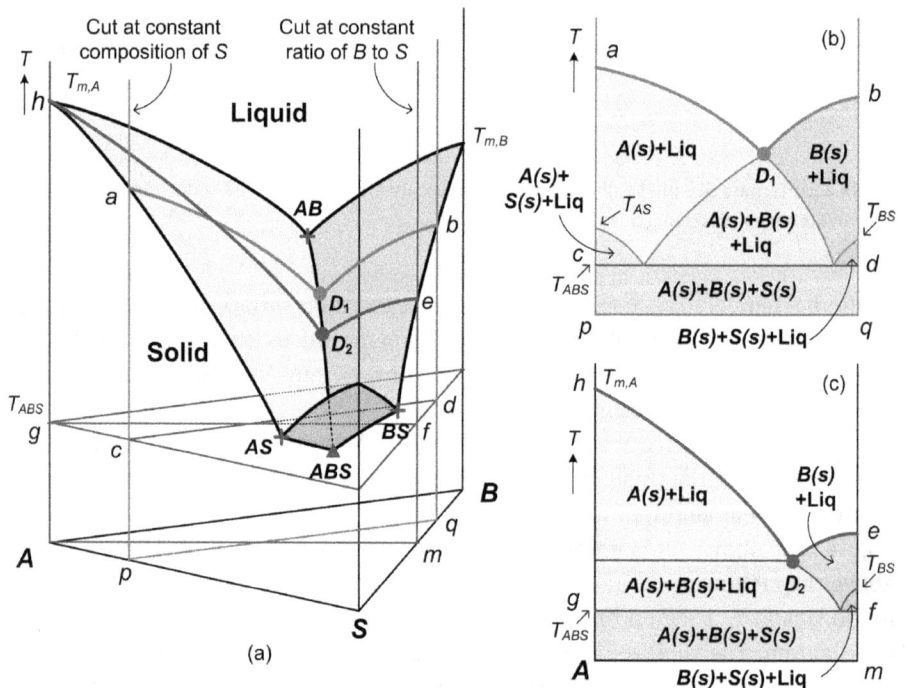

Figure 2.12: Isoplethal cuts of the ternary SLE phase diagram: (a) T–x diagram, (b) cut at constant composition of S, (c) cut at constant ratio of B to S.

The different regions on the cut at a constant composition of S is shown in Figure 2.12b. Between the single saturation regions of A and B, a double saturation region $A(s) + B(s) + L$ appears below the double saturation point D_1. Furthermore, double saturation regions of $A(s) + S(s) + L$ and $B(s) + S(s) + L$ appear below the binary eutectic temperatures of AS and BS, respectively. Line cd lies on the isothermal plane of T_{ABS}. All three double saturation regions end at the ternary eutectic temperature, below which only a solid mixture of $A(s)+B(s) + S(s)$ can exist.

Similarly, the different regions on the cut at a constant ratio of B to S are depicted in Figure 2.12c. Again, the double saturation region $A(s)+B(s)+L$ appears below the double saturation point D_2. The boundary between this region and $A(s)+L$ region is a straight line, which is the tie-line connecting pure A and the double saturation point at D_2. Line fg lies on the isothermal plane of T_{ABS}. There is no double saturation region of $A(s)+S(s)+L$ at any temperature on this cut and the AS binary eutectic temperature has no relevance.

Example 2.3 – Ternary simple eutectic system

The ternary system of sulfonal (sulfonmethane, $C_7H_{16}O_4S_2$, melting point 124.5 °C), β-naphthol (2-naphthol, $C_{10}H_8O$, melting point 122 °C), and salol (phenyl salicylate, $C_{13}H_{10}O_3$, melting point 42.5 °C) exhibits simple eutectic behavior. Experimental solubility data of sulfonal in mixtures of β-naphthol and salol as well as β-naphthol in mixtures of sulfonal and salol at different temperatures, including the binary and ternary eutectic points, have been reported in the literature [34]. By plotting the data on a T–x diagram (Figure 2.13a), the solubility surfaces of the pure components can be identified. The diamonds in the figure indicate data points with pure sulfonal (A) as the solid phase, while the crosses represent data points with pure β-naphthol (B). Except the pure component melting point, no other data point with pure salol (C) as the solid phase is reported, because of the relatively small solubility surface of this component. Figure 2.13b is the polythermal projection of the SLE phase diagram, showing the three compartments where pure components can be obtained as the solid phase. The approximate location of the boundary between compartments A and B is determined from the data points, while the AC and BC boundaries are assumed to be straight lines connecting the eutectic points. Isotherms, lines connecting data points having the same temperature, are also plotted on the polythermal projection.

Isothermal cuts can also be taken to identify the various regions of phase equilibrium at constant temperature. An isothermal cut at 80 °C (Figure 2.13c), which is below the melting points of A and B but above the AB binary eutectic temperature (67 °C), features two separate regions where pure A and pure B, respectively, are in equilibrium with a liquid phase. A cut taken at 60 °C (Figure 2.13d), which is below the AB binary eutectic temperature, features a third region where a mixture of solid A and B coexists with a liquid phase. The boundaries of this region are defined by the double saturation point, which is the intersection of the two solubility curves. The two-solid region also appears at the isothermal cut at 40 °C (Figure 2.13e) and its size gets larger as the double saturation point moves away from the AB edge. Since this temperature is lower than the melting point of pure C, a small region where solid C is in equilibrium with the liquid appears near the C vertex.

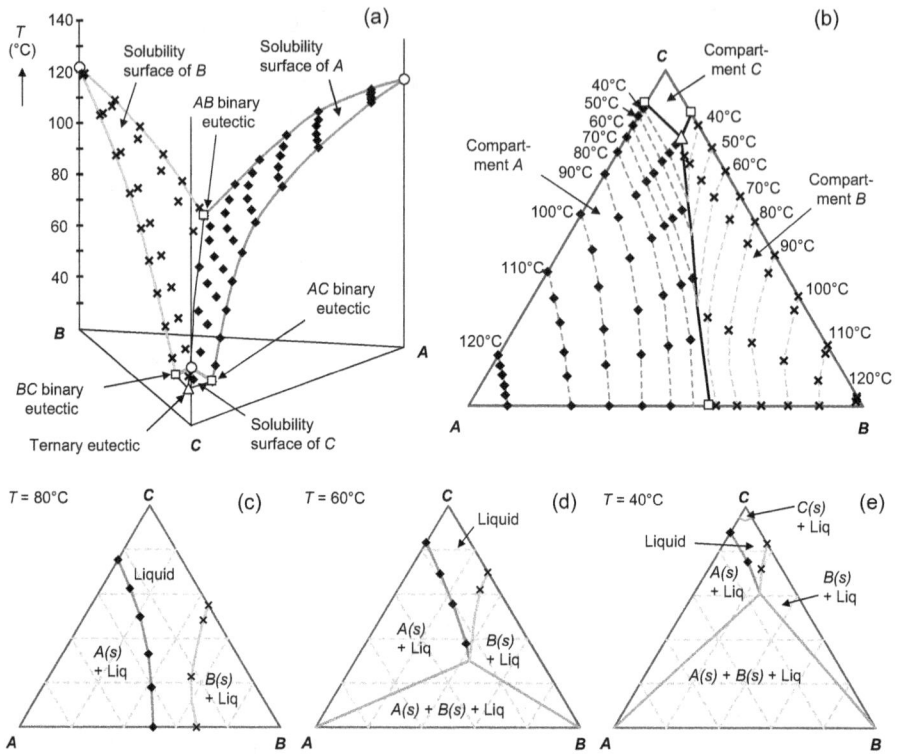

Figure 2.13: Ternary SLE phase diagram of sulfonal (*A*)/β-naphthol (*B*)/salol (*C*) system (data from [34]): (a) T–x diagram, (b) polythermal projection with isotherms, (c) isothermal cut at 80 °C, (d) isothermal cut at 60 °C, and (e) isothermal cut at 40 °C.

2.2.4 Quaternary systems

For a four-component system, representing composition requires a 3D space, which is the composition tetrahedron shown in Figure 2.14. Indicated in the diagram are three planes corresponding to S content of 25%, 50%, and 75%. Point *M* is located on the plane of 50% S. The composition of *A*, *B*, and *C* can be read in the same way as for the ternary system. The T–x diagram for a quaternary system would be four-dimensional, which is obviously impossible to draw on a piece of paper. However, by analogy with systems with fewer components, it is possible to obtain 3D polythermal projection or isothermal cut, which would take the shape of a tetrahedron.

The polythermal projection is shown in Figure 2.15a. There are four 3D compartments, one for each of the components, separated from each other by the eutectic surfaces. For example, compartment *A* is bounded by eutectic surfaces *AB-ABC-ABCS-ABS*,

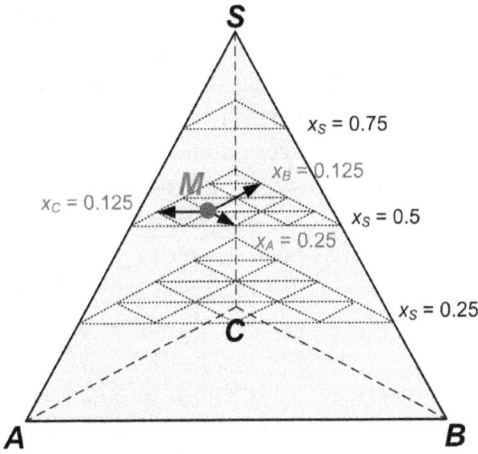

Figure 2.14: Composition tetrahedron for representing quaternary systems.

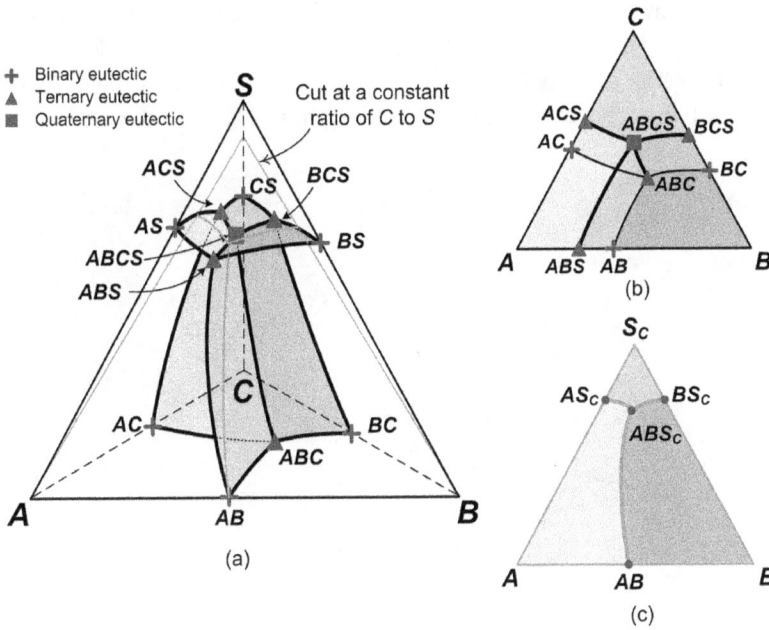

Figure 2.15: Polythermal projection of SLE diagram of a quaternary system: (a) three-dimensional figure, (b) Jänecke projection, and (c) constant composition cut.

AC-ABC-ABCS-ACS, and AS-ABS-ABCS-ACS. The six eutectic surfaces meet along four ternary eutectic troughs (ABC-ABCS, ABS-ABCS, ACS-ABCS, and BCS-ABCS) and at a quaternary eutectic point (ABCS). The dimensionality of this 3D figure can be further reduced by taking a projection or cut. One common approach is to take a *Jänecke projection*, which normalizes the composition with respect to a reference component. Since the solvent is normally not crystallized out in the process, a Jänecke projection from the solvent apex (point S) as shown in Figure 2.15b is particularly useful for process design. The compartments of A, B, and C in the projection are marked with different shades. Compartment S, located above the other three compartments and covering the entire triangle, does not appear in the plot. Another way to reduce dimensionality is to take a cut at constant composition. For example, a cut at a constant C to S ratio (Figure 2.15c) is useful when considering a quaternary system in which C is a soluble impurity that does not crystallize out. In other words, S_c can be viewed as the solvent in a system where only A and B may precipitate out. The subscript C indicates the presence of C as an unsaturated component.

The isothermal cut of the quaternary phase diagram is shown in Figure 2.16. Analogous to the isothermal cut of a ternary phase diagram, it features cone-shaped regions bounded by saturation surfaces, one each for A, B, and C. There is no saturation surface for S because this cut is taken above the melting point of S. Points D_1, D_2, and D_3 are double saturation points. The saturation surfaces intersect each other along double saturation curves (D_1T, D_2T, and D_3T) and all three of them meet at the triple saturation point (T), which is a solution saturated with A, B, and C. To reduce dimensionality, a Jänecke projection can be taken from the solvent apex (point S), as shown in Figure 2.16b. It should be noted that although, at a glance, this projection

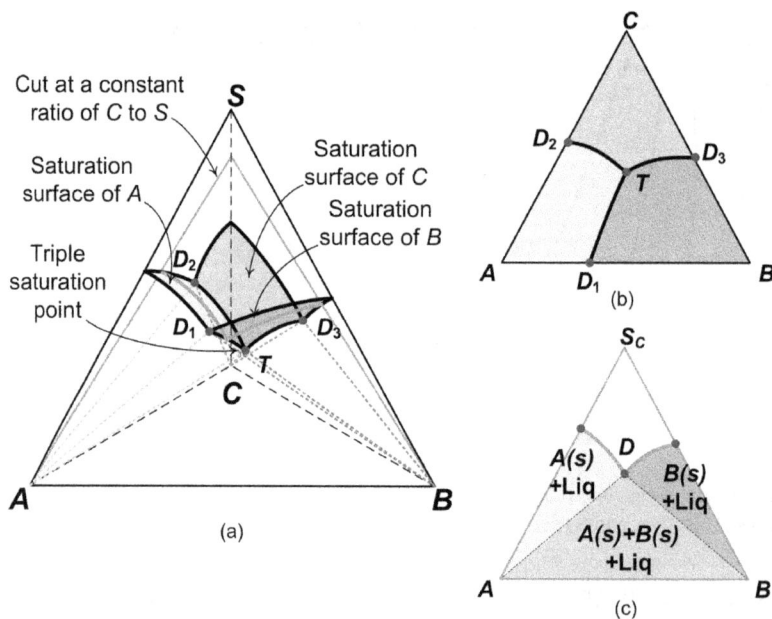

Figure 2.16: Isothermal cut of SLE diagram of a quaternary system: (a) three-dimensional figure, (b) Jänecke projection, and (c) constant composition cut.

appears to be the same as a polythermal projection of a ternary phase diagram, there are fundamental differences between the two. In this projection, each point has the same temperature but different content of a fourth component (i.e., solvent S). Meanwhile, each point in a polythermal projection has a different temperature. A constant composition cut (Figure 2.16c) looks similar to the isothermal cut of a ternary T–x diagram, but it should be kept in mind that C is also present in the liquid phase.

Example 2.4 – Isothermal SLE phase diagram of a quaternary system
The SLE phase behavior of amino acids in aqueous solution has been widely studied. For example, Grosse Daldrup et al. [64] measured the solubility of mixtures of L-alanine, L-leucine, and L-valine in water at different temperatures. These amino acids do not form solid solution with each other. Figure 2.17a–c depicts the isothermal cut of the SLE phase diagrams of L-alanine/L-leucine/water, L-alanine/L-valine/water, and L-leucine/L-valine/water, respectively, at 50 °C. The solubility data of Grosse Daldrup et al. are plotted on the diagrams; diamonds, crosses, and plus signs represent the liquid compositions in equilibrium with a solid phase of pure L-alanine, L-leucine, and L-valine, respectively. The solubility curves are estimated based on these data points. Note that due to the relatively low solubility of the amino acids, only the top portion of the triangle (mass fraction $H_2O \geq 0.75$) is shown for clarity.

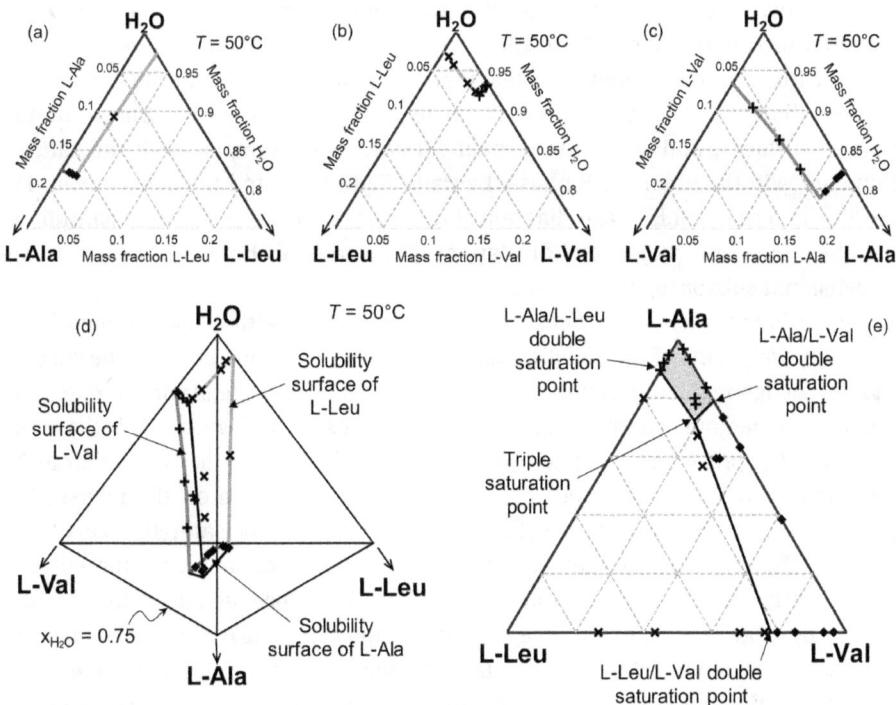

Figure 2.17: Isothermal cut of the SLE phase diagram of L-alanine/L-leucine/L-valine water system at 50 °C (data from [64]): (a) L-alanine/L-leucine/water subsystem, (b) L-alanine/L-valine/water subsystem, (c) L-leucine/L-valine/water subsystem, (d) quaternary system, and (e) Jänecke projection of the quaternary system.

Figure 2.17d shows the isothermal cut of the quaternary SLE phase diagram involving all three amino acids and water. In addition to the ternary data points, six quaternary data points (two each with L-alanine, L-leucine, and L-valine as the solid phase) are also plotted on the diagram. Again, only the top portion of the tetrahedron is displayed. The phase diagram features three solubility surfaces, one for each amino acid, passing through the data points and intersecting along three double satura- tion curves. A clearer view of these double saturation curves and the triple saturation point – where the three curves meet – can be had by taking a Jänecke projection with respect to water, as shown in Figure 2.17e. Note that the location of the double and triple saturation points have been estimated based on the requirement that all data points with the same solid phase must fall within the corre- sponding region. For example, all data points with L-alanine as the solid phase (plus signs) lie within the projected L-alanine saturation surface, which is the shaded region in the figure.

2.2.5 Multicomponent systems

Beyond four components, visualization is difficult because it is impossible to view entities of dimensions four or higher, in their entirety, on a piece of paper. The only practical approach is to take cuts and projections to reduce the di- mensionality, and then examine a set of those lower dimensional images to form a mental picture of the complete space. A comprehensive set of methods is avail- able for this purpose. Generally speaking, any phase diagram can be considered as an n-dimensional object. Geometrical features such as points, lines, curves, planes, surfaces, and so on are termed *varieties*. For example, both the satura- tion curves in the binary SLE phase diagram (Figure 2.4) and the solubility surfa- ces in the ternary SLE phase diagram (Figure 2.11a) are referred to as saturation varieties. Cuts and projections display, at least, parts of these varieties in an m- dimensional subspace, where $m<n$.

A *cut* is the intersection between the original object with a lower dimensional subspace. Only parts of the varieties that lie on the intersection appear in the cut and the rest are ignored. For example, the isothermal cut at T_1 of the 3D prism represent- ing the T–x diagram of a three-component molecular system (Figure 2.11) is the inter- section of the prism with a plane representing $T = T_1$. Parts of the two-dimensional (2D) saturation varieties (A and B saturation surfaces) that lie on the intersecting plane appear as one-dimensional (1D) saturation varieties (the saturation curves) in the cut (Figure 2.11c). The cut does not contain any information on the eutectic points, as they do not lie on the intersecting plane. To reduce the dimensionality of a system by more than one, multiple successive cuts can be taken. For example, the constant composition cut in Figure 2.16c is obtained by first intersecting the 4-D ob- ject representing the T–x diagram of the quaternary system with a 3D hyperplane rep- resenting constant temperature, and then intersecting the resulting tetrahedron (Figure 2.16a) with a plane representing a constant ratio of C to S. If all saturation varieties in the phase diagram as well as the intersecting subspace are represented by mathematical equations, it is straightforward to obtain the cut by solving those equations simultaneously.

A *projection* can be viewed as getting the image of an object on a lower dimensional subspace called a *projective space*, by shining light on the object. The lines connecting each point in the object with its image is termed a *projection ray*. There are three types of projection: *orthogonal projection* (projection rays are parallel and orthogonal to the projective space), *parallel projection* (projection rays are parallel but not necessarily orthogonal to the projective space), and *central projection* (all projection rays go through a point called the center of projection). Figure 2.18 explains the differences between these three types of projection. Note that the polythermal projection is an example of an orthogonal projection, while the Jänecke projection is a central projection with the reference component apex as the center of projection. Although visualization is limited up to 3D, the same concept applies to projections in higher dimensions.

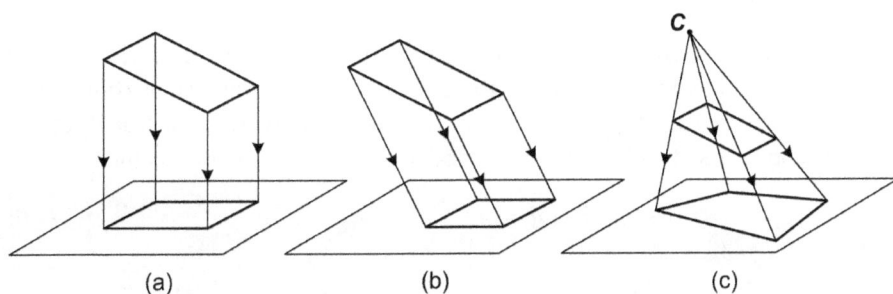

(a)   (b)   (c)

Figure 2.18: Three basic types of projection: (a) orthogonal, (b) parallel, and (c) central.

In performing a projection, the projective subspace and the center of projection (in central projection) or direction of projection rays (in parallel projections) need to be defined. In the case of orthogonal projection, once the projective subspace is given, the direction of projection rays cannot be chosen arbitrarily and vice versa, as they are always orthogonal to each other. Once the projection rays are defined using a set of basis vectors, the coordinates of the projective subspace, referred to as the *canonical coordinates*, naturally appear. Wibowo and Ng [65] showed that the canonical coordinates for multiple orthogonal projections from an n-dimensional space to an m-dimensional subspace are given by

$$Y_i = y_i - \theta_i^T Q_{ref}^{-1} y_{ref}; \; i = 1, \, 2, \, \ldots, m \tag{2.2}$$

where

$$\boldsymbol{\theta}_i = [q_{1,i} \quad q_{2,i} \quad \cdots \quad q_{n-m,i}]^T, \; i = 1,\, 2,\, \ldots,\, m \tag{2.3}$$

$$\boldsymbol{Q}_{\text{ref}} = \begin{bmatrix} q_{1,m+1} & q_{2,m+1} & \cdots & q_{n-m,m+1} \\ q_{1,m+2} & q_{2,m+2} & \cdots & q_{n-m,m+2} \\ \vdots & \vdots & \ddots & \vdots \\ q_{1,n} & q_{2,n} & \cdots & q_{n-m,n} \end{bmatrix} \tag{2.4}$$

$$\boldsymbol{y}_{\text{ref}} = [y_{m+1} \quad y_{m+2} \quad \cdots \quad y_n]^T \tag{2.5}$$

Here, $q_{j,i}$ are the elements of the jth projection vector, $\boldsymbol{q}_j$. There are $(n - m)$ linearly independent projection vectors, each of which defines the direction of projection rays for reducing the dimensionality by one. The projection of any variety of dimension m or higher in the original object (which are mathematically described by functional forms in terms of y_i's) can be obtained by casting the functions in terms of Y_i's.

As an example, consider an orthogonal projection of a right-angled triangular pyramid which lies in a 3D space ($n = 3$) represented by the set of coordinates $\{y_1, y_2, y_3\}$, onto a 2D subspace ($m = 2$), as illustrated in Figure 2.19a. The direction of projection rays shown in the figure is described by the projection vector

$$\boldsymbol{q}_1 = [1 \quad 1 \quad -1]^T \tag{2.6}$$

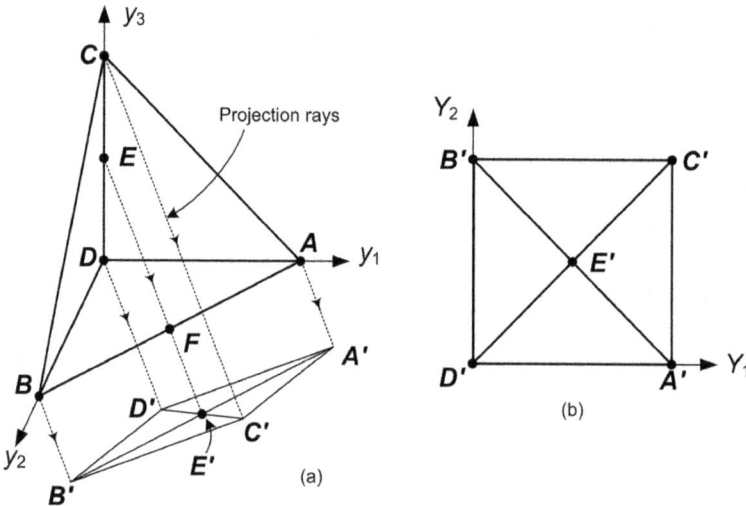

Figure 2.19: Projection of a right-angled triangular pyramid onto a two-dimensional subspace [65]: (a) direction of projection rays, and (b) canonical coordinates defining the projective subspace.

The elements of this vector are $q_{1,1} = 1$, $q_{1,2} = 1$, and $q_{1,3} = -1$. As $n - m = 1$ for this projection, all eqs. (2.3)–(2.5) give 1×1 matrices:

$$\boldsymbol{\theta}_1 = [q_{1,1}] \tag{2.7}$$

$$\boldsymbol{\theta}_2 = [q_{1,2}] \tag{2.8}$$

$$\mathbf{Q}_{\text{ref}} = [q_{1,3}] \tag{2.9}$$

$$\mathbf{y}_{\text{ref}} = [y_3] \tag{2.10}$$

Here, y_3 has been conveniently chosen as the reference coordinate. Substituting into eq. (2.2), the canonical coordinates representing the 2D subspace are obtained:

$$Y_1 = y_1 - (1)(-1)^{-1} y_3 = y_1 + y_3 \tag{2.11}$$

$$Y_2 = y_2 - (1)(-1)^{-1} y_3 = y_2 + y_3 \tag{2.12}$$

Figure 2.19b shows the projection of the pyramid in the 2D subspace, which takes the form of a square. The projections of lines AB and CD form the diagonals of the square and both points, E and F, are projected to their intersection (point E'). This projection has been originally proposed by Cruickshank et al. [66], who found it to be useful for visualizing the liquid–liquid equilibrium behavior of four-component systems.

Since a projection shows only part of the system under consideration, multiple figures are required to obtain a mental image of the phase behavior of the entire system. For a C-component molecular system, the simplest way to generate those figures is by taking Jänecke projections with respect to any $C - 3$ reference components, which is essentially selecting three components at a time and ignoring the mole fractions of the reference components. The total number of such 2D projections that can be generated by selecting different combinations of reference components is

$$N_P = \binom{C}{3} \tag{2.13}$$

However, Samant et al. [67] showed that the minimum number of 2D projections required to completely represent the phase behavior is given by

$$N_{P,\,\text{min}} = \left(\frac{C-1}{2}\right)^2 + k \; ; \; k = \begin{cases} 0 & \text{if } C \text{ is odd} \\ 3/4 & \text{if } C \text{ is even} \end{cases} \tag{2.14}$$

The remaining $N_P - N_{P,\text{min}}$ projections do not offer any additional information, but may be used as supplementary information. Interested readers are referred to the original publication for the derivation of eq. (2.14).

To illustrate the application of projections and cuts in visualizing high-dimensional SLE phase diagrams, consider the isobaric, polythermal phase diagram of a five-component (A, B, C, D, E) molecular system with simple eutectic behavior, which is four dimensional in nature. Figure 2.20 shows one such projection, with D and E as the reference components. Different points on the projection, including pure components and eutectic points, are labeled after the saturated components. For example, vertex AB indicates a binary mixture where both A and B are saturated; that is, the AB binary eutectic. As a consequence of the choice of reference components, saturation varieties involving D and E do not appear in this projection, making the representation of the phase behavior incomplete. For example, the single saturation varieties of D and E are absent in the projection. To view the missing saturation varieties, other projections need to be taken by choosing different sets of reference components. Since eq. (2.14) gives $N_{P,\min} = 4$, three more projections are needed to completely represent the phase behavior of this system.

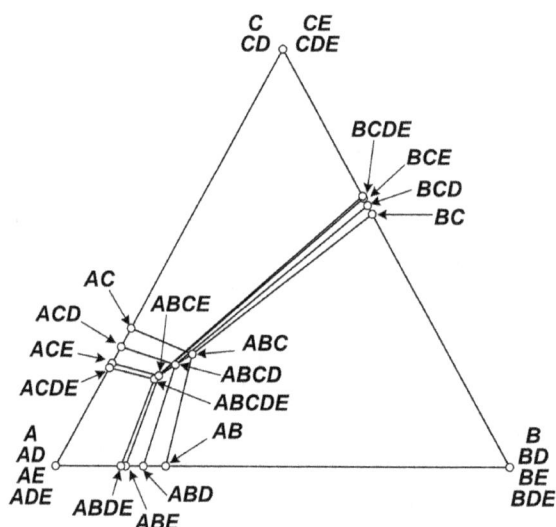

Figure 2.20: Projection of the polythermal SLE phase diagram of a quinary system [65].

Figure 2.21 shows four 2D cuts at constant $x_E = 0.2$ and different values of x_D. The saturation points on the cuts are labeled after the saturated components. Unsaturated components, if present, are written as subscripts. For example, AB_{DE} is a double saturation point at which four components (A, B, D, and E) are present, but only A and B are saturated. At $x_D = 0.1$ (Figure 2.21a), both D and E are unsaturated at any proportion of A, B, and C. Therefore, the cut looks very much like the polythermal phase diagram of the ABC ternary system, with the exception that D and E appear as unsaturated components at all saturation points. The situation is different at $x_D = 0.55$ (Figure 2.21b), where D becomes saturated at the region bounded by ABD_{CE}, ACD_{BE}, and BCD_{AE}. As it is no longer possible for A, B, and C to be co-saturated in the

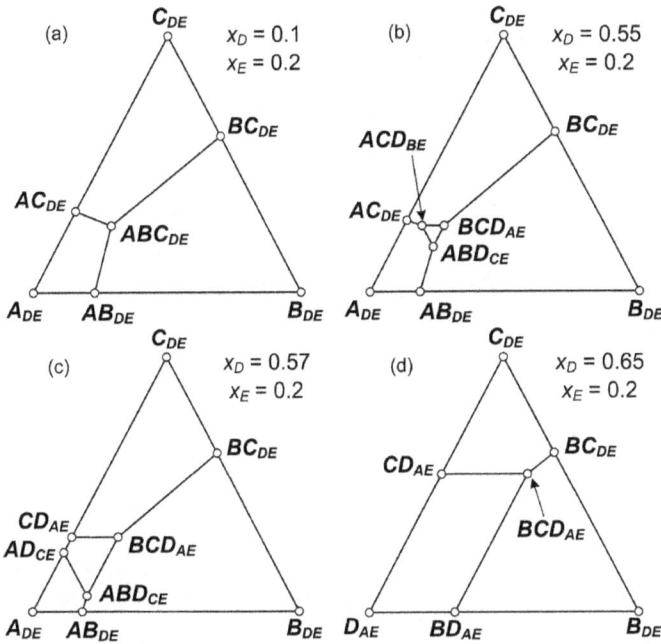

Figure 2.21: Isoplethal cuts of the polythermal projection at different concentrations of D and E.

presence of such a large amount of D, point ABC_{DE} does not appear in this cut. When x_D is further increased to 0.57 (Figure 2.21c), it is not even possible for just A and C to be saturated together. Therefore, point ACD_{BE} disappears and is replaced by AD_{CE} and CD_{AE}. At $x_D = 0.65$ (Figure 2.21d), A can no longer be saturated. Thus, the single saturation variety of A disappears altogether from the cut, and point AD_E is replaced by D_{AE}, signifying the fact that D is now the saturated component at that point.

One important task in visualizing multicomponent SLE phase diagrams using cuts and projections is to identify the different varieties in the high-dimensional space that should appear in a certain cut or projection. To systematize this task, Samant et al. [67] proposed a framework in which the geometry of the phase diagram is captured in a digraph as represented by its associated matrices. Different points on the phase diagram, including pure components, eutectic points, and saturation points are represented by *vertices*. Two vertices are said to be *adjacent* to each other if they are connected by an *edge*. An *adjacency matrix* defines the presence of an edge between any pair of vertices. These vertices and edges define the boundaries of saturation varieties of higher dimensions, inside which one or more components are saturated. A *saturation variety matrix* can be derived from the adjacency matrix to clarify which edges form the bounds of which saturation varieties. The phase diagram is defined by identifying and locating all vertices, edges, and saturation varieties.

Figure 2.22 shows the representation of the polythermal phase diagram of a ternary system with simple eutectic behavior as a digraph, along with the corresponding adjacency and saturation variety matrices. In the adjacency matrix, "1" signifies the pair of vertices that are adjacent to each other, such as A and AB, AB and ABC, and so on. Meanwhile, in the saturation variety matrix, "1" identifies the vertices on the column heading that form the boundaries of the saturation variety on the row heading. For example, the A single saturation variety (A-sat) is the area bounded by vertices A, AB, AC, and ABC, while the AB double saturation variety (AB-double sat) is the line AB–ABC. Arrows on the digraph indicates the direction of decreasing temperature.

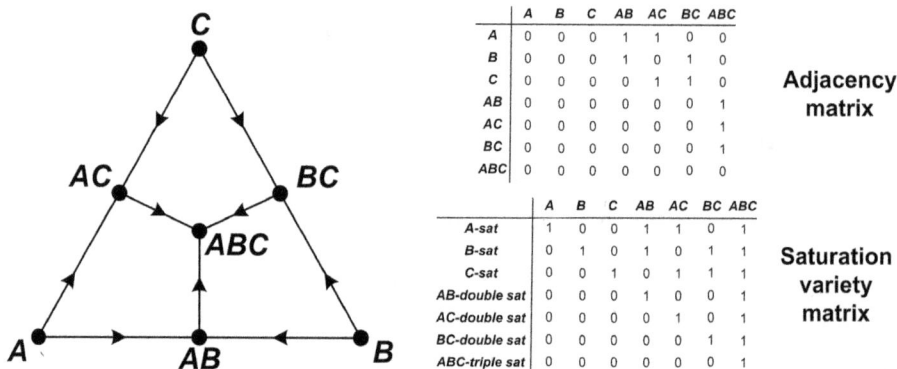

	A	B	C	AB	AC	BC	ABC
A	0	0	0	1	1	0	0
B	0	0	0	1	0	1	0
C	0	0	0	0	1	1	0
AB	0	0	0	0	0	0	1
AC	0	0	0	0	0	0	1
BC	0	0	0	0	0	0	1
ABC	0	0	0	0	0	0	0

Adjacency matrix

	A	B	C	AB	AC	BC	ABC
A-sat	1	0	0	1	1	0	1
B-sat	0	1	0	1	0	1	1
C-sat	0	0	1	0	1	1	1
AB-double sat	0	0	0	1	0	0	1
AC-double sat	0	0	0	0	1	0	1
BC-double sat	0	0	0	0	0	1	1
ABC-triple sat	0	0	0	0	0	0	1

Saturation variety matrix

Figure 2.22: Representation of a polythermal phase diagram of a ternary system as a digraph with the corresponding adjacency and saturation variety matrices.

When taking a projection, the adjacency and saturation variety matrices remain unchanged, as a projection does not affect the connections among vertices. Note that some saturation varieties may not appear on the projection; this happens when any vertex that forms the boundaries of the saturation variety is missing in the projection (e.g., a vertex corresponding to a reference component in a Jänecke projection). On the other hand, a cut only shows parts of the saturation varieties. Therefore, taking a cut has to involve a modification of the adjacency matrix to identify which edges should appear as a new vertex in the cut. In each new vertex, the components common to both original vertices connected by the surviving edge would be saturated and the ones present in only one of them would be unsaturated. The saturation variety matrix is also modified correspondingly to identify the new saturation varieties in the cut. More details and examples can be found in the original publication [67].

2.2.6 Reactive systems

If there are chemical reactions among components in a system and the reactions achieve equilibrium, the compositions are constrained to the equilibrium curve. In other words, the presence of reactions reduces the number of degrees of freedom in the system because the composition has to obey the reaction equilibrium equations. Correspondingly, the possible compositions are confined to a lower dimensional subspace of the phase diagram.

As discussed in Section 2.1.2, an isobaric phase diagram for a mixture with C components can be represented in a C-dimensional space. A complete representation of the isobaric phase behavior of a C-component system with R independent reactions requires a $(C - R)$-dimensional space and the compositions of the reaction mixture can be sufficiently represented using $(C - R - 1)$ new composition variables. It is convenient to define them in such a way that they are reaction invariant, that is, they take the same numerical values before and after the reaction. This can be achieved by taking projections along *stoichiometric lines*, which represent the loci of reaction mixture compositions at different conversions, resulting in *reaction-invariant projections*. Note that there is one set of stoichiometric lines for each independent reaction.

As an illustration, consider a three-component system with one reaction $(A+B=D)$, for which Figure 2.23 shows the isothermal cut of the phase diagram. Since D can exist as a solid at this temperature, but the melting points of A and B are below this temperature, only the saturation region of $D(s)$ appears in this cut. Reaction equilibrium restricts the composition along the reaction equilibrium curve [68], which can be mathematically expressed as

$$\frac{x_D}{x_A x_B} = K \tag{2.15}$$

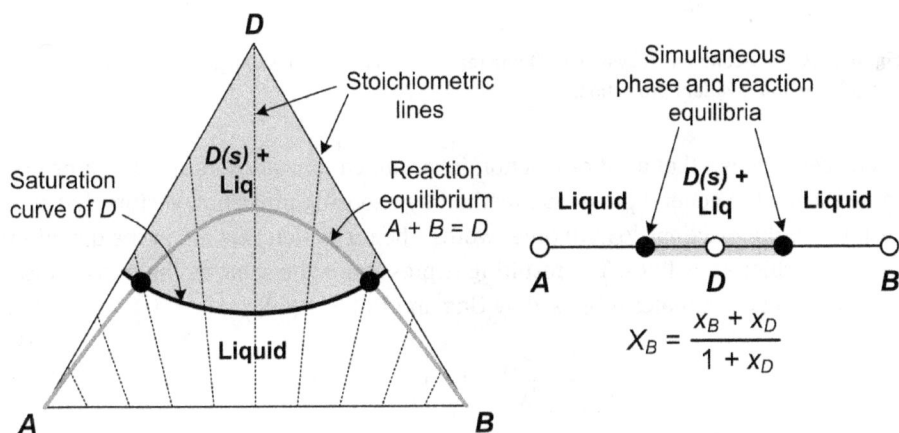

Figure 2.23: Reaction-invariant projection in a three-component system with one reaction.

where K is the equilibrium constant. Projection along the stoichiometric lines gives a 1D figure with compositions expressed in terms of a *transformed mole fraction*, X_B. The projection features a segment of the line as the saturation region of D bounded by two saturation points where phase and reaction equilibria occur at the same time. The choice of the expression for X_B will be explained further.

Figure 2.24a shows an example of a reactive phase diagram for a system of four components: A, B, D, and S, and one reaction: $A+B = D$, at a constant temperature. The reaction equilibrium is represented by the indicated reaction surface. At this temperature, the surface intersects the saturation surface of D, beyond which solid D precipitates. A projection of the phase diagram in which the compositions are expressed in terms of transformed mole fractions X_1 and X_2, are shown in Figure 2.24b. The D saturation curve results from the intersection between reaction equilibrium and SLE surfaces. Inside the region bordered by this curve and the AB edge, the solution is in equilibrium with solid D. Outside of the region, the solution is unsaturated.

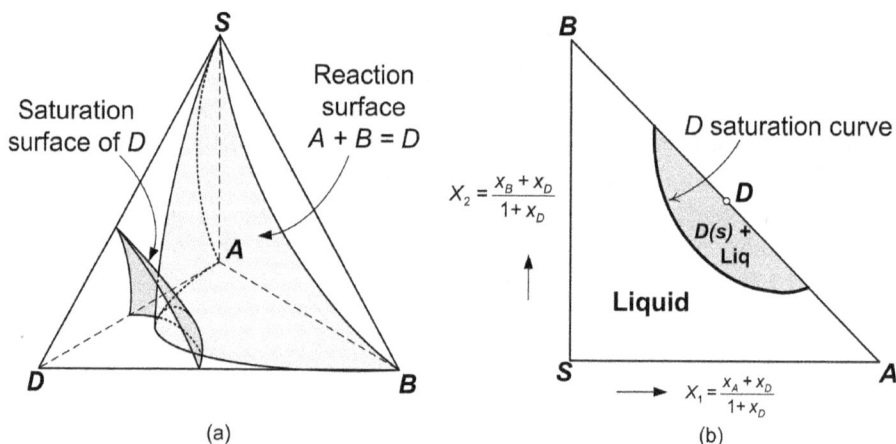

(a)   (b)

Figure 2.24: Four-component system with one reaction: (a) three-dimensional isothermal diagram and (b) reaction-invariant projection.

It has been shown that a set of reaction-invariant canonical coordinates can be obtained using the general projection method by choosing projection vectors that correspond to the reactions [65]. In other words, the projection rays are in the direction of the stoichiometric lines. The resulting expression is the same as the transformed mole fraction coordinates proposed by Ung and Doherty [69]:

$$X_i = \frac{x_i - \boldsymbol{v}_i^T \boldsymbol{V}_{\text{ref}}^{-1} \boldsymbol{x}_{\text{ref}}}{1 - \boldsymbol{v}_{\text{TOT}}^T \boldsymbol{V}_{\text{ref}}^{-1} \boldsymbol{x}_{\text{ref}}} \quad ; i = 1, 2, \ldots, (C - R - 1) \tag{2.16}$$

where

$$v_i^T = [v_{i1} \quad v_{i2} \quad \cdots \quad v_{iR}] ; i = 1, 2, \ldots, (C - R - 1) \tag{2.17}$$

$$v_{TOT}^T = \left[\sum_{i=1}^{C} v_{i1} \quad \sum_{i=1}^{C} v_{i2} \quad \cdots \quad \sum_{i=1}^{C} v_{iR} \right] \tag{2.18}$$

$$V_{ref} = \begin{bmatrix} v_{C-R,1} & v_{C-R,2} & \cdots & v_{C-R,R} \\ v_{C-R+1,1} & v_{C-R+1,2} & \cdots & v_{C-R+1,R} \\ \vdots & \vdots & \ddots & \vdots \\ v_{C-1,1} & v_{C-1,2} & \cdots & v_{C-1,R} \end{bmatrix} \tag{2.19}$$

$$x_{ref} = [x_{C-R} \quad x_{C-R+1} \quad \cdots \quad x_{C-1}]^T \tag{2.20}$$

where v_{ij} represents the stoichiometric coefficient of component i in reaction j, and x_i is the mole fraction of component i. Different transformed coordinates have been proposed in terms of concentration of elements [70]. Despite the seemingly different approaches and resulting expressions, these coordinates actually describe the same projective space.

The expressions for the transformed coordinates in Figures 2.23 and 2.24b can be readily obtained from eq. (2.16) by taking D as the reference component. For the three-component system with one reaction ($C = 3$, $R = 1$) shown in Figure 2.23, eqs. (2.17)–(2.19) give

$$v_B^T = [v_{B,1}] = [-1] \tag{2.21}$$

$$v_{TOT}^T = [v_{A,1} + v_{B,1} + v_{C,1}] = [-1] \tag{2.22}$$

$$V_{ref} = [v_{D,1}] = [1] \tag{2.23}$$

Since $C - R - 1 = 1$, eq. (2.16) leads to only one reaction-invariant canonical coordinate for this system:

$$X_B = \frac{x_B - (-1)(1)^{-1}x_D}{1 - (-1)(1)^{-1}x_D} = \frac{x_B + x_D}{1 + x_D} \tag{2.24}$$

On the other hand, the reactive projection of the four-component system in Figure 2.24a ($C = 4$, $R = 1$, hence $C - R - 1 = 2$) is represented by two canonical coordinates, with one more expression obtained from eq. (2.17),

$$v_A^T = [v_{A,1}] = [-1] \tag{2.25}$$

resulting in the expressions for X_1 and X_2 as shown in Figure 2.24b.

The phase behavior of systems forming adducts, hydrates, or racemic compounds can be obtained using the same approach by treating the compound formation as a reaction [71]. All pure components as well as the compounds are considered as separate components and the compounds are taken as reference components in deriving the transformed coordinates. For example, consider a system consisting of components A and B that can form a compound C. Figure 2.25a shows the SLE phase behavior of

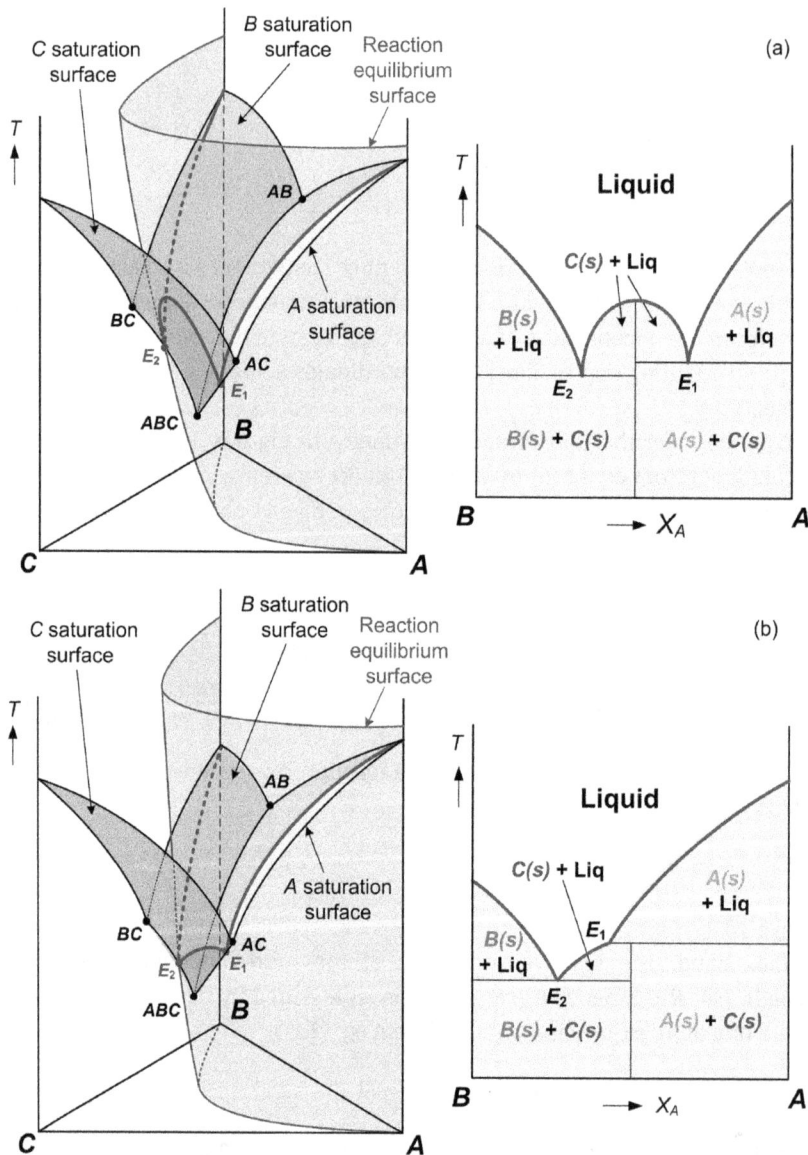

Figure 2.25: SLE phase diagram of a ternary system with reaction: (a) resulting in compound formation behavior and (b) resulting in peritectic behavior.

the ternary system of A, B, and C. Also shown is a reaction equilibrium surface corresponding to the reaction $A+B = C$. The shape of the surface accounts for the fact that the equilibrium constant K is a function of temperature. Under equilibrium condition, which can be assumed to be achieved quickly, possible compositions would be restricted to the reaction equilibrium surface which intersects the saturation surfaces of A, B, and C as well as various solid-liquid regions in the triangular prism. Projecting the intersection to a flat surface in the direction of stoichiometric lines gives a reaction-invariant projection that is equivalent to the phase diagram of a binary system with compound formation (Figure 2.5a). Taking compound C as the reference component, the transformed coordinate for this projection is

$$X_A = \frac{x_A + x_C}{1 + x_C} \tag{2.26}$$

However, since C can be considered as an equimolar mixture of A and B rather than a separate component, it is common to express the composition of the mixture in terms of effective mole fractions of A and B. The expression of eq. (2.26) actually represents the effective mole fraction of A in the binary mixture of A and B. It should also be noted that chemical analysis often gives the concentration of a compound in terms of its constituents, making the effective mole fraction more practical for representing the composition of the mixture.

Figure 2.25b shows a similar ternary system involving a reaction but with different values of melting points (particularly, lower melting point of B) and a reaction equilibrium surface with a slightly different shape (representing a different function of equilibrium constant K with respect to temperature). When projected to a flat surface along the stoichiometric lines, the intersection between the reaction equilibrium surface and the saturation surfaces gives an image that is equivalent to the phase diagram of a binary system with peritectic behavior (Figure 2.5b). Therefore, the binary SLE phase diagrams involving compounds can be considered as a reactive projection of a higher dimensional phase diagram which includes the compounds as separate components. Depending on the relative positions of the reaction equilibrium and saturation varieties, the resulting projection can exhibit either compound formation or peritectic behavior.

2.3 Electrolyte Systems

Electrolyte systems are encountered in many industrially important crystallization processes. In contrast to molecular species, electrolytes partially or completely dissociate into ions in solution. For example, when NaCl is dissolved in water, it dissociates into Na^+ and Cl^- ions. When a mixture of two salts having no common ion (referred to as a *conjugate salt system*) is dissolved in water, the salts lose their original identity. For example, a solution containing an equimolar amount of KNO_3 and

NaCl is indistinguishable with another solution containing the same number of moles of KCl and NaNO$_3$, as both mixtures exist as a mixture of K$^+$, Na$^+$, Cl$^-$, and NO$_3^-$ ions in water.

Because of this special characteristic, mole fraction or weight fraction coordinates are not suitable to represent the composition of an electrolyte system. Instead, ionic coordinates are used for plotting the phase diagram [72]. For a solution containing m simple cations (such as proton, sodium ion, potassium ion, and so on), n simple anions (such as hydroxide and chloride), and s nonelectrolytes (such as water or organic solvents), the solution composition can be adequately represented using $m - 1$ cationic coordinates, $n - 1$ anionic coordinates, and s nonelectrolyte coordinates, as given by

$$R(M_i) = \frac{z_{M_i}[M_i]}{\sum_{i=1}^{m} z_{M_i}[M_i]} \tag{2.27}$$

$$R(N_j) = \frac{z_{N_j}[N_j]}{\sum_{j=1}^{n} z_{N_j}[N_j]} \tag{2.28}$$

$$R(S_k) = \frac{[S_k]}{\sum_{i=1}^{m} z_{M_i}[M_i]} = \frac{[S_k]}{\sum_{j=1}^{n} z_{N_j}[N_j]} \tag{2.29}$$

Concentrations of the cations $[M_i]$, anions $[N_j]$, and $[S_k]$ are expressed in molality, and z is the ionic charge.

As an example of how this coordinate system is used, consider a solution containing an acid HX and a base MOH. In solution, these species dissociate into H$^+$, M$^+$, OH$^-$, and X$^-$ ions, which can also be obtained from a mixture of salt MX and water. Since both H$^+$ and OH$^-$ are present in this system, water should be considered as one of the conjugate salts instead of a nonelectrolyte solvent. Because of the reaction-invariant nature of the projection, the actual degree of water dissociation into H$^+$ and OH$^-$ is irrelevant. Therefore, in plotting a composition using the coordinate system, water should always be assumed to completely dissociate into ions. In this system, $m = 2$ and $n = 2$, so all compositions can be plotted using a coordinate system consisting of 1 cationic coordinate and 1 anionic coordinate, as shown in Figure 2.26a. Using eqs. (2.27 and 2.28), the coordinates are

$$R(M^+) = \frac{[M^+]}{[H^+] + [M^+]} \tag{2.30}$$

$$R(OH^-) = \frac{[OH^-]}{[OH^-] + [X^-]} \tag{2.31}$$

Note that the choice of M$^+$ and OH$^-$ for the numerator is arbitrary. Replacing them by H$^+$ and X$^-$ does not change the features on the phase diagram. A mixture containing 1 mol HX and 1 mol MOH, which is equivalent to a mixture of 1 mol MX and

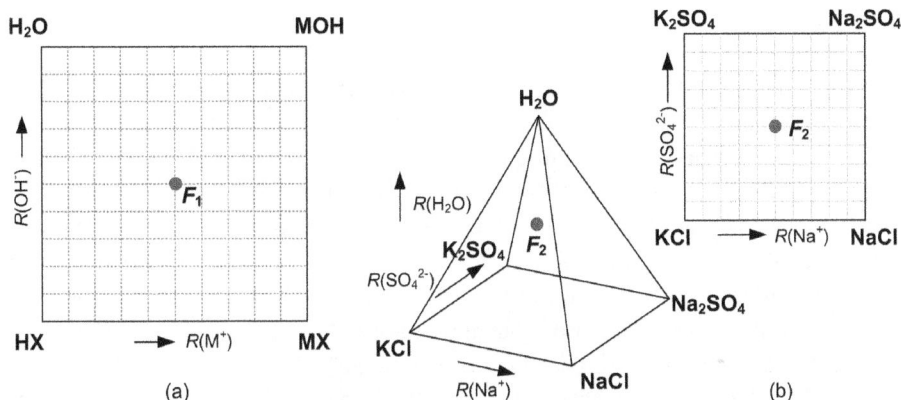

Figure 2.26: Coordinate systems for representing composition in electrolyte systems: (a) an acid-base system and (b) a conjugate salt system.

1 mol H_2O, has a value of $R(M^+) = 0.5$ and $R(OH^-) = 0.5$. Therefore, it is represented by point F_1 in Figure 2.26a.

Another example is a conjugate salt system containing NaCl, KCl, Na_2SO_4, and K_2SO_4 in aqueous solution. Since there are two simple cations and two simple anions in this system, one cationic coordinate, $R(Na^+)$, and one anionic coordinate, $R(Cl^-)$ can be chosen to describe the ionic compositions. In this system, water is assumed to be a nonelectrolyte, although in reality it dissociates to a small extent to form H^+ and OH^-. These two ions are not taken into account since their amounts are negligible. Thus, one more coordinate representing water concentration extends to the third dimension, as shown in Figure 2.26b. Using eqs. (2.27)–(2.29), the coordinates are

$$R(Na^+) = \frac{[Na^+]}{[Na^+] + [K^+]} \tag{2.32}$$

$$R(Cl^-) = \frac{[Cl^-]}{[Cl^-] + 2[SO_4^{2-}]} \tag{2.33}$$

$$R(H_2O) = \frac{[H_2O]}{[Cl^-] + 2[SO_4^{2-}]} \tag{2.34}$$

In the Jänecke projection with respect to water, $R(H_2O)$ is omitted and a 2D figure is obtained. For a mixture containing 2 mol of NaCl and 1 mol of K_2SO_4 in water, R (Na^+) and $R(Cl^-)$ are both 0.5 according to eqs. (2.32) and (2.33). This mixture is represented by point F_2 in Figure 2.26b, appearing at the center of the Jänecke projection.

2.3.1 Acid–base systems

The $T-x$ diagram of an acid-base system involving acid HX and base MOH is shown in Figure 2.27a. The base of this prism is the composition square shown in Figure 2.26a. The three prominent solubility surfaces (MOH, MX, and HX) are identified. There is also a fourth solubility surface near the H_2O vertex, belonging to water. Since there is little water for the dissociation of salts near the HX/MX and MX/MOH faces of the rectangular prism, the phase behavior there can be drastically different from that in the interior of the prism. Also, the melting points of the salts tend to be outside the temperature range of interest. For these reasons, the top portion of the diagram has been omitted in Figure 2.27a. The dimensionality can be reduced by taking an isothermal cut. Figure 2.27b features regions where solid HX, MX, MOH, or their mixtures can be found in equilibrium with a liquid phase. Another useful view is a diagonal cut shown in Figure 2.27c. Such a cut shows the existence of a subsystem along the diagonal plane, which contains a solution of a

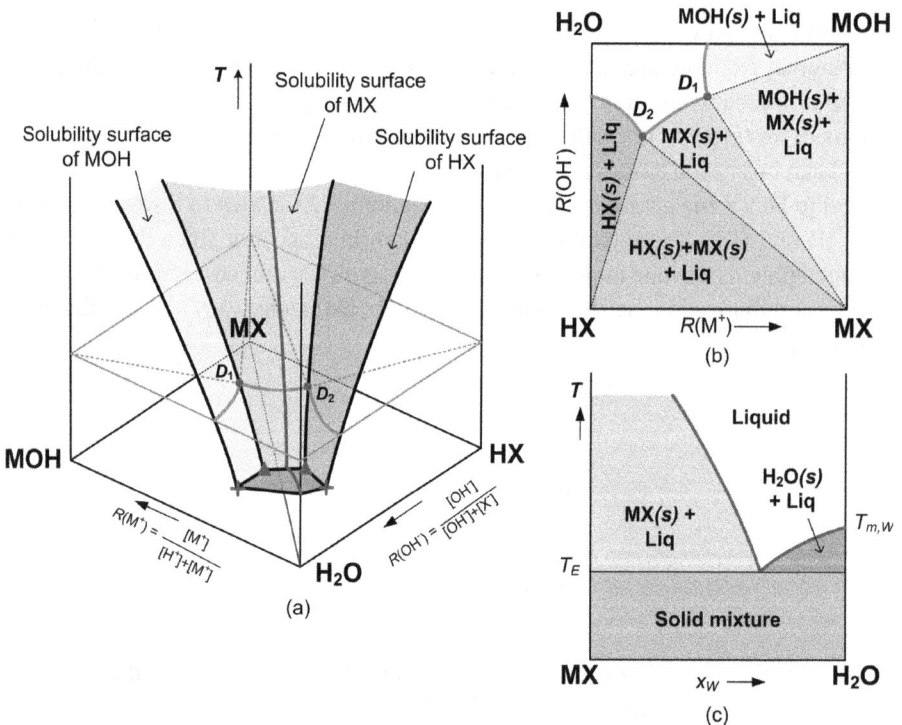

Figure 2.27: SLE phase diagram of an acid-base system: (a) three-dimensional figure, (b) isothermal cut, and (c) diagonal cut.

single salt MX in water with no excess hydrogen or hydroxide ions. This 2D T–x diagram is similar to that of a molecular system, with solubility curves of the salt and water defining the regions in the phase diagram.

Figure 2.28a shows the isothermal cut of an SLE phase diagram involving two acids and one base. H_2A is a dibasic acid because it has two replaceable hydrogen ions. Upon reaction with a monovalent base MOH, either a mono-salt MHA or a di-salt M_2A can be formed. To precipitate a certain acid or salt form, the pH is often controlled by adding the base as well as another acid, such as HX. This results in a multicomponent system involving two simple cations (H^+ and M^+) and three simple anions (A^{2-}, X^-, and OH^-). The three coordinates defining the triangular prism are

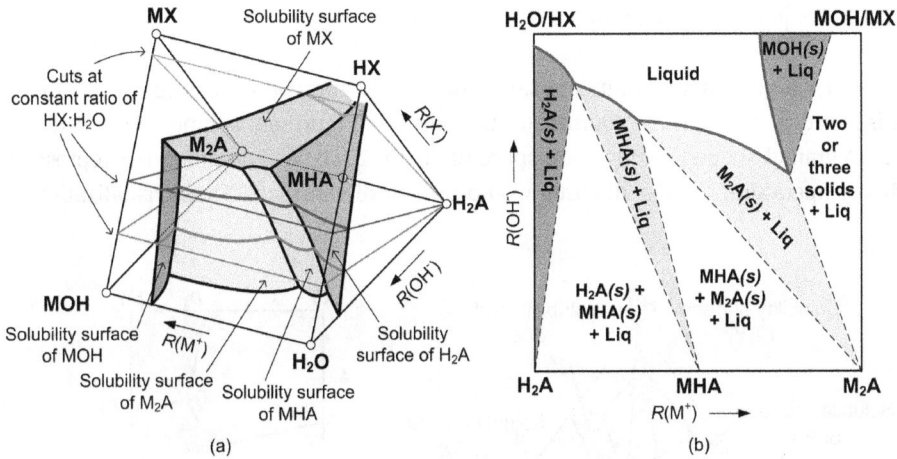

Figure 2.28: Isothermal SLE phase diagram of a system involving two cations and three anions: (a) three-dimensional figure and (b) cut at a constant ratio of HX: H_2O.

$$R(M^+) = \frac{[M^+]}{[H^+] + [M^+]} \tag{2.35}$$

$$R(X^-) = \frac{[X^-]}{[OH^-] + [X^-] + 2[A^{2-}]} \tag{2.36}$$

$$R(OH^-) = \frac{[OH^-]}{[OH^-] + [X^-] + 2[A^{2-}]} \tag{2.37}$$

The various solubility surfaces are marked on Figure 2.28a. The solubility surfaces do not extend all the way to the back face of the prism, at which the phase behavior is irrelevant for process design due to the absence of water.

It is often convenient to take cuts at constant ratios of HX to water as indicated on the triangular prism because it gives a clear cross-sectional view of the regions where a single component of interest (acid H_2A, monosalt MHA, or disalt M_2A) or a mixture of components can be obtained. One such cut is shown in Figure 2.28b, featuring the single saturation regions of H_2A, MHA, and M_2A, with double saturation regions in between. The single saturation region of MOH also appears in the cut. However, since the cut does not include the MOH vertex, from which all tie-lines in that region must originate, those tie-lines do not lie on the cut. On the right side of the figure, there are several regions where two or three solids (including MOH or MX) are in equilibrium with the liquid. They are not detailed in the figure as they are generally irrelevant.

2.3.2 Conjugate salt systems

Consider a quaternary conjugate salt system involving two cations and two anions dissolved in water. Three coordinates are needed to represent composition, resulting in a pyramid-shaped composition space of Figure 2.26b. With another axis representing temperature, the $T–x$ diagram would be four dimensional. Thus, visualization of

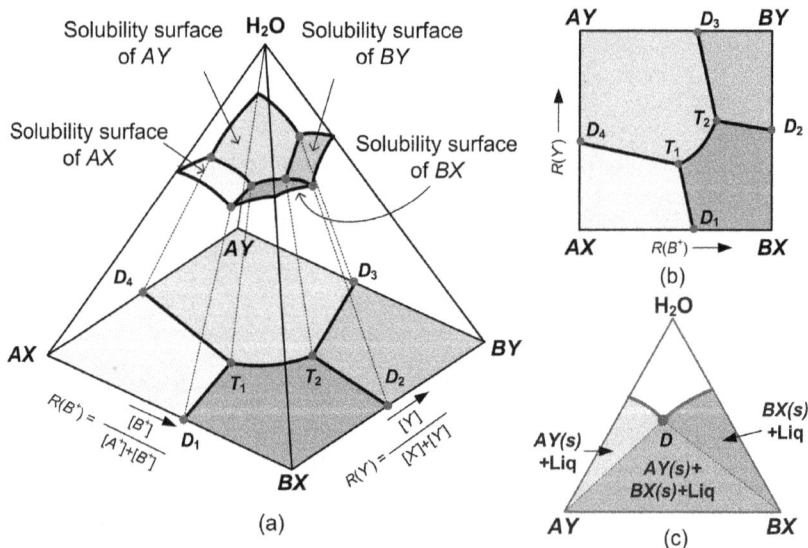

Figure 2.29: Isothermal SLE phase diagram of a conjugate salt system: (a) three-dimensional pyramid; (b) Jänecke projection, and (c) diagonal cut.

the phase diagram often focuses on a 3D isothermal cut such as the one shown in Figure 2.29a.

The solubility surfaces for the four salts intersect each other along the five double saturation curves (D_1T_1, D_2T_2, D_3T_2, D_4T_1, and T_1T_2) and at two triple saturation points (T_1 and T_2). The Jänecke projection (Figure 2.29b) displays the projection of these saturation curves as compartments. For example, the compartment of AX is the four-sided region defined by the AX vertex, D_1, T_1, and D_4. AY and BX are said to be *compatible salts* because their compartments share a common boundary (double saturation curve T_1T_2) so that they can coexist in equilibrium in a solution. On the contrary, AX and BY are *incompatible salts*, as they cannot coexist in equilibrium with a liquid phase [73].

A diagonal cut including the compatible salts (Figure 2.29c) can sufficiently represent the SLE phase behavior of an aqueous solution of compatible salts. This cut sits on top of the line connecting AY and BX, which can be mathematically expressed as

$$R(B^+) + R(Y^-) = 1 \tag{2.38}$$

Since the electrolyte mixture must always be neutral, $[A^+] + [B^+] = [X^-] + [Y^-]$. Hence, eq. (2.38) implies that $[B^+] = [X^-]$ and $[A^+] = [Y^-]$. Consequently, any composition in the triangle can be simply expressed in terms of mole fractions of BX, AY, and water, despite the fact that BX and AY physically exist as ions in solution. In other words, the diagonal cut is essentially equivalent to a ternary molecular system phase diagram. Similarly, a cut along a triangular face of the pyramid is sufficient to represent the SLE behavior of a subsystem consisting of salts with a common ion. For example, consider a cut along BX-BY-H_2O, which corresponds to $[A^+] = 0$. Since B^+ is the only cation in the resulting subsystem, any X^- must originate from BX and any Y^- must originate from BY. Therefore,

$$R(Y^-) = \frac{[Y^-]}{[X^-] + [Y^-]} = \frac{x_{BY}}{x_{BX} + x_{BY}} \tag{2.39}$$

where x denotes mole fraction. As the other coordinate, $R(H_2O)$, is related to the mole fraction of water, it is possible to express compositions in this triangle in terms of mole fractions of BX, BY, and water. Thus, this cut is also equivalent to a ternary molecular system phase diagram.

Example 2.5 – Conjugate salt system with simple SLE behavior

The conjugate salt system Na^+, K^+/Cl^-, NO_3^- is known to only involve simple salts and does not form any hydrate above 0 °C. The solubility data of the pure salts ($NaCl$, KCl, $NaNO_3$, and KNO_3) as well as the double and triple saturation points at different temperatures are available in the literature [34]. Figure 2.30a shows a plot of the isothermal SLE phase diagram of this system at 25 °C, featuring the pure component saturation points (open circles), double saturation points (open squares), triple saturation point (open triangle), and the solubility surfaces of four salts, obtained by connecting the saturation points. In the absence of other data, the connections are drawn as straight lines for illustration purposes, although this may not be the case in reality. Only the top of the pyramid (mol fraction $H_2O \geq 0.6$) is shown for clarity. The Jänecke projection of the phase diagram, which is depicted on the right-hand side, shows the compartments (projected solubility surfaces) of the salts with double saturation curves forming their boundaries. $NaCl$ and KNO_3 are the compatible salts in this system, while $NaNO_3$ and KCl form an incompatible pair as they cannot be saturated together at this temperature.

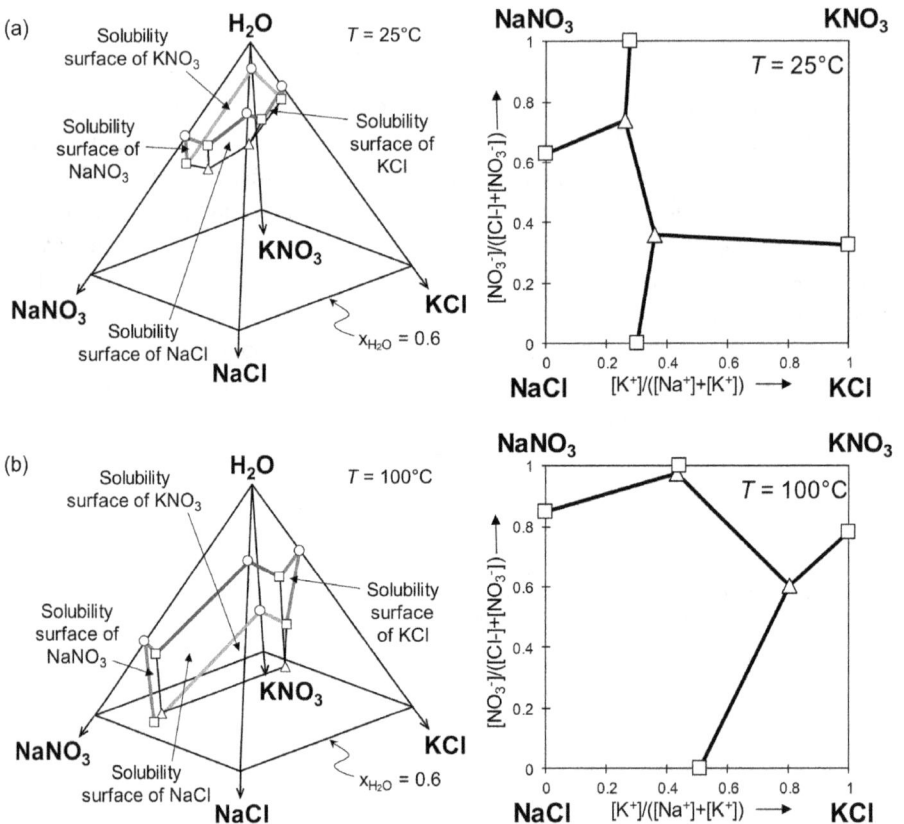

Figure 2.30: Isothermal SLE phase diagram of Na^+, K^+/Cl^-, NO_3^- system (data from [34]) and its Jänecke projection: (a) at 25 °C and (b) at 100 °C.

Figure 2.30b shows the isothermal SLE phase diagram of the same system at 100 °C along with its Jänecke projection. The salt solubilities are higher at this temperature, as reflected by the position of the solubility surfaces; they are farther away from the H_2O vertex compared to their positions at 25 °C. Furthermore, because the relative solubilities of the salts also change with temperature, the size of the compartments on the Jänecke projection also becomes different. For example, the solubility of NaCl only increases slightly from 111.0 mol/1,000 mol H_2O at 25 °C to 120.6 mol/1,000 mol H_2O at 100 °C, while that of KNO_3 increases sharply from 68.3 to 438 mol/1,000 mol H_2O over the same temperature range. As a result, the double saturation curve between the two salts moves closer to KNO_3 as temperature increases, leading to a much larger NaCl compartment and much smaller KNO_3 compartment at 100 °C compared to those at 25 °C. While the NaCl–KNO_3 pair remains compatible at this temperature, this does not have to be case in general; the compatible and incompatible pairs may alter depending on the change in relative solubilities of all salts.

Example 2.6 – Conjugate salt system with complex SLE behavior

The phase behavior of salt systems can be complicated by the presence of compounds such as double salts and hydrates. A classic example of such a system is the sodium-potassium-chloride-sulfate system, whose phase behavior has been thoroughly studied since the early twentieth century [74]. Besides the simple salts NaCl, Na_2SO_4, KCl, and K_2SO_4, it is also possible to crystallize sodium sulfate decahydrate ($Na_2SO_4.10H_2O$), also known as Glauber's salt, and a double salt $NaK_3(SO_4)_2$, commonly known as glaserite. Therefore, the phase diagram features more solubility surfaces, as apparent from the Jänecke projections shown in Figure 2.31.

Figure 2.31a shows the phase behavior at 0 °C, which features compartments of KCl, NaCl, $Na_2SO_4.10H_2O$, K_2SO_4, and the double salt. At this low temperature, there is no compartment for anhydrous Na_2SO_4, as it always forms the decahydrate. Note that the vertex labelled Na_2SO_4 in the Jänecke projection also represents $Na_2SO_4.10H_2O$. The double salt compartment appears in the middle of the diagram and does not include the vertex representing the salt itself. This indicates incongruent dissolution behavior, where the double salt would decompose into single salts before being completely dissolved. At 10 °C (Figure 2.31b) the compartment of the double salt becomes larger and extends all the way to the bottom edge, while the compartment of the hydrate becomes smaller. When the temperature is further increased to 25 °C (Figure 2.31c), the hydrate compartment shrinks further and a new compartment corresponding to anhydrous sodium sulfate begins to appear. At 100 °C (Figure 2.31d), the anhydrous compartment becomes much larger, while the hydrate region completely disappears. It can be seen that the compartment boundaries change with temperature and certain products can only be obtained at a certain range of temperature.

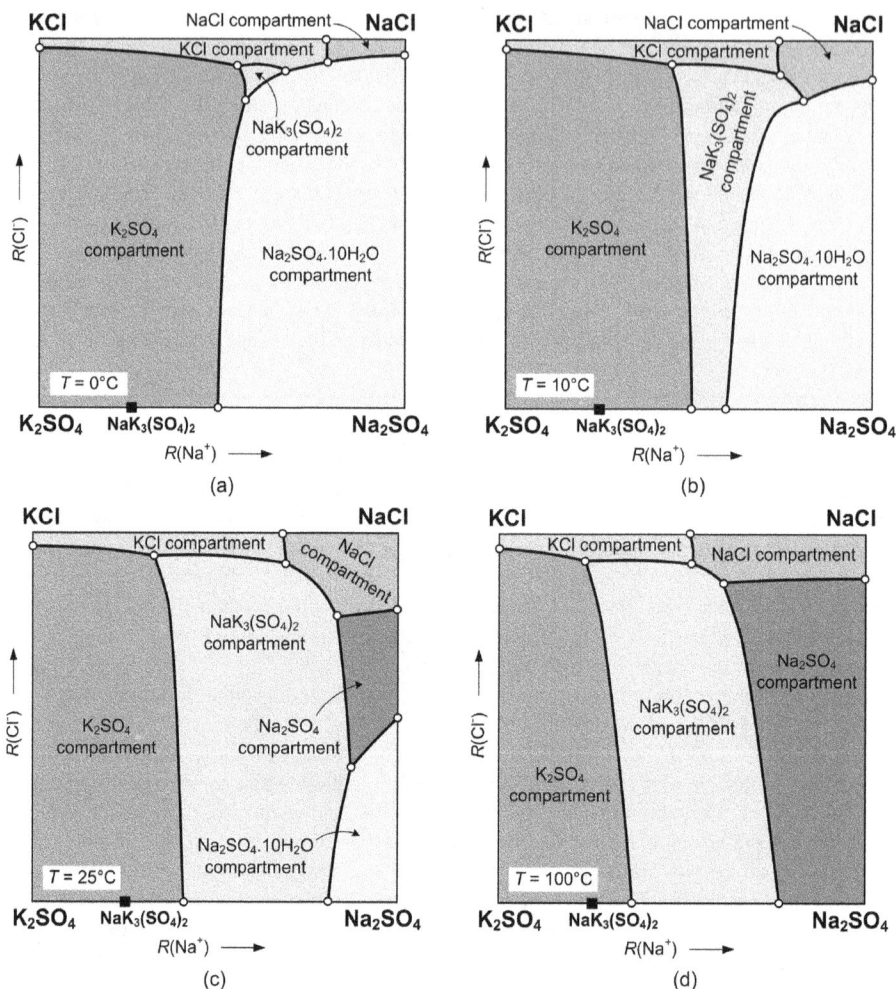

Figure 2.31: Jänecke projection of isothermal SLE phase diagrams of Na^+, K^+/Cl^-, SO_4^{2-} (data from [72]): (a) at 0 °C, (b) at 10 °C, (c) at 25 °C, and (d) at 100 °C.

2.3.3 Multicomponent salt system

When there are more ions involved in the system, the phase diagram quickly becomes high dimensional and visualization becomes increasingly difficult. In this situation, a series of cuts and projections can be used to form a mental picture of the system in its entirety. For example, Kwok et al. [75] presented the isothermal–isobaric phase diagram of the aqueous system comprising Li^+, Na^+, K^+, Mg^{2+}, Cl^-, and SO_4^{2-} ions at 25 °C and 1 atm, which is characteristic of salt lake brines containing relatively high content of lithium. This aqueous system of four cations and two anions can form 21 stable salts in various forms (simple salts, multiple salts, and hydrates) at this temperature and pressure, any of which can crystallize from the aqueous solution depending on the composition, operating temperature, and pressure. Following eqs. (2.27)–(2.29), the phase diagram has five independent coordinates. Correspondingly, its Jänecke projection, which omits the solvent coordinate, is four dimensional. Several cuts and projections of this Jänecke projection are considered in the forth-coming paragraphs.

A 2D cut at $R(K^+) = 0$ and $R(Mg^{2+}) = 0$, shown in Figure 2.32a, is simply the Jänecke projection of the phase diagram of a subsystem consisting of Li^+, Na^+, Cl^-, and SO_4^{2-} ions. This cut features the compartments of various single salts, double salts, and hydrates as marked on the figure. Due to its high solubility relative to other salts, the compartment of $LiCl.H_2O$ only consists of a very small area near the LiCl ver-tex. The area is so small that $LiCl.H_2O$ is labelled but no area is shown in Figure 2.32a. Figure 2.32b shows another 2D cut, this time at $R(Na^+) = 0$ and $R(Mg^{2+}) = 0$, which signifies the Jänecke projection of the phase diagram of a subsystem consisting of Li^+, K^+, Cl^-, and SO_4^{2-} ions. Compartments of various salts and hydrates can also be ob-served in this figure.

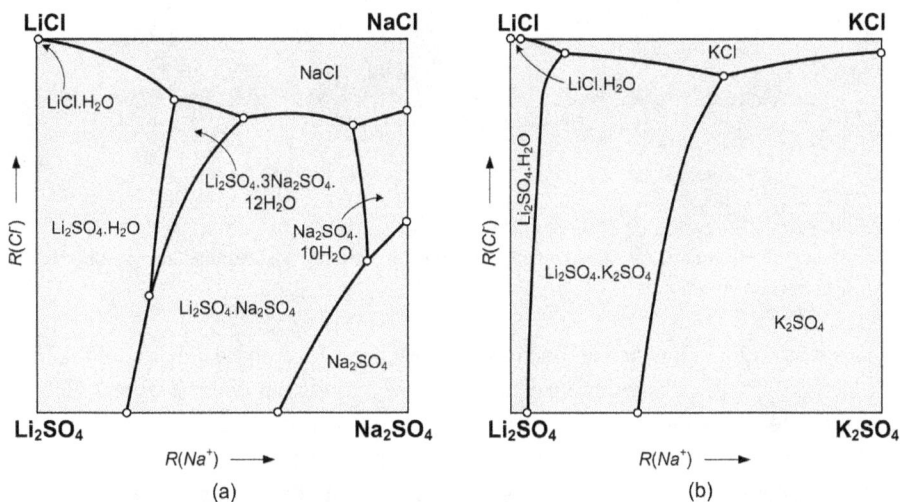

Figure 2.32: Jänecke projection of isothermal-isobaric SLE phase diagrams at 25 °C and 1 atm of lithium salt systems [75]: (a) Li^+, Na^+/Cl^-, SO_4^{2-} subsystem and (b) Li^+, K^+/Cl^-, SO_4^{2-} subsystem.

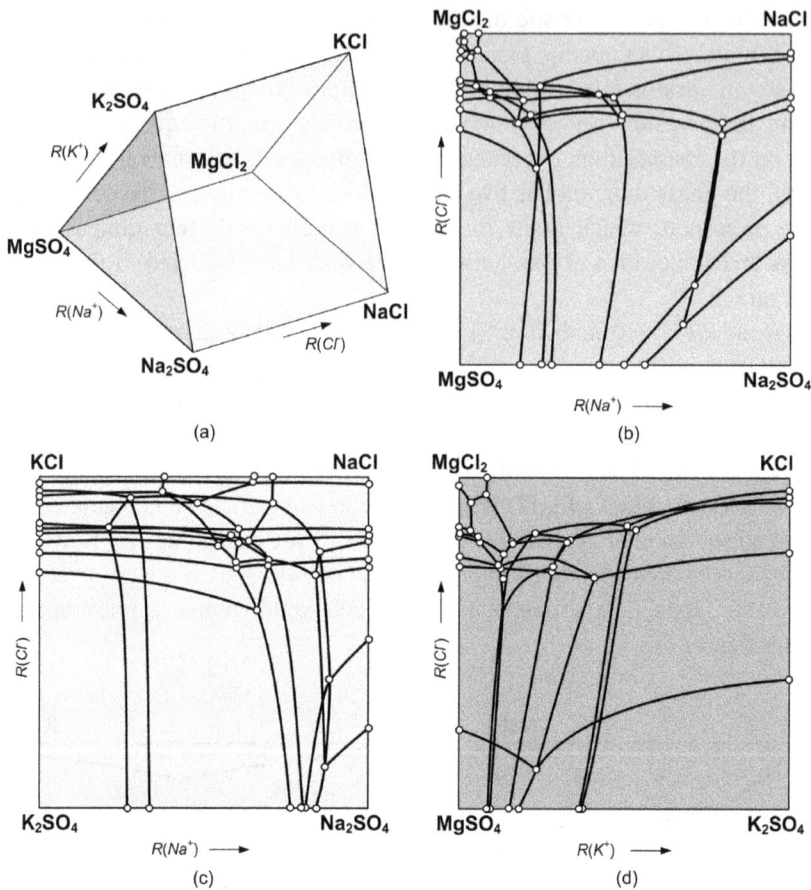

Figure 2.33: Jänecke projection of isothermal SLE phase diagrams of Na^+, K^+, Mg^{2+}/Cl^-, SO_4^{2-} subsystem [72]: (a) three-dimensional prism, (b) projection ignoring K^+, (c) projection ignoring Mg^{2+}, and (d) projection ignoring Na^+.

A 3D cut at $R(Li^+) = 0$, which is the Jänecke projection of the phase diagram of the Na^+, K^+, Mg^{2+}/Cl^-, SO_4^{2-} subsystem, takes the shape of a triangular prism (Figure 2.33a). To visualize the various saturation varieties that are present in the interior of the prism, three projections are presented. In plotting these projections, compositions of both anions are considered, but the composition of one of the three cations is ignored: K^+ in Figure 2.33b, Mg^{2+} in Figure 2.33c, and Na^+ in Figure 2.33d. The resulting images are basically the projection of the interior of the phase diagram onto the three rectangular faces of the prism. Although it is difficult to see the overlapping compartments in the projections, each compartment can be located by identifying the vertices and edges that form its boundaries. Together, these projections provide a mental picture of the interior of the prism in Figure 2.33a.

2.4 Summary

The basics of SLE phase behavior of relevance to conceptual design of crystallization processes have been discussed in this chapter. Of particular importance is the
phase diagram that shows the presence of solid or liquid phase, compounds, hydrates, and so on under different conditions of temperature, pressure, and composition. For separation of a multicomponent mixture, the phase diagram is essentially
a navigation map that guides the process designer to the proper region where the
desired component can be recovered. The way in which these phase diagrams should
be plotted depends on whether the components are molecular or ionic and whether
reactions are involved. Well-established techniques such as polythermal projections
and isothermal cuts can be used to facilitate the visualization of these phase diagrams. As more components are involved, the phase diagram becomes high dimensional. Visualization of such phase diagrams can be achieved using the generalized
projections and cuts discussed in this chapter. By visualizing the phase behavior in
lower dimensions, a mental picture of the phase diagram in high dimensions can be
formed. A large number of terms have been introduced. The key terms related to SLE
phase behavior are summarized in Table 2.3, while the special terminologies that
have been used throughout the discussion on high-dimensional SLE phase diagrams
are summarized in Table 2.4.

Table 2.3: Glossary of key terms related to SLE phase behavior.

Electrolyte system: A mixture consisting of electrolytes that partially or completely dissociate into
ions in solution.

    Conjugate salt system: A quaternary electrolyte system formed by two salts that have no
    common ion and inter-react in aqueous solution.

    Compatible salt pair: The salt pair of a conjugate salt system that can coexist in equilibrium in
    a solution.

    Incompatible salt pair: The salt pair of a conjugate salt system that cannot coexist in
    equilibrium in a solution.

Eutectic: A eutectic of order n (**binary eutectic** if $n = 2$, **ternary eutectic** if $n = 3$, and so on) is a
liquid mixture of n components, all of which are saturated.

Melting point: The temperature at which the transformation from solid phase to liquid phase occurs.

    Congruent melting point: The temperature at which a solid compound melts to form a liquid
    phase of the same composition.

    Incongruent melting point: The temperature at which a solid compound decomposes into
    another solid (pure component or another solid compound) and a liquid phase of a different
    composition.

Molecular system: A mixture consisting only of molecules that do not dissociate to form ions.

Table 2.3 (continued)

Peritectic: A type of SLE phase behavior characterized by incongruent melting of a solid compound, during which the compound decomposes into another solid and a liquid.

Simple eutectic behavior: The type of SLE behavior for which the solid phase is always pure regardless of the liquid composition.

Solid solution behavior: The type of SLE behavior featuring region(s) where a solid solution can be found in equilibrium with a liquid phase.
    Liquidus: The upper boundary of a solid+liquid region in the $T–x$ phase diagram of a system with solid solution behavior, adjacent to the unsaturated liquid region.
    Solidus: The lower boundary of a solid+liquid region in the $T–x$ phase diagram of a system with solid solution behavior, adjacent to the solid region.
    Solid solution: A solid mixture of multiple components where molecules of different components are interspersed in the same crystal lattice.

Triple point: The temperature and pressure at which solid, liquid, and gas phases coexist and are in equilibrium with each other.

Table 2.4: Glossary of special terms used in high-dimensional SLE phase diagrams.

Canonical coordinates: A set of coordinates used to plot a projective subspace, which appears naturally once the projection rays have been defined.
    Transformed mole fraction coordinates: A set of canonical coordinates for plotting the composition of reactive mixtures in a reaction-invariant projection.

Compartment: A subspace in a projection of an SLE phase diagram within which a component can be crystallized in pure form.

Cut: A picture obtained by viewing a part of the system and ignoring the rest. An m-dimensional cut of an n-dimensional object is an intersection between the object with a variety of dimension m.
    Isoplethal cut: A cut of an isobaric phase diagram or its projection taken at a constant composition of one or more components.
    Isothermal cut: A cut of an isobaric temperature-composition phase diagram taken at a constant temperature.

Digraph: A discrete mathematical model representing the phase diagram of a system consisting of a set of **vertices** connected to each other by a set of **edges,** where each edge has a direction from an initial vertex to a terminal vertex.
    Adjacency matrix: A matrix whose row and column headings correspond to the vertices of a digraph; its elements are either 1 (if an edge is present between a pair of vertices) or 0 (if an edge is absent).
    Saturation variety matrix: A matrix whose row and column headings correspond to the saturation varieties and vertices of a digraph, respectively; its elements are either 1 (if the vertex lies on the corresponding saturation variety) or 0 (if it does not).

Table 2.4 (continued)

Projection: An image obtained by shining light on an object. An m-dimensional projection of an n-dimensional object is the image of the object on an m-dimensional projective subspace (which can be a plane or any other geometrical feature). The line connecting the light source, a point in the object, and the image of the point in the projection is called a *projection ray*.

    Central projection: A projection in which all projection rays pass through a point called the center of projection.

    Jänecke projection: A central projection with a vertex representing a pure component in the system (usually the solvent) being the center of projection.

    Parallel projection: A projection in which all projection rays are parallel.

    Polythermal projection: A parallel projection along the temperature axis.

    Reaction-invariant projection: A special projection along the stoichiometric lines resulting in transformed mole fractions that have the same numerical values regardless of the extent of reaction.

Stoichiometric lines: Lines on the phase diagram describing composition changes due to a reaction.

Tie-line: The line connecting the compositions of two different phases (such as a solid and a liquid) that are in equilibrium with each other.

Variety: A general term for an m-dimensional geometrical object in space. The variety is a *point* if $m = 0$, a *line, curve* or *trough* if $m = 1$, a *plane* or *surface* if $m = 2$, a *volume* if $m = 3$, or a *hypervolume* if $m > 3$.

    Saturation variety: A variety on the phase diagram, bounded by vertices and edges, inside which one or more components are saturated. A saturation variety of order n (*single saturation variety* if $n = 1$, *double saturation variety* if $n = 2$, and so on) has n saturated components.

Exercises

2.1. The feed to a crystallizer in an adipic acid plant consists of adipic acid, succinic acid, glutaric acid, and a solvent (which can be considered as a single component). Based on Gibbs phase rule, what would be the dimensionality of the isobaric SLE phase diagram of the system? What shape would the phase diagram take?

2.2. The binary system o-nitrochlorobenzene and p-dibromobenzene is known to exhibit simple eutectic behavior. The melting points of the pure components are 33 °C and 86 °C, respectively. The eutectic temperature is 25 °C and the eutectic composition is 21 wt% p-dibromobenzene. Sketch the SLE phase diagram of this binary system based on the given information.

2.3. The composition of a ternary system can be plotted on a composition triangle. On the composition triangle for methanol/ethanol/water system shown in Figure 2.34, locate mixtures with the following compositions:
a. Mixture A: 30% methanol, 70% water
b. Mixture B: 90% ethanol, 10% water
c. Mixture C: 20% methanol, 60% ethanol, 20% water
d. Mixture D: 40% methanol, 30% ethanol, 30% water

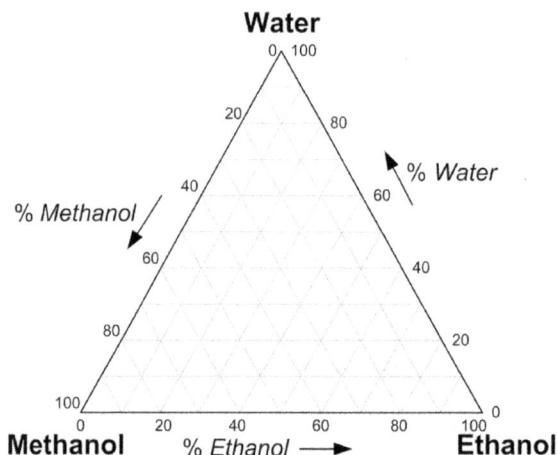

Figure 2.34: Composition triangle for methanol/ethanol/water system (problem 2.3).

2.4. The eutectic points for the ternary system p-xylene/m-xylene/o-xylene are given as follows:

Eutectic point	E1	E2	E3	E4
Temperature (K)	220.8	238.32	211.84	209.59
p-Xylene (mol fraction)	0.118	0.235	0	0.072
m-Xylene (mol fraction)	0.882	0	0.675	0.629
o-Xylene (mol fraction)	0	0.765	0.325	0.299

Based on this information, sketch the polythermal projection of the SLE phase diagram of the ternary system and identify the compartments of p-, m-, and o-xylene.

2.5. Consider an isobaric, polythermal SLE phase diagram of a five-component molecular system with simple eutectic behavior, a projection of which appears in Figure 2.20. The composition of the eutectic points as well as their adjacency information is given below.

Type	Eutectic	x_A	x_B	x_C	x_D	x_E	Adjacent to
Binary	AB	0.757	0.242	0	0	0	A, B, ABC, ABD, ABE
	AC	0.670	0	0.330	0	0	A, C, ABC, ACD, ACE
	AD	0.246	0	0	0.754	0	A, D, ABD, ACD, ADE
	AE	0.078	0	0	0	0.922	A, E, ABE, ACE, ADE
	BC	0	0.395	0.605	0	0	B, C, ABC, BCD, BCE
	BD	0	0.080	0	0.920	0	B, D, ABD, BCD, BDE
	BE	0	0.016	0	0	0.984	B, E, ABE, BCE, BDE
	CD	0	0	0.124	0.876	0	C, D, ACD, BCD, CDE
	CE	0	0	0.028	0	0.972	C, E, ACE, BCE, CDE
	DE	0	0	0	0.231	0.770	D, E, ADE, BDE, CDE
Ternary	ABC	0.564	0.168	0.268	0	0	AB, AC, BC, ABCD, ABCE
	ABD	0.231	0.055	0	0.715	0	AB, AD, BD, ABCD, ABDE
	ABE	0.076	0.014	0	0	0.910	AB, AE, BE, ABCE, ABDE
	ACD	0.221	0	0.088	0.690	0	AC, AD, CD, ABCD, ACDE
	ACE	0.075	0	0.024	0.000	0.900	AC, AE, CE, ABCE, ACDE
	ADE	0.056	0	0	0.216	0.728	AD, AE, DE, ABDE, ACDE
	BCD	0	0.067	0.113	0.820	0	BC, BD, CD, ABCD, BCDE
	BCE	0	0.015	0.027	0	0.958	BC, BE, CE, ABCE, BCDE
	BDE	0	0.010	0	0.228	0.762	BD, BE, DE, ABDE, BCDE
	CDE	0	0	0.018	0.226	0.756	CD, CE, DE, ACDE, BCDE
Quaternary	ABCD	0.209	0.049	0.083	0.659	0	ABC, ABD, ACD, ABCDE
	ABCE	0.074	0.013	0.024	0	0.889	ABC, ABD, ACD, ABCDE
	ABDE	0.055	0.009	0	0.214	0.722	ABD, ABE, ADE, ABCDE
	ACDE	0.055	0	0.017	0.212	0.717	ACD, ACE, ADE, ABCDE
	BCDE	0	0.010	0.018	0.223	0.749	BCD, BCE, BDE, ABCDE
Quinary	ABCDE	0.054	0.009	0.017	0.210	0.711	ABCD, ABCE, ABDE, ACDE, BCDE

a. How many other 2D projections are needed to completely represent the SLE phase behavior of this system?

b. Plot the 2D projection obtained by taking B and C as reference components.

2.6. To design a process to obtain magnesium and calcium compounds from calcined dolomite (a mixture of MgO and CaO), the SLE phase diagram of a system involving MgO (A), CaO (B), CO_2 (C), $MgCO_3$ (D), $CaCO_3$ (E), and water (S), and two reactions ($A+C \rightarrow D$ and $B+C \rightarrow E$) is to be constructed.

a. What is the dimension of the isobaric, isothermal, and reaction-invariant phase diagram?

b. Taking D and E as the reference components, determine the canonical coordinates used to plot the phase diagram.

2.7. The composition of an aqueous solution containing K^+, Na^+, Cl^-, and SO_4^{2-} can be represented on a Jänecke projection, with the coordinate system shown in Figure 2.35.

a. Provide the appropriate expressions for $R(Na^+)$ and $R(SO_4^{2-})$ in terms of ionic concentrations.

b. A mixture is obtained by dissolving 5 mol KCl, 2 mol NaCl, and 1.5 mol K_2SO_4 in water. Locate this mixture on the Jänecke projection.

c. Can the same mixture be obtained by dissolving KCl, Na_2SO_4, and K_2SO_4 instead? If yes, how many moles of each salt should be dissolved?

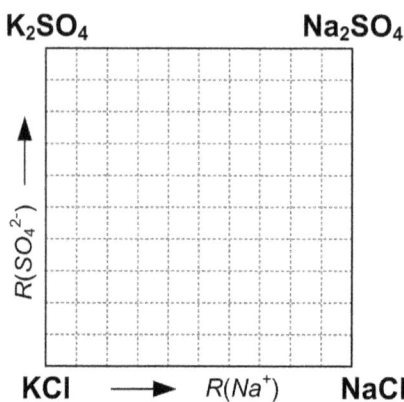

Figure 2.35: Jänecke projection for K^+, Na^+/Cl^-, SO_4^{2-} system (problem 2.7).

2.8. The SLE data of the aqueous system containing Na^+, NH_4^+, NO_3^-, and HCO_3^- at 30 °C, as reported in the International Critical Tables [34] are given as follows:

Solid phase	Concentration in liquid phase, mol/1,000 mol H_2O			
	$NaNO_3$	$NaHCO_3$	NH_4NO_3	NH_4HCO_3
$NaNO_3$	203.4			
$NaHCO_3$		23.6		
NH_4NO_3			523.1	
NH_4HCO_3				61.38
$NaNO_3+NaHCO_3$	201.4	3.78		
$NaNO_3+NH_4NO_3$	186.8		496.4	
$NaHCO_3+NH_4HCO_3$		15.5		58.14
$NH_4NO_3+NH_4HCO_3$			521.5	28.6
$NaHCO_3+NH_4NO_3+NH_4HCO_3$	45.2	24.6	538.9	
$NaNO_3+NaHCO_3+NH_4NO_3$	173.7	11.9	505.3	

a. Plot the Jänecke projection of the isothermal SLE phase diagram of this system. Assume straight lines connecting the saturation points.

b. Based on the phase diagram, identify the compatible and incompatible salt pairs.

3 Thermodynamic-Based Conceptual Design of Crystallization Processes

3.1 Movements in Composition Space

Phase diagrams are a useful visualization tool because both the phase behavior and the basic operations in a chemical process can be represented together in composition space. There are three types of basic operations in crystallization processes: changing temperature, changing pressure, or changing composition (Figure 3.1). As mentioned previously, SLE behavior is normally insensitive to pressure. For this reason, pressure swing is seldom used in crystallization processes. Temperature swing can either be cooling or heating. Composition change can be further classified into addition or removal of a component, which can be a solvent or a mass separating agent (MSA), and simple stream combination or splitting. An MSA is an external component added to the system to effect separation, such as an antisolvent or precipitating reactant. Stream combination is simply mixing two or more streams in the plant. Stream splitting refers to the physical separation of a stream, such as the use of a filter for a solid–liquid stream or a flash tank for a vapor–liquid stream. All these operations can be represented as movements on the phase diagram.

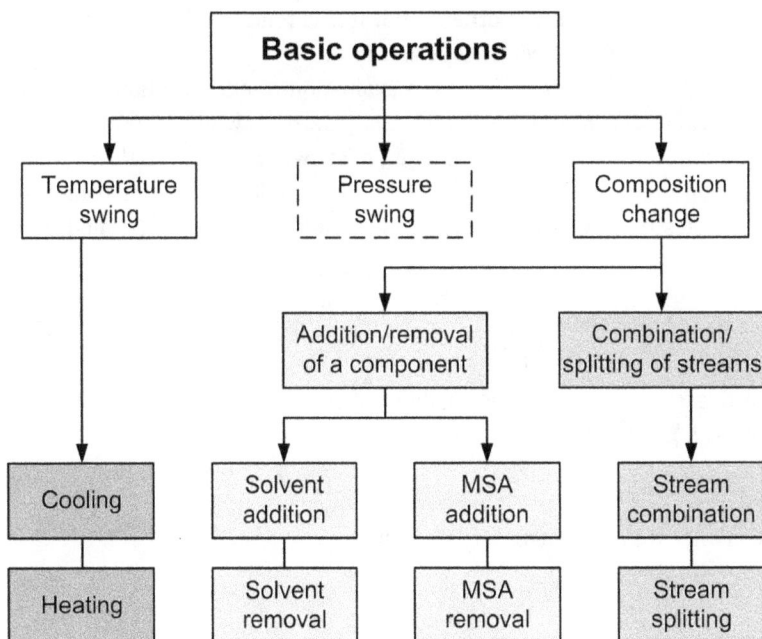

Figure 3.1: Basic operations in crystallization processes.

https://doi.org/10.1515/9781501519901-003

3.1.1 Lever rule

One important concept in dealing with phase diagrams and movements in composition space is the so-called lever rule. The rule enables the determination of the relative amounts of two masses A and B in their mixture M, from the phase diagram. Appropriately, the name "lever rule" comes from the physics principle of a lever. If A and B were placed on the ends of a lever with the pivot point (or fulcrum) at the composition of mixture M, as illustrated in Figure 3.2, the lever would balance. Mathematically, this requires that the weight or amount of A multiplied by the distance, a, equals the amount of B multiplied by the distance, b. In other words, the location of M depends on the relative amounts of A and B.

**Weight of A x a =
Weight of B x b**
                         Figure 3.2: Illustration of the lever rule.

In the context of phase diagrams, it is more important to understand the qualitative implication of the lever rule. If more B is added to the mixture, point M moves toward B as it gets richer in B. On the other hand, if B is removed from the mixture, point M moves away from B as it becomes leaner in B.

    The lever rule is directly applicable in the evaluation of an SLE phase diagram. Consider a mixture M with composition x_M and temperature T_c, which is located inside the two-phase region $A(s)+Liq$ in the phase diagram shown in Figure 3.3a. At equilibrium, mixture M splits into solid S, which is pure A, and liquid L with composition x_L. The lever rule dictates that the mass ratio of solid to liquid phases in the mixture is equal to the ratio of segment lengths ML to SM:

$$\frac{F_S}{F_L} = \frac{ML}{SM} = \frac{x_L - x_M}{x_M} \tag{3.1}$$

The validity of this relationship can be easily proven by performing a simple mass balance. The overall mass balance is given by

$$F_S + F_L = F_M \tag{3.2}$$

and the mass balance for component B is given by

$$F_S x_S + F_L x_L = F_M x_M \tag{3.3}$$

Since $x_S=0$, substituting eq. (3.2) into eq. (3.3) gives

$$F_L x_L = (F_S + F_L) x_M \tag{3.4}$$

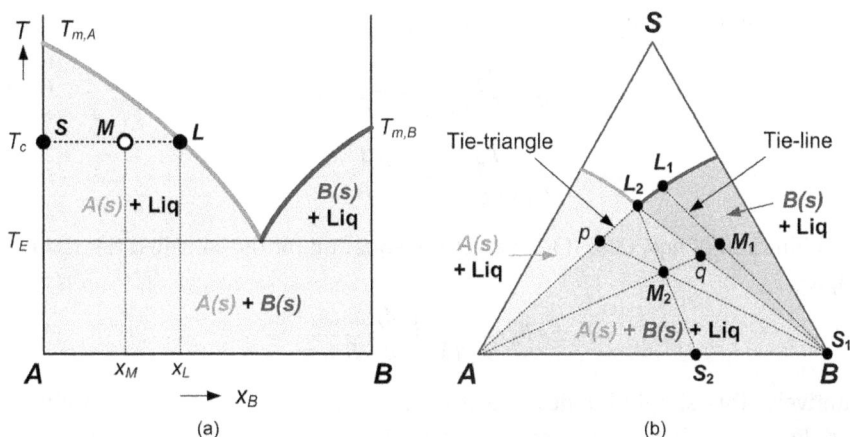

Figure 3.3: Application of lever rule on SLE phase diagrams: (a) binary system and (b) ternary system.

which is equivalent to eq. (3.1). The more qualitative interpretation is that the closer M is to S, the more solid is present in the mixture. Conversely, if M is closer to L, the mixture contains more liquid, or the yield of crystals is lower.

Figure 3.3b illustrates the application of lever rule in the SLE phase diagram of a ternary system. Point M_1 is located inside the two-phase region $B(s)+$Liq. At temperature T_c, this mixture splits into a solid phase (pure B) and a liquid phase with a composition given by point L_1 on the saturation curve of B. By the lever rule:

$$\frac{F_S}{F_{L_1}} = \frac{M_1 L_1}{BM_1} \tag{3.5}$$

Meanwhile, point M_2 is inside the three-phase region $A(s)+B(s)+$liquid. At T_c, this mixture splits into two solid phases (A and B) and a liquid having a composition of the double saturation point L_2. Application of the lever rule gives

$$\frac{F_{S_2}}{F_{L_2}} = \frac{M_2 L_2}{S_2 M_2} \tag{3.6}$$

$$\frac{F_A}{F_B} = \frac{S_2 B}{A S_2} \tag{3.7}$$

Combining eqs. (3.6) and (3.7),

$$\frac{F_{L_2}}{F_A + F_B + F_{L_2}} = \frac{S_2 M_2}{S_2 L_2} \tag{3.8}$$

In a similar manner, it can be obtained that

$$\frac{F_A}{F_A + F_B + F_{L_2}} = \frac{qM_2}{qA} \tag{3.9}$$

$$\frac{F_B}{F_A + F_B + F_{L_2}} = \frac{pM_2}{pB} \tag{3.10}$$

The combination of eqs. (3.8)–(3.10) gives the equation for the so-called tie-triangle principle:

$$F_A{:}F_B{:}F_{L_2} = \frac{qM_2}{qA} : \frac{pM_2}{pB} : \frac{S_2M_2}{S_2L_2} \tag{3.11}$$

Qualitatively, this simply implies that if M_2 is closer to A, there is more solid A; if M_2 is closer to B, there is more solid B; and if M_2 is closer to L, there is more liquid in the mixture.

3.1.2 Representation of basic operations on SLE phase diagram

The representation of basic operations in a cooling crystallization process as movements on the phase diagram is illustrated in Figure 3.4a. The starting point is an unsaturated solution with a composition represented by point 1. Cooling causes the temperature to decrease, but does not change the composition. Thus, it is represented on the phase diagram by a vertical movement, parallel to the temperature axis, toward point 2. At this point, which lies on the saturation curve of A, the solution becomes saturated in A. Further cooling brings the composition into the two-

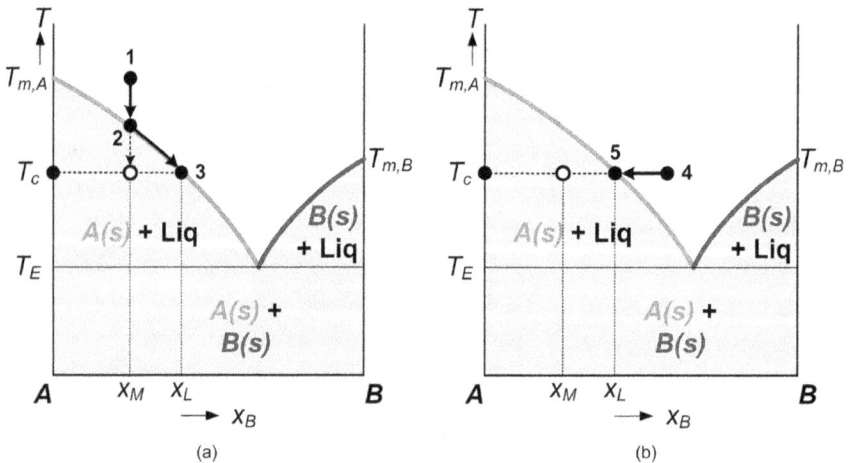

Figure 3.4: Representation of basic operations as movement on SLE phase diagrams: (a) cooling crystallization and (b) evaporative crystallization.

phase region, and solid A would crystallize out. As A leaves the solution, the liquid composition moves along the saturation curve from point 2 to point 3. Notice that as temperature decreases, the distance between the open circle (which represents the overall mixture) and the point representing the liquid increases. In accordance with the lever rule, more solids would come out.

The operations involved in an evaporative crystallization process can also be readily represented as movements on the phase diagram. Again, the starting point is an unsaturated solution, the composition of which is given by point 4 in Figure 3.4b. If B is evaporated from the solution under constant temperature, the composition would move horizontally to the left (closer to A), because the remaining solution becomes richer in A. When the composition reaches point 5, the solution becomes saturated in A. Further evaporation would cause the composition to move into the two-phase region and induces the crystallization of solid A. As more B is removed by evaporation, the overall composition (indicated by the open circle) would move further inside the two-phase region and more A crystallizes out, while the liquid composition stays at point 5.

The representation of basic operations as movements in composition space also applies to high-dimensional phase diagrams. Consider an unsaturated solution containing three components A, B, and S, which is represented by point 1 in the SLE phase diagram in Figure 3.5a. Cooling of this solution is represented by a vertical movement down to point 1a in the three-dimensional prism. On the polythermal projection, which is shown in Figure 3.5b, points 1 and 1a appear as a single point, since they have the same composition. The projection also gives a clear view that

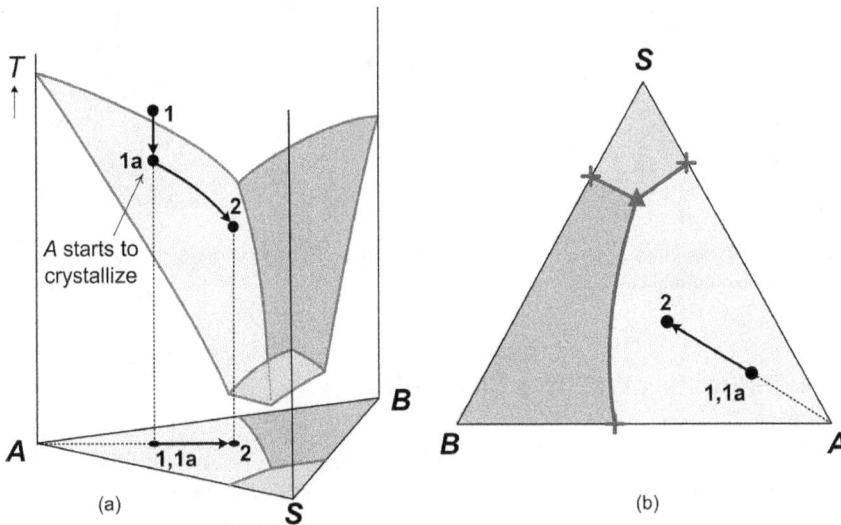

Figure 3.5: Representation of polythermal movements on ternary SLE phase diagram: (a) *T–x* diagram and (b) polythermal projection.

point 1 lies within compartment A, which implies that cooling beyond point 1a would lead to crystallization of A.

Since A is removed from the solution during crystallization, the liquid composition has to move away from A, as indicated by the movement from point 1a to point 2 in the polythermal projection. As shown in Figure 3.5a, these movements are in the direction of decreasing temperature.

Representing the basic operations on an isothermal cut is also straightforward, as shown in Figure 3.6. Again, the starting point is an unsaturated solution, represented by point 1. If S is removed from the solution, such as by evaporation, the composition moves away from S, toward point 2. Since the proportion of A and B in the solution remains unchanged during evaporation, the correct direction of the movement is not vertical, but along the extension of a line connecting vertex S and point 1. At point 2, the solution becomes saturated with A, and further removal of S would cause A to crystallize out. The solution composition, which would be enriched in B, as both S and A leave the solution, would follow the saturation curve of A toward point 3.

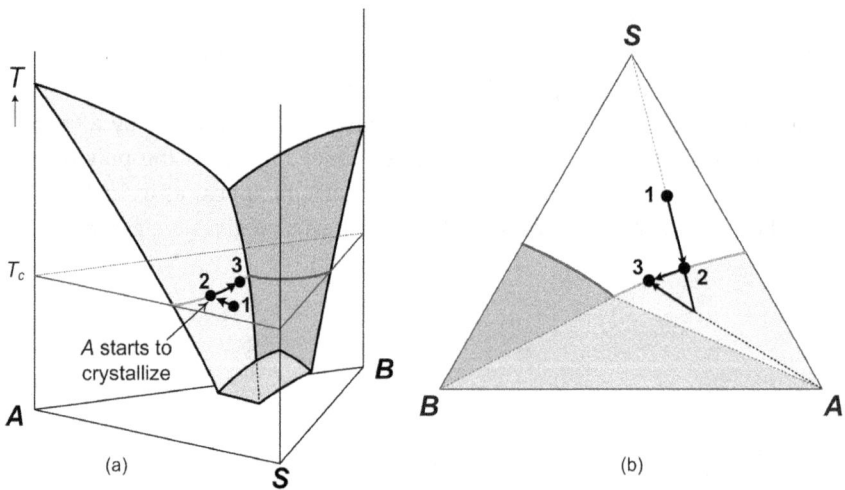

Figure 3.6: Representation of isothermal movements on ternary SLE phase diagram: (a) T–x diagram and (b) isothermal cut at T_c.

3.2 Maximum Recovery of a Pure Solid

A typical application of SLE in the conceptual design of crystallization processes is the determination of the maximum recovery of a pure solid. The presence of eutectics (or multiple saturation varieties in general) impose a limit on how much of a pure component can be crystallized before another component also starts to crystallize. To maximize the per-pass yield in a crystallization process while keeping the solid pure, identifying the conditions corresponding to the maximum recovery of a pure product is crucial.

Consider again the cooling crystallization process in a binary system, as shown in Figure 3.7. Starting with an A-rich mixture represented by point 1, cooling beyond point 3 results in more crystallization of pure A. However, once the liquid composition reaches the eutectic point E, further cooling would cause B to crystallize together with A, as the mixture enters the solid mixture region of $A(s)+B(s)$. Therefore, point E represents a *thermodynamic boundary* that defines the maximum recovery of pure A. The maximum recovery Y_A is given by

$$Y_A = \frac{F_S}{F_1 x_{A,1}} \tag{3.12}$$

where F_1 and F_S are the mass of feed mixture (stream 1) and the mass of solid A obtained at maximum recovery, respectively, and $x_{A,1}$ is the mass fraction of A in the feed.

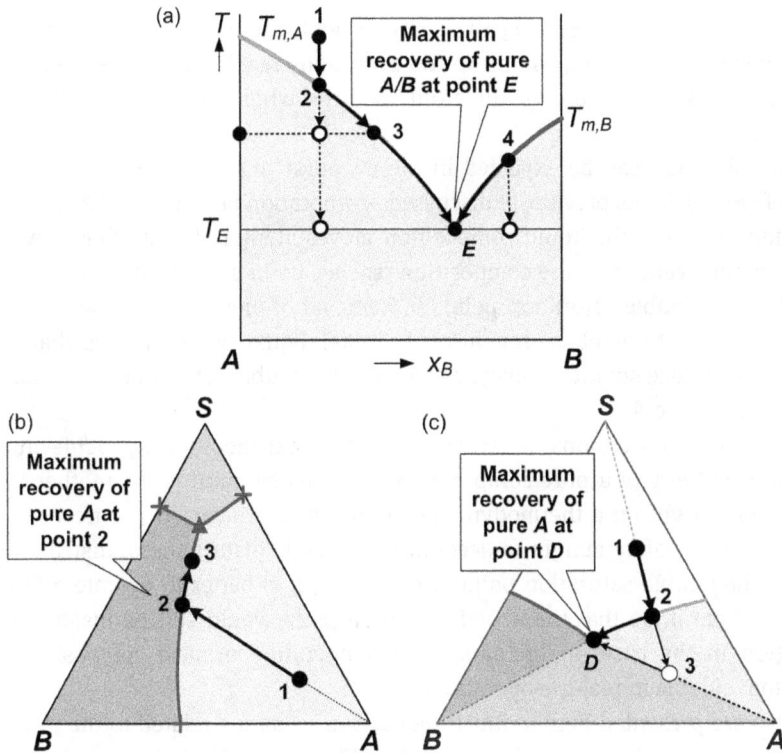

Figure 3.7: Concept of thermodynamic boundary: (a) binary system, (b) polythermal movements in ternary system, and (c) isothermal movements in ternary system.

Since the liquid composition at maximum recovery would be the eutectic composition, $x_{A,E}$, lever rule gives

$$\frac{F_S}{F_1 - F_S} = \frac{x_{A,1} - x_{A,E}}{1 - x_{A,1}} \tag{3.13}$$

It follows that

$$Y_A = \frac{x_{A,1} - x_{A,E}}{x_{A,1}(1 - x_{A,E})} \tag{3.14}$$

Similarly, if the starting point is a B-rich mixture as represented by point 4, cooling leads to crystallization of pure B. Once again, the eutectic point limits the recovery of pure B, as further cooling leads to co-crystallization of A.

For a ternary system, starting with a composition shown by point 1 in Figure 3.7b, cooling leads to crystallization of pure A, until the composition reaches point 2 on the eutectic trough, which is the boundary with compartment B. If cooling is continued beyond point 2, additional crystallization of A would move the composition into compartment B. Consequently, B crystallizes out together with A, and the liquid composition would move along the compartment boundary toward point 3. Upon further cooling, the composition would eventually reach the ternary eutectic point, at which S would also crystallize out, and the whole mixture would turn into a block of solid.

A similar situation can be expected in the evaporative crystallization process depicted in Figure 3.7c. As previously discussed, evaporation beyond point 2 causes crystallization of A, and the liquid composition moves along the saturation curve. Eventually, as the overall mixture composition reaches point 3, the solution composition reaches the double saturation point, D. Removal of more S would cause the mixture to enter the three-phase region of $A(s)+B(s)+$liquid, which means that A and B would crystallize simultaneously. Therefore, the double saturation point limits the recovery of pure A.

From the above discussions, it can be concluded that theoretically, achieving the maximum recovery of a pure component would involve putting the mother liquor composition right on a thermodynamic boundary. However, in practice, it is necessary to have a safety margin to account for possible disturbances. Instead of operating at the double saturation point temperature, it is better to operate a few degrees away from it, so that the actual operation point would still be inside the desired region in the case of fluctuations in temperature or feed composition, which are unavoidable in real-life operation.

Examples are presented here to illustrate various scenarios related to the issue of maximum recovery of a pure solid, where the SLE phase behavior is used in the conceptual design of the process.

Example 3.1 – Crystallization of *p*-xylene
Para-xylene (PX) is one of the precursors for the production of polyethylene terephthalate (PET), which is widely used as polyester fiber, film, and resin for various applications. The conventional process to produce PX involves methylation of benzene and toluene, followed by separation of a mixture of xylene isomers, which is one of the earliest known applications of crystallization as a separation process for organic commodity chemicals. Typically, the PX content in a mixed feed to the crystallizer is only about 20–25 wt%, with the rest being *m*-xylene (MX) and *o*-xylene (OX), along with some toluene and ethylbenzene.

The SLE phase diagram of the PX/MX/OX ternary system is depicted in Figure 3.8a. Based on available SLE data in the literature [76], this system exhibits near-ideal behavior. The feed composition, assumed to contain only the three xylene isomers (point 1), is also plotted on the diagram. Crystallization of PX upon cooling of this feed mixture causes the liquid composition to move toward point 2, at which it reaches the boundary between PX and MX compartments. Cooling should be stopped just before this point is reached, to avoid any co-crystallization of MX. Consequently, the maximum recovery of PX in the crystallizer is limited by the eutectic troughs.

Figure 3.8b indicates the component recovery profiles as functions of crystallization temperature, which is calculated based on the SLE phase behavior. PX starts to crystallize at −33 °C, and reaches a recovery of about 74% at −61 °C, before MX starts to co-crystallize. If cooling is continued to about −63.5 °C, OX would also crystallize, and the entire mixture would turn into solid. Therefore, in order to recover pure PX, the crystallizer has to be operated above −61 °C, which is the temperature corresponding to the maximum recovery of pure PX. Note that these results are specific to a given feed composition (point 1). A different composition would lead to different values of minimum temperature and maximum recovery.

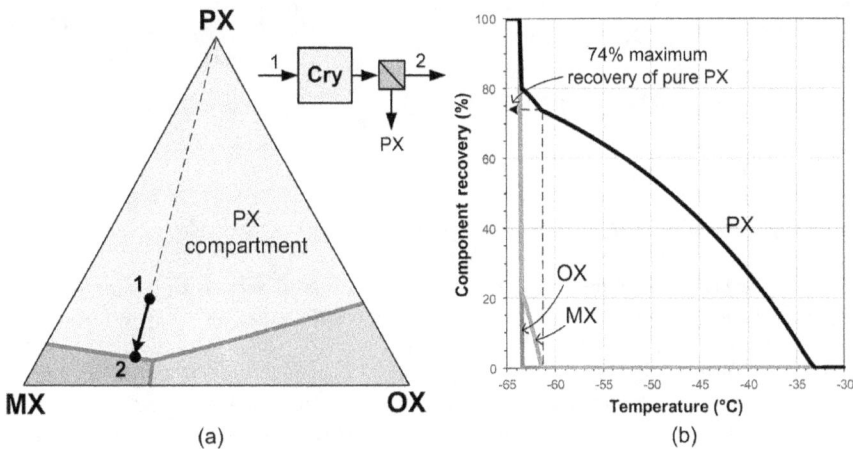

Figure 3.8: Recovery of pure *p*-xylene by crystallization: (a) process path on SLE phase diagram and (b) component recovery profile as a function of temperature.

Example 3.2(I) – Maximum purity of a secondary product
An industrially important dicaboxylic acid is produced via a hydrogenation reaction, which produces a mixture of *cis* and *trans* isomers in an aqueous solution. With the typical catalyst and reaction conditions used in commercial processes, the *trans* content in the mixture is about 75%. While the *trans* isomer is usually the desirable one for downstream applications, there is a new potential market for the *cis* isomer. Therefore, a specialty chemical company was interested in producing high purity *cis* isomer via a crystallization-based process. One idea was to utilize the mother liquor, after crystallizing out the *trans* isomer from the reaction mixture.

To address this issue, an experimental investigation of the SLE phase behavior of the ternary system involving the two isomers and water as the solvent was conducted to identify the location of the *cis/trans* double saturation point at different temperatures. The results are plotted on the polythermal projection of the phase diagram, together with the binary eutectic trough drawn through the data points, as depicted in Figure 3.9. Note that only the top portion of the phase diagram (mass fraction $H_2O \geq 0.5$) is displayed, because this is the region where the data points are located. The experimental data points indicate that the compartment of *trans* is much larger than that of *cis*, mostly owing to the large difference in the melting point of the pure isomers.

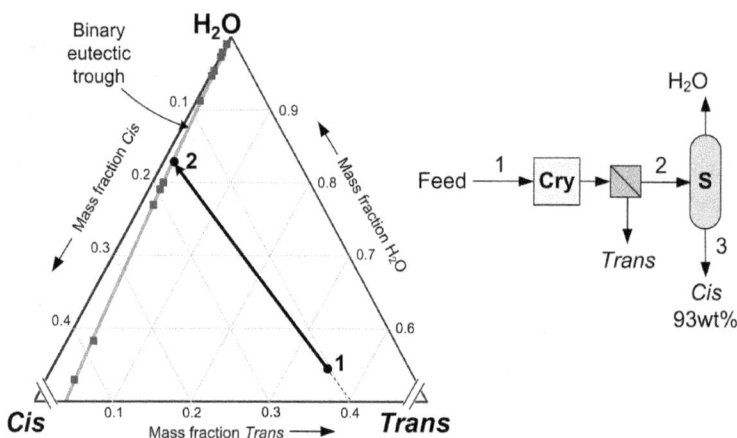

Figure 3.9: A process alternative to obtain high purity cis from a trans-rich feed.

Starting with a *trans*-rich feed (point 1), pure *trans* can be crystallized out by cooling until the liquid composition hits the binary eutectic trough at point 2. Note that apart from achieving the maximum recovery of pure *trans*, operation at point 2 gives mother liquor with the highest purity of *cis*. Therefore, the simplest process alternative to obtain high purity *cis* is to completely evaporate water from this mother liquor, yielding a solid product containing 93 wt% *cis* (stream 3). Since the binary eutectic trough is almost linear (the *cis/trans* ratio at the double saturation point stays nearly constant around 7:93 as temperature changes), 93 wt% represents the maximum purity of *cis* product that can be obtained from any feed in the *trans* compartment. A *cis* product of higher purity can be obtained by performing the hydrogenation reaction at a different condition or with a different catalyst, or performing an isomerization reaction after hydrogenation, so that a mixture with a *cis/trans* ratio less than 7:93 can be produced.

Example 3.3(I) – Crystallization of a specialty monomer

A specialty monomer A is produced in a batch reaction in solvent S, in which two major impurities, B and C, are also generated. After the reaction, most of the solvent is removed by evaporation, so that its final concentration is only about 1%. As shown in Figure 3.10, the mixture is then sent to a batch crystallizer operating at 10 °C. The mixture is cooled down to crystallize out pure A, which is separated by centrifugation. Due to batch-to-batch variation in reaction conditions, the composition of the crystallization feed is not constant. As shown in Table 3.1, the content of A can vary from 55% to 80%, and the ratio of C to B can be anywhere between 0.7 and 3. As a result, the recovery of A also varies from batch to batch, and sometimes the product becomes off-spec, if the impurities also crystallize out at 10 °C. It is desirable to increase the recovery of A and, at the same time, prevent co-crystallization of the impurities.

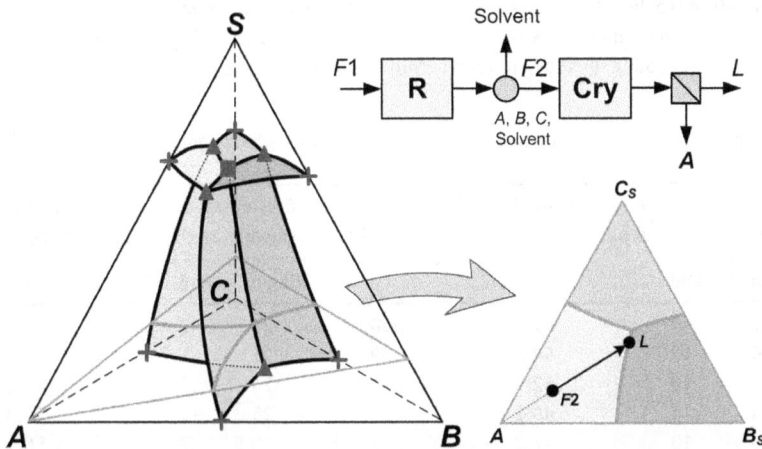

Figure 3.10: Crystallizing a pure product from a reaction mixture.

Table 3.1: Batch-to-batch variation in crystallization feed composition.

Batch number	Normalized crystallization feed composition (wt%)			
	S	A	B	C
001	0.89	71.31	16.76	11.04
002	0.78	70.74	15.81	12.66
003	0.58	62.33	23.00	14.08
004	0.56	62.65	21.82	14.96
005	1.93	69.08	12.69	16.31
006	1.51	67.87	19.28	11.34
007	1.20	68.05	20.65	10.10
008	1.17	62.63	26.00	10.20
009	0.77	57.36	31.26	10.61
010	1.51	63.24	24.51	10.74
011	1.07	59.42	30.20	9.30
012	0.18	64.58	24.51	10.74
013	0.64	55.45	32.83	11.08

To address this problem, the SLE phase behavior of the system was established from experimental solubility data. Since this is a four-component system, the polythermal projection takes the shape of a tetrahedron, as shown in Figure 3.10. Taking a cut at 1% S, which corresponds to the crystallization condition, an image similar to the polythermal projection of a ternary system phase diagram is obtained. If the feed composition to the crystallizer is given by point $F2$, the maximum recovery of pure A is achieved when the composition reaches point L. Cooling beyond this point would lead to co-crystallization of impurity B and, eventually, also impurity C.

Based on the phase diagram, it is possible to determine the crystallization temperature that results in the maximum recovery of A for different feed compositions, instead of fixing the crystallizer temperature at 10 °C. The results for the 13 batches are summarized in Table 3.2. For batch 001, A starts to crystallize at 31 °C and B begins to co-crystallize at 9 °C, when the recovery of A reaches about 61%. Therefore, in order to get the maximum recovery of pure A, cooling must be stopped before the temperature reaches 9 °C. However, with the feed composition of batch 005, A does not start to crystallize until the temperature reaches 26.5 °C, and co-crystallization already occurs at 13 °C, which means that cooling must be stopped before reaching this temperature.

Table 3.2: Maximum yield and crystallization temperatures for various feed compositions.

Batch number	Crystallization temperature (°C)		Maximum yield of pure A (%)	Batch number	Crystallization temperature (°C)		Maximum yield of pure A (%)
	Begin	End			Begin	End	
001	31	9	61.0	008	24.5	2	56.8
002	30	12	54.8	009	21	1.5	49.8
003	24	10	42.2	010	24	2	56.3
004	24	11	40.1	011	23	−1.5	57.3
005	26.5	13	42.2	012	28.5	7	56.1
006	27.5	6	57.5	013	20	2	45.5
007	28.5	4	62.2				

As can be seen from the table, the maximum recovery of A, as well as the corresponding crystallization ending temperature, does not depend solely on the concentration of A in the feed, but on the overall composition. Analysis of the relevant SLE phase behavior has resulted in an operation guideline for the crystallization process, which allows the determination of the conditions corresponding to the maximum recovery, without the risk of overshooting the compartment boundary and co-crystallizing the impurities.

Example 3.4(I) – Purification by selective dissolution

A specialty chemical company manufactures a product (P) via a substitution reaction that also produces a byproduct (X), in approximately an equal amount as P. As shown in the flowsheet in Figure 3.11, after a series of preliminary separation steps, a slurry containing crystals of P and X in solvent S, in which P has a moderate solubility but X is only slightly soluble, is obtained (point 1). The amount of solvent is then increased to completely dissolve P. The solution (point 2), which contains only a small amount of X, is taken as the product (point 3), and the undissolved solid is filtered out.

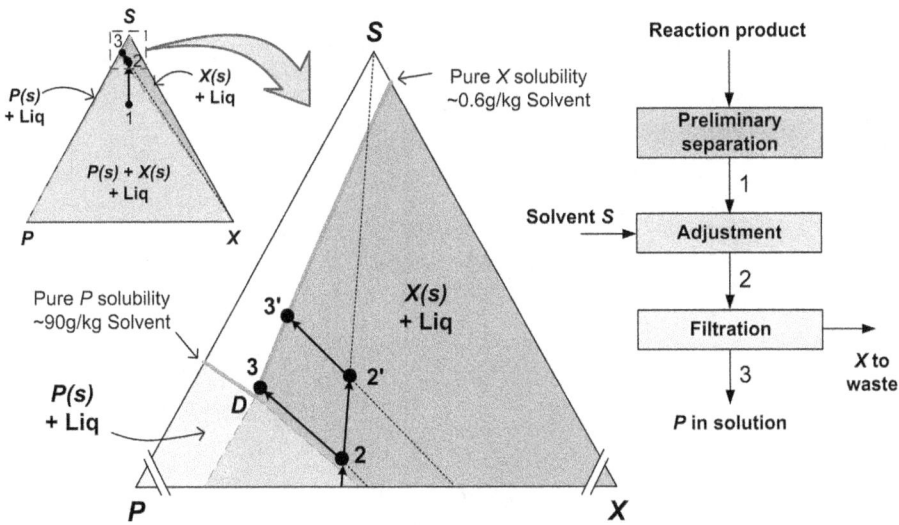

Figure 3.11: Selective dissolution to remove an insoluble impurity.

To better understand the process, consider the relevant isothermal SLE phase diagram (at solvent adjustment temperature) of the ternary system involving P, X, and S, as shown in Figure 3.11. As the operating points tend to cluster near the solvent apex due to the relatively low solubility of the components at this temperature, the top part of the phase diagram has been enlarged to provide a clearer view. As the solubility of X in S is much smaller than that of P in S, the region where pure X can be crystallized out (dark-shaded area) is much larger than the region where pure solid, P, can be obtained (light-shaded area), and the double saturation point, D, is located much closer to P than to X. The process path is plotted on the phase diagram. After the preliminary separation steps (point 1), S is added to the slurry to move the composition into the X region (point 2). This causes all P in the solid phase to dissolve, leaving behind crystals of pure X. The composition of the liquid phase, which is the product, is given by point 3.

The company would like to optimize the separation process based on the knowledge of SLE phase behavior. In particular, they wanted to maximize the product purity and minimize the product loss to the waste stream. The phase diagram suggests that there is a tradeoff between these two objectives. If more solvent is added such that the overall composition is at point 2′ instead of point 2, more X dissolves, causing the purity of P in the liquid product to decrease (that is, comparing point 3′ to point 3). On the other hand, if not enough solvent is added, the overall composition would fall into the mixed solids region ($P(s)+X(s)$+Liq), which means some P would still exist in the solid phase and get purged along with X, causing a loss to the waste stream. Therefore, operation at the double saturation point would give the highest product purity, while at the same time minimizing the loss. Consequently, knowing the location of the double saturation point is essential in determining the optimum condition of the solvent adjustment step.

Figure 3.12a shows the tradeoff between product purity and yield at 20 °C, based on the SLE phase diagram established from experimental data. Note that although all P is dissolved when the solvent to solid ratio is above 4.42 L/kg, there is some residual liquid attached to the crystals (assumed to be 0.2 kg/kg solid), which goes to the waste stream; this is why the product yield is less than unity. Clearly, operating as close as possible to the double saturation point would be optimal. The corresponding solvent to solid ratio can be readily obtained from the composition of the double saturation point.

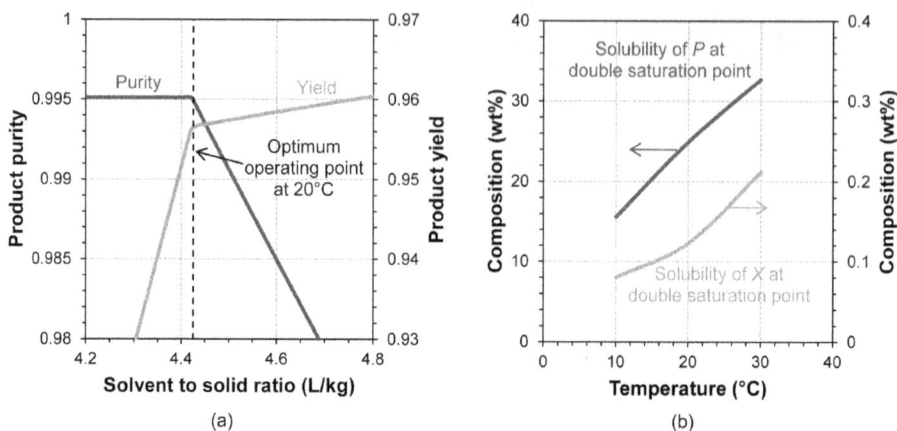

Figure 3.12: Selective dissolution process: (a) effect of solvent to solid ratio on product purity and yield at 20 °C and (b) double saturation point composition as a function of temperature.

Since the ratio of P to X at the double saturation point can vary with temperature, the product purity and yield are also affected by the choice of temperature in the solvent adjustment step. Figure 3.12b shows the composition of the double saturation point in the temperature range of 10 to 30 °C. These data indicate that the relative amount of P and X in the double saturation point is almost the same at all three temperatures; hence the maximum product purity would not be significantly affected by the choice of temperature. However, less solvent is required to dissolve P at higher temperature (thus, a higher solid to solvent ratio), and on the other hand, the higher solubility of P at higher temperature leads to more loss to the residual liquid. Based on this observation, operation at lower temperature is more desirable.

Example 3.5(I) – Purity improvement by adjusting operating conditions
An intermediate chemical for manufacturing a pharmaceutical product is synthesized via a hydrogenation reaction, which produces a mixture of *cis* and *trans* isomers in solid form, along with some minor impurities. A batch crystallization process is used to get the desirable *trans* isomer from the solid mixture. The mixture is first dissolved in a 14:86 B-to-D mixture of two solvents – solvent B in which both isomers are highly soluble and solvent D in which the isomers are less soluble. The solution is then cooled down to 32 °C, so that the *trans* isomer crystallizes out. The product crystals are separated by centrifugation. The crystals are found to contain about 6.5 wt% *Cis*, as well as a small amount of other impurities.

Figure 3.13a shows the current situation with an isothermal SLE phase diagram of the pseudoternary system involving *cis*, *trans*, and solvent. The impurities are lumped to the solvent, because only a very small amount crystallized out under the current operating conditions. The overall feed to the crystallizer (point F) is located in the double saturation region, which is consistent with the fact that the solid (point S) is a mixture of *cis* and *trans*.

Since the presence of *cis* in this intermediate product leads to losses in the next processing step due to side reactions, it is desirable to increase the *trans* purity in the crystals. This can be achieved by operating the crystallizer in the *trans* single saturation region, so that the crystals would be pure *trans*. An experimental study was performed to understand how to avoid the co-crystallization of *cis*.

Based on the SLE phase behavior, three ideas were proposed to achieve this goal. The first idea is to add more solvent to the mixture without changing the proportion of B and D in the solvent, as illustrated in Figure 3.13b. The final temperature also remains the same at 32 °C. By adding about 10% more solvent, the overall crystallization feed moves to point *P* in the *trans* single saturation region, resulting in crystals of pure *trans* (point *S'*). The second idea, illustrated in Figure 3.13c, is to use more *B* which offers higher solubility, while keeping the same amount of *D*. The overall feed moves up slightly to point *P''*, but at the same time, the solubility curves shift away from the solvent vertex, as the solubility of both isomers increases with increasing *B* content. With a *D* content of 84 wt% in the solvent, point *P''* falls inside the *trans* single saturation region, and the crystals (point *S''*) would again be pure *trans*. Finally, another alternative is to increase the final temperature without changing the composition or amount of solvent, as depicted in Figure 3.13d. By increasing the temperature to 35 °C, the solubility curves shift away from the solvent vertex, so that the overall feed *F* is located in the *trans* single saturation region at this temperature.

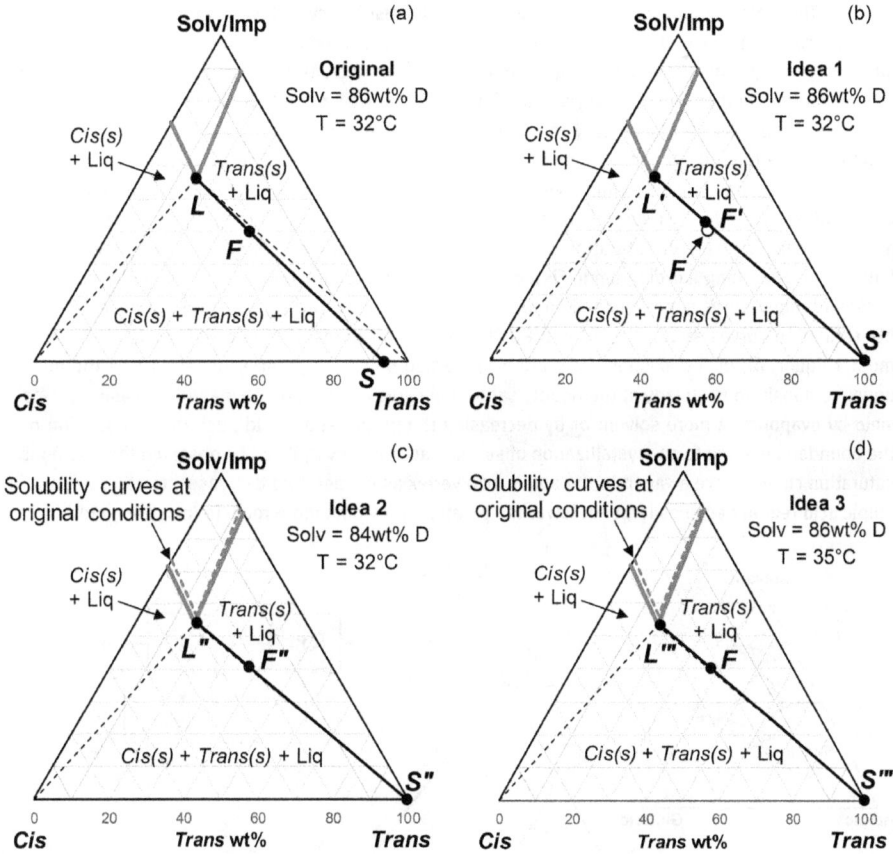

Figure 3.13: SLE phase diagram of *cis*/*trans*/solvent pseudo-ternary system: (a) original conditions, (b) adding more solvent, (c) changing solvent composition, and (d) increasing temperature.

Combinations of the three ideas can also be implemented; for example, the solvent amount can be increased, while at the same time increasing the final temperature. The basic concept is to place the overall feed composition in the region where pure *trans* isomer can be crystallized, instead of in the two-solid region where the *cis* isomer can co-crystallize. With a model of the SLE phase behavior, the impact of various improvement ideas on the crystallization yield can be visualized and estimated, and the best crystallization conditions can be identified.

Example 3.6 – Retrofitting of an adipic acid plant

This example is based on a basic study of a real-life problem: the retrofit of an existing adipic acid plant with a production rate of 60 ton/day, which had been operating at 98.6% overall recovery [77]. Adipic acid, an important dicarboxylic acid produced on a large scale worldwide, is produced from a mixture of cyclohexanol and cyclohexanone called "KA oil." The reactor outlet, which is fed to the crystallizer, contains only 6.5 wt% adipic acid. It also contains 27.5 wt% glutaric acid, 8 wt% succinic acid, and the rest is solvent, which is a mixture of water and nitric acid. Due to the relatively low solubility of adipic acid compared to the other components, crystals of adipic acid are obtained at the crystallizer temperature of 50 °C. The mother liquor M is partially purged, and the rest is recycled back to the reactor.

Using common sense, it is possible to come up with retrofit ideas to increase the overall recovery, such as reducing temperature, evaporating more solvent, or decreasing the purge ratio (defined as the ratio of purged mother liquor to recycled mother liquor). However, an understanding of the SLE phase behavior of the system is necessary to identify the feasible ideas. Figure 3.14 shows the SLE phase diagram of the four-component system consisting of adipic acid, glutaric acid, succinic acid, and solvent. The isothermal phase diagram takes the shape of a tetrahedron and features the saturation surfaces of adipic, glutaric, and succinic acid. At different temperatures, the location of these surfaces is also different. Two sets of saturation surfaces at different temperatures are shown in the figure for illustration. The triangular diagram in Figure 3.14 is the Jänecke projection, with four isothermal cuts at different temperatures superimposed on top of one another. The four sets of curves represent the double saturation curves at 10, 20, 50, and 80 °C. The compositions of the crystallizer feed, F, and the mother liquor, M, of the existing plant are also marked on the diagram. It turns out that the mother liquor composition is already at the double saturation curve at 50 °C, which means increasing per-pass yield by evaporating more solvent or by decreasing the purge ratio would push the composition over the boundary, leading to co-crystallization of succinic acid. However, it can be observed that the double saturation curves move away from the adipic acid vertex as temperature decreases. In other words, the adipic acid region becomes larger at lower temperature, providing more room to increase yield.

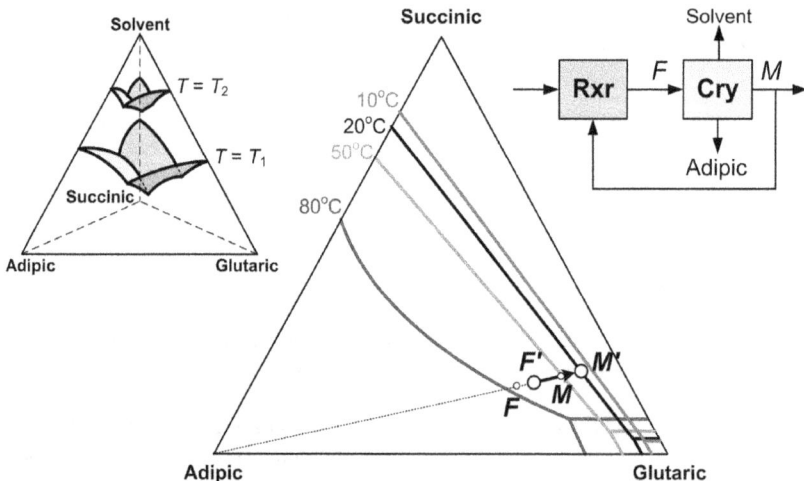

Figure 3.14: SLE phase diagram of adipic acid/glutaric acid/succinic acid/solvent system.

Based on the SLE phase behavior, it is possible to increase the overall recovery by reducing the crystallization temperature and, at the same time, decreasing the purge ratio. By purging less, more succinic and glutaric acid are returned to the reactor, and their concentration in the reactor outlet also increases. This is why the crystallizer feed composition shifts from point *F* to point *F'*. Reducing the crystallizer temperature to 20 °C and operating close to the double saturation curves, the new mother liquor composition is represented by point *M'*. Based on the new operating points, the overall recovery would increase by 0.6 percentage point to 99.2%. Because of economies of scale, such a relatively small increase actually corresponds to a potential economic merit of nearly USD 1.5 million per year [77].

3.3 Crystallization of Desirable Product

Another typical application of SLE phase behavior in designing a crystallization process is the identification of suitable conditions for crystallizing the desirable product. Starting from a given mixture, it is often possible to obtain different products such as compounds, solvates or hydrates, polymorphs, acid form, or salt form (for electrolytes), under different crystallization conditions. Depending on the objective, it may be desirable to crystallize only a certain product and not the others.

The SLE phase behavior, as visualized on a phase diagram, provides a map of the composition space that shows where the regions for different products are located. As such, it helps guide the determination of the suitable conditions (composition, temperature, and sometimes pressure) under which the desirable product can be obtained. The crystallization process can then be designed by selecting the appropriate movements in the right direction, in order to get from the feed point to the desired operating point. The examples here illustrate how this strategy is applied in various situations.

Example 3.7 – Crystallization of sodium perborate

Sodium perborate is a bleaching agent that is widely used in detergents and other cleaning products, due to its good stability and high speed of oxygen delivery. It can crystallize as an anhydrate ($NaBO_3$) as well as three different hydrates: monohydrate ($NaBO_3 \cdot H_2O$), trihydrate ($NaBO_3 \cdot 3H_2O$), and tetrahydrate ($NaBO_3 \cdot 4H_2O$). It is manufactured in the tetrahydrate form by reacting borax pentahydrate with hydrogen peroxide and sodium hydroxide, but is mainly sold in monohydrate form as it dissolves better and has higher heat stability compared to the tetrahydrate [78]. Furthermore, the spherical monohydrate crystals are more desirable than the rod-shaped anhydrate crystals, since the latter tend to break and form small fines during filtration [79]. For this reason, it is crucial to pick a crystallization condition under which the monohydrate would be obtained. Figure 3.15 is a sketch of the SLE phase diagram of the sodium perborate/water system, showing the regions where various solid forms can be crystallized out, based on general information available in the literature. Regions of solid mixtures are not marked on the diagram, but can be easily identified by comparison with Figure 2.5.

A mixture of tetrahydrate crystals and water (point *F*) is the starting point of the crystallization process. If this mixture is heated to temperature T_1 (as represented by the vertical movement from *F* to point 1), followed by removal of water (represented by the horizontal movement from 1 to 2), the composition would enter the two-phase region where the liquid phase is in equilibrium with crystals of monohydrate. Consequently, the monohydrate crystallizes out.

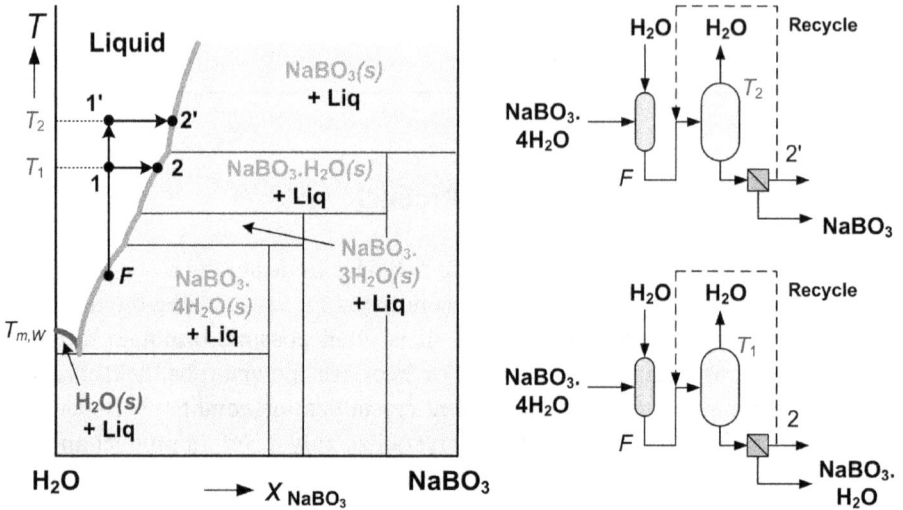

Figure 3.15: Process alternatives for crystallizing sodium perborate.

If the mixture is heated to temperature T_2 (point 1′) instead, water removal would move the composition to point 2′, which is located in the two-phase region where the solid phase is the anhydrate. Accordingly, anhydrate crystals would be obtained. The corresponding flowsheets for these two process alternatives are also shown in Figure 3.15. Clearly, the selection of crystallizer temperature is the key in producing the desirable monohydrate crystals. With the knowledge of the SLE phase diagram, the proper crystallizer temperature can be readily selected.

Example 3.8 – Monosodium glutamate from glutamic acid

Monosodium glutamate (MSG) is widely used in the food industry as a flavor enhancer. MSG can be produced from glutamic acid, which involves addition of sodium hydroxide (NaOH) and crystallization of MSG in its monohydrate form. Figure 3.16 shows the SLE phase diagram of the glutamic acid (H_2Glu/NaOH, with the solubility curves of H_2Glu and the monosodium salt hydrate (NaHGlu.H_2O) obtained from literature data at 35 °C [14]. The crystallization regions of H_2Glu, NaHGlu.H_2O, as well as mixed solids region, are marked on the diagram. A zoomed view of the upper left corner of the diagram, in which the process points are concentrated, is shown as an inset.

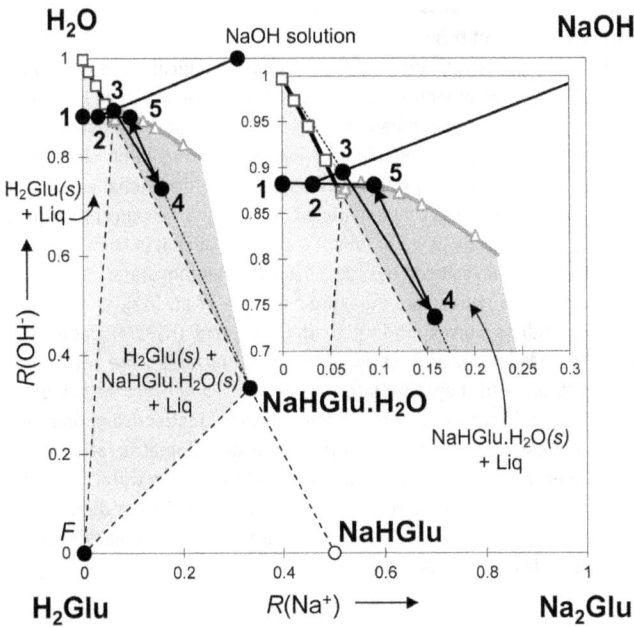

Figure 3.16: SLE phase diagram of glutamic acid/NaOH system.

Glutamic acid is first mixed with water to produce a mixture indicated by point 1 in Figure 3.16. This mixture is further combined with a recycle stream (point 5) to give a mixture of point 2. Since point 2 is inside the H_2Glu crystallization region, it represents slurry (solid–liquid mixture). Aqueous NaOH solution is then added to bring the composition to point 3, which is located in the unsaturated liquid region, thereby completely dissolving H_2Glu. Next, the solution is fed to an evaporative crystallizer, in which water is removed and the composition moves to point 4 inside the $NaHGlu.H_2O(s)$+Liq region. Consequently, the hydrate crystallizes out, leaving mother liquor with the composition of point 5, which is recycled. The corresponding process flowsheet is shown in Figure 3.17 [80].

The knowledge of the SLE phase diagram is useful in determining proper process conditions such as how much water and NaOH to add such that point 3 falls in the liquid region, and how much evaporation is required to place point 4 in the $NaHGlu.H_2O$ crystallization region.

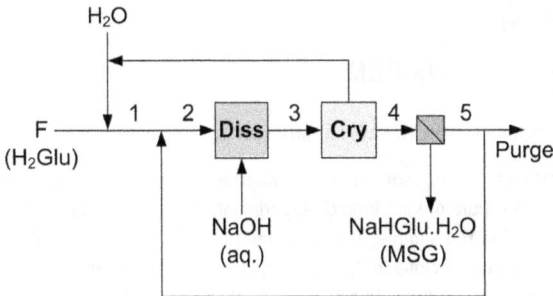

Figure 3.17: Process for crystallizing monosodium glutamate.

Example 3.9 – Selection of pharmaceutical salt form
Salt formation is the most common and effective method employed in pharmaceutical industry to mod-ify and optimize the physicochemical properties of ionic compounds [81]. Properties such as solubility, dissolution rate, hygroscopicity, stability, impurity profiles, and crystal habit can be influenced by using a variety of pharmaceutically acceptable counterions. In addition to alterations in physicochemi-cal properties, salt formation can alter organoleptic properties such as taste and, occasionally, pharma-cological response and toxicity. This example focuses on pemetrexed, a drug for lung cancer treatment post-chemotherapy, which has a very low solubility in water (about 0.0455 mg/mL). It is known to form monosodium or disodium salt, and the disodium salt can be crystallized as a heptahydrate.

A sketch of the relevant phase diagram, which was measured by Lam et al. [82], is shown in Figure 3.18. There are solid–liquid regions corresponding to the free acid (H_2PEM), mono-salt (NaHPEM), di-salt heptahydrate ($Na_2PEM.7H_2O$), and NaOH. The remaining areas are regions of solid mixtures. The target is to crystallize the heptahydrate, which has a much higher solubility than the free acid, making it more suitable for intravenous administration. Instinctively, one may think that a mixture with an $NaOH/H_2PEM$ molar ratio of 2, which is the stoichiometric ratio, would be suitable to obtain the di-salt. However, the phase diagram suggests that adding water to such a mixture, represented by point A in Figure 3.18, would result in a movement along the direction of the arrow. Regardless of the amount of water added to the mixture, it would be impossible to ob-tain a composition inside the region of $Na_2PEM.7H_2O$.

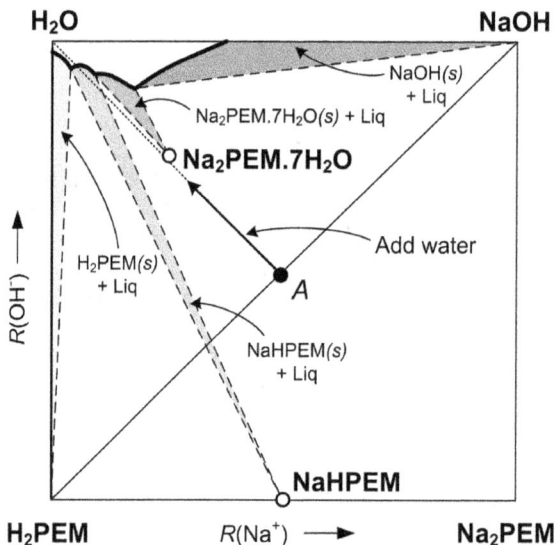

Figure 3.18: Sketch of the SLE phase diagram of $H_2PEM/NaOH$ binary system.

Indeed, a series of verification experiments with varying molar ratio, summarized in Table 3.3, shows that a molar ratio of about 2.5 is required to obtain pure di-salt. In each experiment, the Na^+ and PEM^{2-} con-tents in the solids are determined by analysis of the sample. The molar ratio of Na^+/PEM^{2-} in the solids serves as an indication of its identity. A molar ratio of close to 0, 1, or 2 signifies the acid, mono-salt, and di-salt, respectively, while ratios between those values indicate a mixture of solids. The feed compositions are plotted in Figure 3.19 along with the solubility data (plotted as open squares) that are used to deter-mine the saturation curves. It can be seen that the feed points for data sets 1–4 (solid triangles), with a

NaOH/H_2PEM molar ratio of around 2, lie inside the mono-salt/di-salt solid mixture region. Compositions with a NaOH/H_2PEM molar ratio greater than 2.5 (data sets 5–8) lie within the region of pure di-salt.

In summary, NaOH should be added to get the desired di-salt heptahydrate from pemetrexed, but the appropriate NaOH/H_2PEM molar ratio does not correspond to the stoichiometry. With the knowledge of the SLE phase diagram, it is possible to determine, in advance, the required NaOH/H_2PEM molar ratio to get the desired product.

Table 3.3: Feed composition and solid analysis results of the verification experiments.

Set	Mass in feed (g)			NaOH/H_2PEM molar ratio	Content in solids (mmol/g)		Na$^+$/PEM^{2-} molar ratio
	H_2PEM	NaOH	H_2O		Na$^+$	PEM^{2-}	
1	0.4273	0.0805	1.0255	2.0134	3.433	2.024	1.70 (mixture)
2	0.4221	0.0879	1.2223	2.2245	3.474	1.970	1.76 (mixture)
3	0.4213	0.0889	1.0961	2.2549	3.159	1.708	1.85 (mixture)
4	0.4326	0.0961	1.2269	2.3746	3.649	1.927	1.89 (mixture)
5	0.4310	0.1046	1.3348	2.5931	3.621	1.879	1.93 (di-salt)
6	0.4207	0.1043	1.0694	2.6493	3.335	1.749	1.91 (di-salt)
7	0.4204	0.1046	0.9947	2.6588	3.789	1.874	2.02 (di-salt)
8	0.4305	0.1080	1.0910	2.6808	3.794	1.831	2.07 (di-salt)

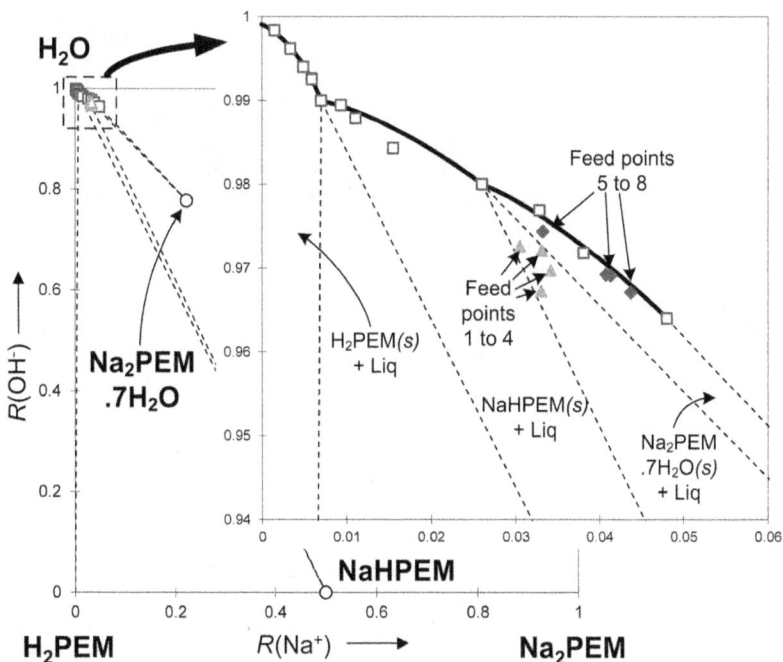

Figure 3.19: Experimental solubility data and feed compositions of verification experiments plotted on the SLE phase diagram of H_2PEM/NaOH system.

Example 3.10 – Production of succinic acid

Succinic acid (SA) is widely used in the synthesis of various polymers as well as pharmaceutical products such as vitamin, expectorant, and sedative [83]. Nowadays, SA is mainly produced by microbial fermentation of glucose from renewable feedstocks. One such process is described in U.S. Patent 6,265,190 [84] and is illustrated in Figure 3.20. Ammonia is added to the fermenter to keep the broth at near-neutral pH, so as to prevent injuries to the microorganisms, as SA is produced during fermentation. After filtering out the cells and any other insolubles, a solution of ammonium succinate salts (stream 1) is obtained. This solution is then concentrated in an evaporator, before being sent to a crystallizer. The pH in the crystallizer is adjusted by adding an acid such as sulfuric acid, so that SA is recovered by crystallization. The mother liquor (stream 5) is sent to a recovery unit to recover ammonium sulfate, which can be thermally cracked to produce ammonia and ammonium bisulfate (NH_4HSO_4). Ammonia is recycled to the fermenter, and ammonium bisulfate (stream 11) is sent back to the crystallizer. SA is further purified by recrystallization in methanol.

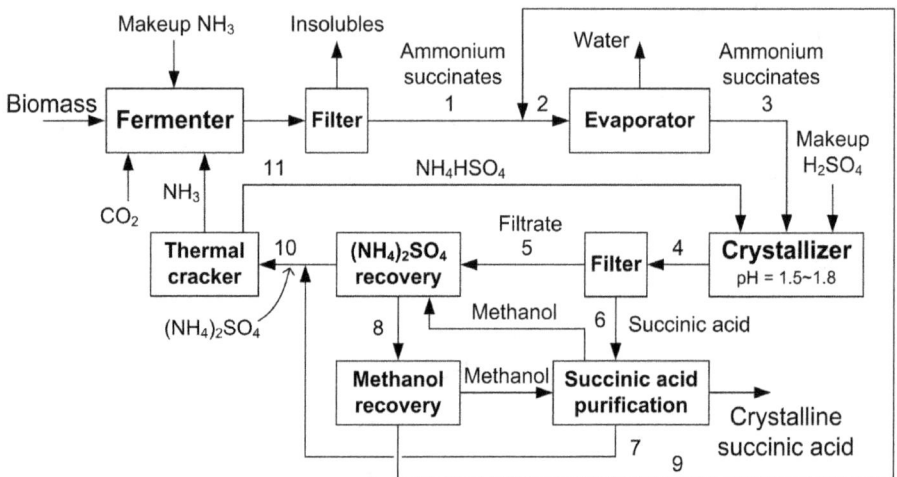

Figure 3.20: Succinic acid production process according to U.S. Patent 6,265,190.

In order to determine the suitable condition to crystallize SA instead of its mono- or diammonium salt, knowledge of the SLE phase behavior of SA/sulfuric acid/ammonium hydroxide system is important. The SLE phase diagram of this system is similar to that in Figure 2.28, with SA, NH_4OH, and H_2SO_4 in the place of H_2A, MOH, and HX, respectively. Figure 3.21a shows the saturation curves at different cuts, based on experimental data from the literature [85]. The cut at $SO_4^{2-}/OH^-=0$ (no sulfate) features three curve segments, corresponding to SA, monoammonium succinate (MAS), and diammonium succinate (DAS), respectively. Three other cuts at different values of SO_4^{2-}/OH^- are also displayed. Due to the limited availability of experimental data at these other cuts, only the solubility curves of SA and MAS are indicated on the figure. It is clear from the cuts that as SO_4^{2-} is added to the system, the saturation curves shift closer to the top of the diagram, indicating decreasing solubility. Furthermore, the location of the double saturation points also shifts such that the single saturation region of SA is larger at higher SO_4^{2-} content.

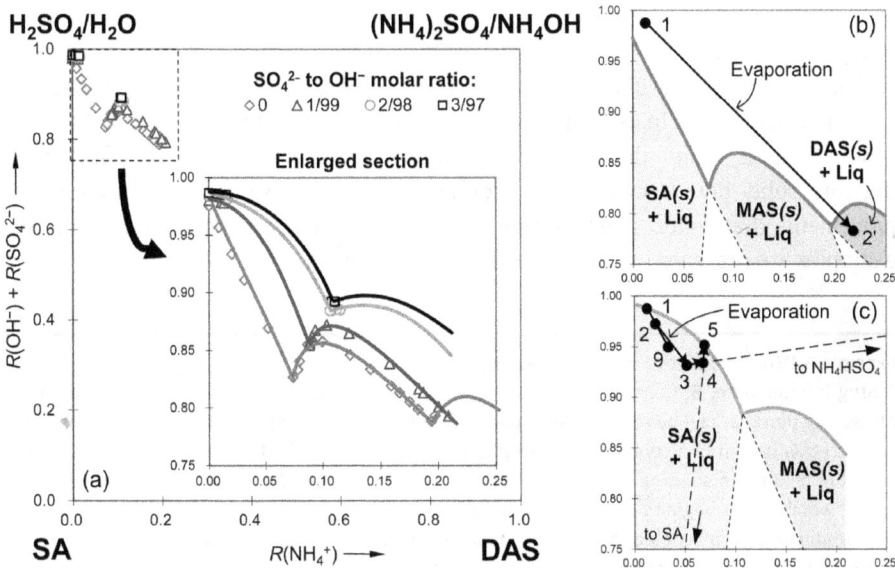

Figure 3.21: SLE phase diagram of succinic acid/H_2SO_4/NH_4OH system: (a) different cuts at various SO_4^{2-}/OH^- ratios, (b) cut at SO_4^{2-}/OH^- = 0, and (c) cut at SO_4^{2-}/OH^- = 2/98.

Figure 3.21b shows more details of the cut at SO_4^{2-}/OH^-=0. The single saturation region of SA, MAS, and DAS are marked on the diagram. If the broth from the fermenter (point 1) is only subjected to evaporation without any pH adjustment, the resulting mixture (point 2′) is located inside the saturation region of DAS, which implies that DAS would crystallize out. Without acid addition, it is not possible to move the composition into the saturation region of SA.

The saturation regions of SA and MAS in the cut at SO_4^{2-}/OH^-=2/98 are featured in Figure 3.21c. Composition points corresponding to process streams in Figure 3.20 are also marked on the diagram, although not all of them actually lie on this cut. The broth from fermenter (point 1) is mixed with the recycle stream (point 9) to give the evaporator feed (point 2). Evaporation brings the composition to point 3, and addition of ammonium bisulfate produces the crystallizer feed given by point 4. SA crystallizes out, and the mother liquor composition is given by point 5. The key point in designing the process is to ensure that point 4 lies within the SA saturation region for the cut, at the corresponding ratio of SO_4^{2-}/OH^-.

3.4 Complete Dissolution

Sometimes it is desirable to identify the suitable conditions under which crystallization can be avoided, or complete dissolution can be achieved. This is important because formation of solid crystals may be undesirable during a processing step, as it may lead to problems such as plugging or fouling. Complete dissolution of solid reactants is desirable to minimize mass transfer limitations during reaction, and reaction products often need to be completely dissolved prior to the subsequent processing step.

As the SLE phase diagram maps out the regions where different phases can exist, it is also useful in determination of the suitable conditions for complete dissolution – that is, inside the region where only a liquid phase can be present, avoiding any solid–liquid region. The process should be designed with appropriate movements that would steer the process toward the desired operating region. Two examples are presented to illustrate this application.

Example 3.11(I) – Production of a drug intermediate
A drug intermediate, B, is produced from a solution of raw material, A, in solvent, S, via a liquid-phase reaction that can be roughly described as $A \rightarrow B$. The reaction product, which contains some unreacted A, is sent to a crystallization step to obtain pure B in solid form. Since it is necessary to keep the mixture in solution during the course of reaction, it was desirable to know the temperature at which solids begin to crystallize out at different reaction extents. In addition, since the reaction product would be cooled to 25 °C, it was also desirable to determine the minimum conversion at which pure B can be obtained at that temperature. To answer these questions, the SLE phase diagram for the ternary system of A, B, and S, similar to the one shown in Figure 2.11, is of interest. Figure 3.22 shows a series of isothermal cuts, which is essentially a contour map of the three-dimensional phase diagram. The saturation regions at 25 °C are also marked on the figure. Note that in order to maintain clarity, only the enlarged top portion of the diagram is shown.

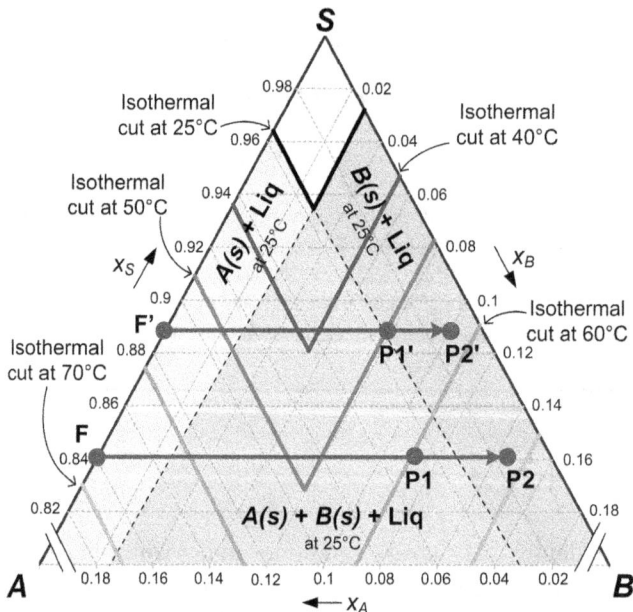

Figure 3.22: SLE phase diagram of the drug intermediate system.

Two reaction paths are shown on the diagram. The first one starts with a feed, F, which contains 16 wt% A. Since the solvent does not participate in the reaction, the total content of A+B remains constant at 16 wt%, as signified by a movement in horizontal direction, as the reaction progresses. The second path starts with a more dilute feed F' (11 wt% A), and follows the horizontal line F'-P2'. Both paths traverse different isotherms, so the dissolution temperature of the reaction mixtures changes with the extent of reaction, as depicted in Figure 3.23a. The profile goes through a minimum, which corresponds to a double saturation point. This plot indicates that the first reaction mixture (starting with 16 wt% A) can always be maintained as a single liquid phase, regardless of conversion, by keeping the reaction temperature above 72 °C. Since the dissolution temperatures are lower for the second reaction mixture (starting with 11 wt% A), it is sufficient to keep the reaction temperature above 60 °C or so.

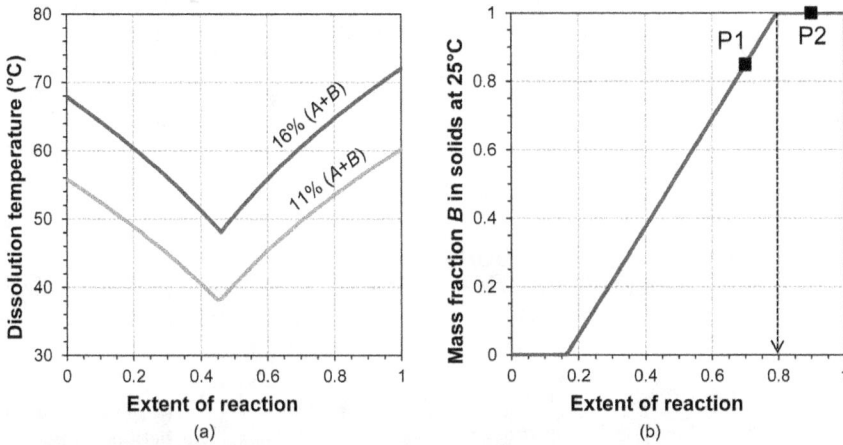

Figure 3.23: Insights from SLE phase diagram: (a) dissolution temperature evolution as reaction progresses and (b) effect of reaction extent on product purity upon cooling to 25 °C.

Figure 3.22 also reveals that the reaction path passes through different regions. For example, consider the path starting with the feed containing 16 wt% A (point F). When the conversion of A reaches 70%, the composition of the reaction mixture (point P1) lies within the region of two solid phases at 25 °C. Consequently, both A and B crystallize out when this mixture is cooled to 25 °C; in other words, the product is not pure B. However, the composition at 90% conversion (point P2) falls within the single solid phase region, which means solids of pure B can now be obtained at 25 °C. Obviously, there is a minimum conversion above which it is possible to obtain pure B. Figure 3.23b, which shows the plot of product purity against reaction extent for this feed, suggests that the minimum conversion is about 80%.

Example 3.12(I) – Dissolution of a fermentation product
A biotechnology company has developed the technology and process involving fermentation to produce a dicarboxylic acid (denoted as H_2D) from natural feedstocks. Due to its low solubility in water, H_2D precipitates out during fermentation. To purify the product, the company considered a process alternative that involves the addition of a strong base such as sodium hydroxide (NaOH) at the end

of fermentation, to raise the pH and dissolve H_2D into solution. At high pH, H_2D is soluble due to the formation of mono- and disodium salts:

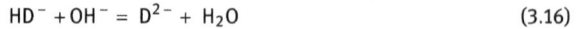

$$H_2D + OH^- = HD^- + H_2O \tag{3.15}$$

$$HD^- + OH^- = D^{2-} + H_2O \tag{3.16}$$

After removing cells and macromolecules from the fermentation broth, an acid such as sulfuric acid (H_2SO_4) is added to neutralize the base and precipitate H_2D out of solution. Crude H_2D crystals are then separated by centrifugation and recrystallized in an organic solvent to obtain high purity crystals as the final product.

An important issue in developing the process is the amount of NaOH that needs to be added to the fermentation broth to completely dissolve H_2D. An initial recipe calls for addition of NaOH until the pH value reaches 8.5, but it is not clear if it is really necessary to add that much NaOH. Obviously, adding less NaOH means reducing the required amount of H_2SO_4 that has to be added to neutralize it as well as the amount of salt waste generated. To address this issue, the SLE phase behavior involving H_2D, NaOH, and H_2O was examined.

Figure 3.24a depicts the conceptual SLE phase diagram of the $H_2D/NaOH/H_2O$ system at a constant pressure and temperature. A zoomed view of the section near the H_2O vertex (upper left corner) is shown in Figure 3.24b. Various regions where one or more species can crystallize out and exist in equilibrium with a saturated liquid are marked on the diagram.

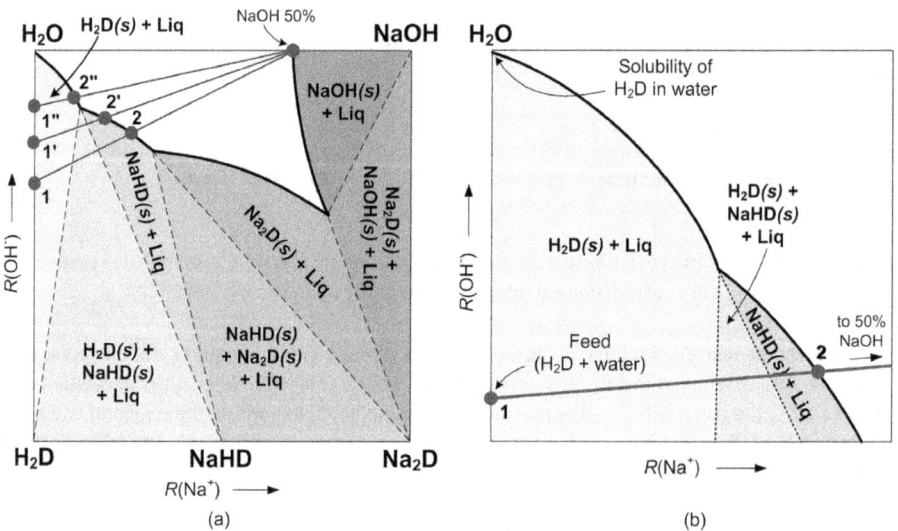

Figure 3.24: SLE phase behavior of $H_2D/NaOH$ system: (a) conceptual image and (b) zoomed section near H_2O corner.

Due to the reactive nature of the salt formation, ionic coordinates are used in plotting the phase diagram. The two coordinates are chosen to be

$$R(Na^+) = \frac{[Na^+]}{[H^+] + [Na^+]} \tag{3.17}$$

$$R(OH^-) = \frac{[OH^-]}{[OH^-] + [HD^-] + 2[D^{2-}]} \tag{3.18}$$

Starting with a mixture of solid H_2D and water (point 1, which is similar to the broth at the end of fermentation), adding NaOH in the form of a 50% solution causes the overall composition to move toward point 2, where it becomes completely dissolved as the mono-salt. Thus, the required amount of NaOH that needs to be added can be determined, once the solubility curves have been identified. Depending on the initial concentration of H_2D in the broth (which can be adjusted by adding or removing water), the required amount of NaOH for complete dissolution varies. For example, if the broth is diluted so that the composition moves to point 1', complete dissolution is achieved at point 2'. If even more water is added to obtain a composition at point 1", complete dissolution occurs at point 2", which lies on the solubility curve of H_2D – implying that H_2D dissolves without being converted to its mono-salt form.

Figure 3.25 shows the required amount of NaOH for dissolution at various dilution ratios (defined as the ratio of the broth amount after dilution to the original broth amount), which is determined based on the solubility curves obtained from experimental data.

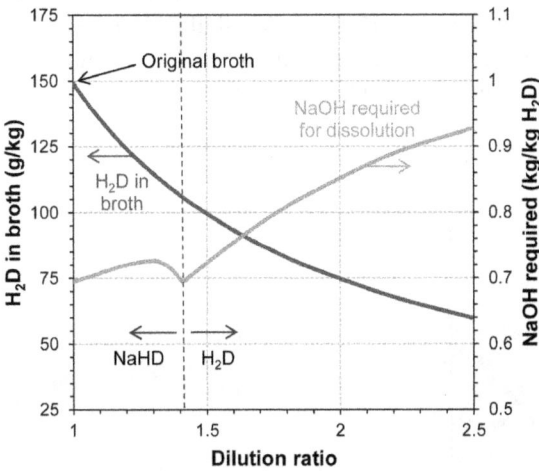

Figure 3.25: Required amount of NaOH for complete dissolution.

Each kg of the original broth (dilution ratio=1) requires about 0.7 kg of NaOH for complete dissolution. The unit consumption of NaOH reaches a maximum and starts to decrease when the dilution ratio is larger than about 1.3. But, beyond a dilution ratio of 1.4, the point of complete dissolution lies on the solubility curve of H_2D, and the required amount of NaOH increases sharply with increasing dilution ratio. From these results, there appears to be no advantage in diluting the broth prior to adding NaOH, as it does not result in a lower amount of NaOH for dissolution.

3.5 Summary

Understanding the SLE phase behavior is crucial in designing a crystallization process. The phase behavior can be visualized using phase diagrams, and operations such as cooling, solvent removal, crystallization, and so on can be represented as movements on the phase diagram. Various applications of this concept, from ensuring maximum recovery of a pure product to determining the conditions under which a particular product can be crystallized, have been illustrated using academic and industrial examples. The key is to identify the location and boundaries of the compartment or region where the desired product can be obtained in pure form. In this chapter, the focus is on how the knowledge of SLE phase behavior, especially the thermodynamic boundaries, can be used in the conceptual design of a crystallization process. There are ways to overcome the limitation of thermodynamic boundaries, as discussed in chapter 4.

Exercises

3.1. The feed to a crystallization process contains 60% A and 40% B as indicated by point M in the T-x diagram shown in Figure 3.26. At the crystallizer temperature, the mixture splits into solid A (point S) and a saturated liquid (point L, containing 50% B). According to the lever rule, how many kilograms of solid A are obtained from 100 kg of feed?

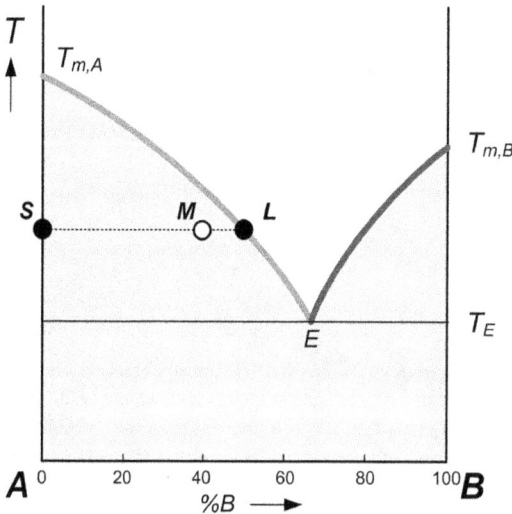

Figure 3.26: SLE phase diagram for a binary system (problem 3.1).

3.2. The phase diagram of the binary system consisting of A and B is shown in Figure 3.27. A feed mixture containing A and B is represented by point F on the diagram.
a. What is the composition of this mixture?
b. Which component would crystallize out if this mixture is cooled down to 50 °C?
c. How much solid would be obtained from 1 kg of mixture upon cooling to 50 °C?

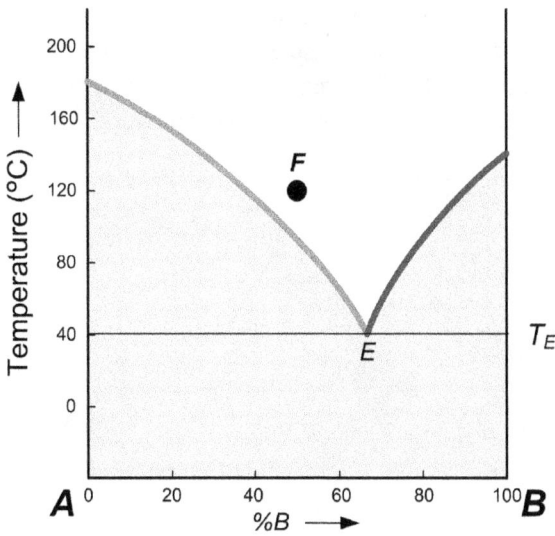

Figure 3.27: SLE phase diagram for a binary system (problem 3.2).

3.3. If pure A is separated from mixture X such that the composition of the remaining mixture is given by point Y, as shown in Figure 3.28, how many kg of A would be obtained from 100 kg of mixture X?

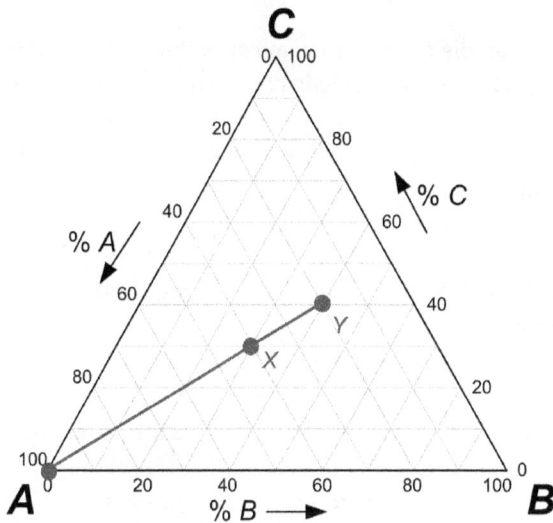

Figure 3.28: Composition triangle (problem 3.3).

3.4. An isothermal cut of the SLE phase diagram involving three components A, B, and S is shown in Figure 3.29. Point M is a mixture of solid A, solid B, and a liquid phase that is saturated in both A and B. If the total mass of the mixture is 100 kg, how many kg each of solid A, solid B, and liquid are present in the mixture?

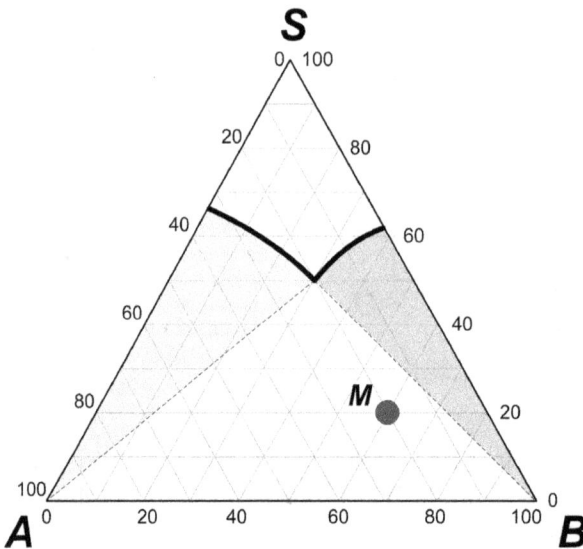

Figure 3.29: Isothermal cut of the SLE phase diagram of a ternary system (problem 3.4).

3.5. The phase diagram of a binary simple eutectic system consisting of A and B is similar to the one shown in Figure 3.26, but the eutectic composition is 79% A and 21% B.
a. Starting from a feed containing 30% A and 70% B, what pure solid can be obtained as a crystallization product?
b. What is the maximum recovery of the pure solid?

3.6. The polythermal projection of the SLE phase diagram involving three components A, B, and S is shown in Figure 3.30.
a. Starting with a saturated solution with an initial composition of point 1, draw the direction of liquid composition change upon cooling.
b. What is the liquid composition that corresponds to the maximum recovery of a pure solid?
c. What is the maximum recovery of the pure solid? (*Hint*: taking as a basis, 1 kg of the initial mixture, calculate the amount of the crystallizing component in the initial mixture and in the solid).

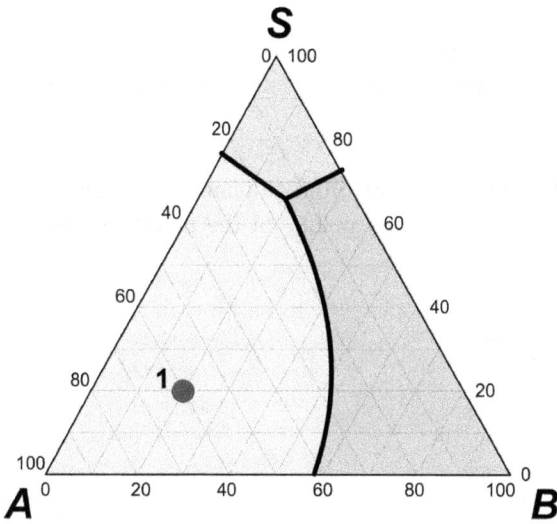

Figure 3.30: Polythermal projection of the SLE phase diagram of a ternary system (problem 3.6).

3.7. A reaction produces a solid mixture of product B and byproduct A in a 50:50 proportion. One hundred kilograms of this mixture is dissolved in 100 kg of solvent C, and then cooled down to 25 °C in a batch crystallizer. The isothermal cut of the phase diagram involving A, B, and C at 25 °C is shown in Figure 3.31. It is also known that the solubility of both A and B in C increases with increasing temperature.

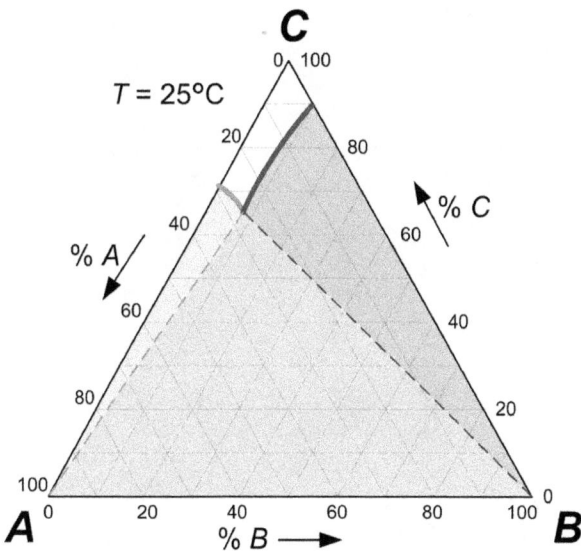

Figure 3.31: Isothermal cut of the SLE phase diagram of a ternary system (problem 3.7).

a. Locate the mixture on the phase diagram.
b. What solid, if any, crystallizes out at this condition?
c. Based on understanding of the phase behavior, come up with two ideas to crystallize pure product B from the given mixture.

3.8. A process alternative for manufacturing MSG from glutamic acid is discussed in Example 3.8. The coordinates of the process points on the phase diagram in Figure 3.16 are given below.

Point	R(Na$^+$)	R(OH$^-$)
1	0.0000	0.8824
2	0.0315	0.8819
3	0.0629	0.8952
4	0.1579	0.7368
5	0.0952	0.8810

Taking a basis of 100 kg/h glutamic acid feed and assuming zero purge (i.e., the recycle stream is the same as stream 5), determine the flow rate of the recycle stream. NaOH is added as a 50 wt% aqueous solution. Use the following values for molecular weights: glutamic acid = 147.1, water = 18, NaOH = 40.

Hint: For easier calculations, first convert the composition data into mass fraction.

4 Synthesis of Crystallization-Based Separation Processes

4.1 Crystallization as a Separation Process

Crystallization is widely used for the purification and separation of various organics, inorganics, and biochemicals in chemical processes. The main focus is to recover the product of interest in high purity and with maximum recovery. The idea of exploiting the solid–liquid equilibrium (SLE) behavior of a multicomponent system in separating a mixture into individual components in pure form by crystallization has been documented in the literature, as early as the 1940s [37]. Over the following decades, various crystallization-based separation and purification techniques have been invented, some of which are listed in Table 4.1. Design methods were subsequently proposed, with consideration of the SLE behavior of the relevant systems.

Despite the different names and terms used to describe these techniques, the underlying principles governing them are the same. Therefore, a unified procedure that guides the user to generate flowsheet alternatives in an evolutionary manner has been developed [100]. The discussion in this chapter will focus on the use of SLE phase diagrams in synthesizing crystallization-based separation processes.

4.2 Bypassing the Thermodynamic Boundary

As thermodynamic boundaries impose a limitation on the recovery of a pure component in a crystallization process, the basic idea behind a crystallization-based separation process is to bypass the boundaries to reach the compartment or crystallization region of the desired component. The same idea is applicable for molecular systems and conjugate salt systems, despite the different ways of representing their phase diagrams.

4.2.1 Ternary system

Figure 4.1a shows the SLE phase diagram of a three-component system consisting of A, B, and S, with a set of basic operations plotted as movements on the phase diagram. The polythermal projection is shown in Figure 4.1b. To make it easier to follow, the explanation starts with an unsaturated liquid denoted by point 3 in the phase diagram. Upon cooling beyond point 3a, which lies on the solubility surface, pure A starts to crystallize out. In the polythermal projection (Figure 4.1b), it is clear that point 3 (as well as point 3a, which coincides with point 3 but is not identified in the projection) is located in compartment A. Due to crystallization of A, the liquid composition moves

https://doi.org/10.1515/9781501519901-004

Table 4.1: Crystallization-based separation and purification techniques.

Technique	Description	Reference
Cooling crystallization	Decreasing temperature to reduce the solubility of the target component so as to cause crystallization	–
Evaporative crystallization	Evaporating the solvent to cause crystallization of the target component	–
Reactive crystallization	chemical reactions which lead to crystallization of one or more products	Söhnel and Garside [86], Berry and Ng [68]
Drowning-out, salting-out, or antisolvent crystallization	Addition of another solute or an antisolvent to reduce the solubility of the target component so as to cause crystallization	Hanson and Lynn [87], Weingaertner et al. [88], Berry et al. [89]
Fractional crystallization	Utilization of heating, cooling, dilution, and evaporation to effect separation of individual components from a mixed solution	Fitch [90], Cisternas and Rudd [91], Cisternas et al. [92], Dye and Ng [93]
Extractive crystallization	Addition of solvent to overcome the eutectic limitation in achieving complete separation of pure components from a mixture	Findlay and Weedman [94], Dale [95], Dye and Ng [96]
Adductive crystallization	Adding a reactant that can form solid adduct with one or more components in the system to effect crystallization of the adduct	Tare and Chivate [45], Dale [95]
Eutectic freeze crystallization	Simultaneous crystallization of solute and solvent (usually salt and water) at the eutectic point, so that solvent is removed by freezing instead of evaporation	van der Haam [97], Himawan et al. [98], Reddy et al. [99]

from point 3 to point 4 at the boundary between compartments *A* and *B*. If cooling is continued beyond point 4, *B* crystallizes together with *A*. Instead, the mixture at point 4 is heated up to an appropriate temperature for evaporating the solvent *S*, as indicated by the vertical movement to point 4a in Figure 4.1a. By removing *S*, the composition moves to point 5, which is located above the saturation surface of *B* in Figure 4.1a. In the polythermal projection (Figure 4.1b), it is clear that point 5 is located inside compartment *B*. Therefore, cooling of this mixture allows the crystallization of pure *B*. Again, cooling is stopped before the liquid composition reaches point 6 at the boundary between compartments *A* and *B*, so as to prevent co-crystallization of *A*. Mixing this final mother liquor with a fresh feed (point 1) results in a composition given by

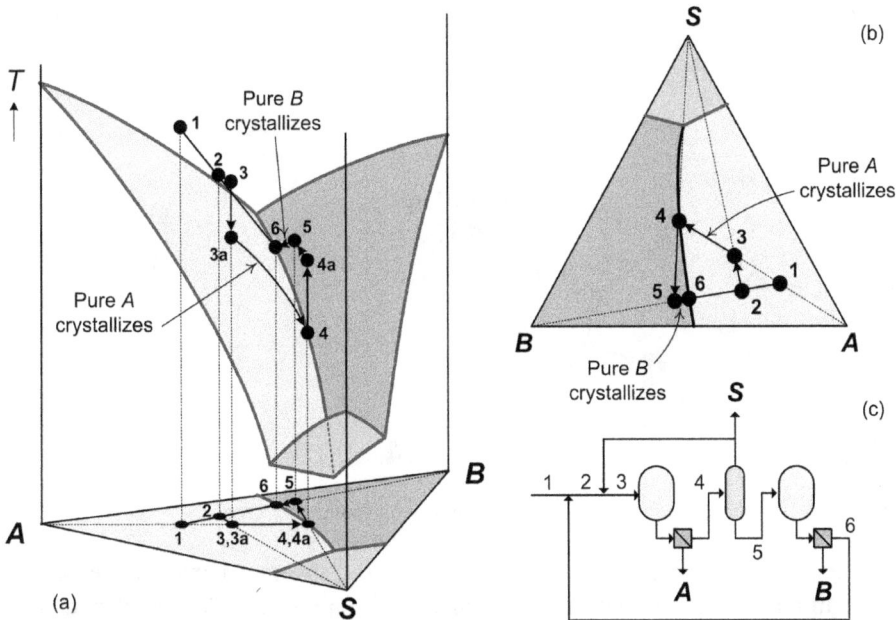

Figure 4.1: Bypassing thermodynamic boundary to crystallize both *A* and *B*: (a) representation on *T–x* diagram, (b) representation on polythermal projection, and (c) process flowsheet.

point 2, which is back in compartment *A* (Figure 4.1b). Adding some *S* to this mixture gives back point 3, and the cycle repeats itself.

As summarized in the flowsheet in Figure 4.1c, this sequence of movements allows complete separation of *A* and *B* from the mixture. In order to crystallize out pure *A* and pure *B*, the respective compartments must be "visited" at least once. In order to move from one compartment to the other, the boundary between the compartments has been bypassed. This is achieved by using another driving force (in this case evaporation) to remove *S* and move from point 4 to point 5, and by mixing to reach point 2 from point 6.

Isothermal cuts, instead of polythermal projection, can also be used for synthesizing the separation process. Figure 4.2 shows two isothermal cuts at temperatures, T_1 and T_2 (with T_2 higher than T_1), which are drawn on the same triangle, and a set of movements corresponding to the same separation process is depicted in Figure 4.1. The process starts with the combination of the feed (point 1) with a recycle stream (point 6) to form a mixture (point 2); it is followed by addition of solvent to give a composition inside the saturation region of *A* at temperature T_1 (point 3). Since this region corresponds to pure solid *A* at temperature T_1, crystals of *A* are obtained. The mother liquor composition is given by point 4. Removal of solvent from this stream produces a composition of point 5, which is located inside the crystallization region of pure solid *B* at T_2. Therefore, crystals of *B* can be obtained by bringing the solution to temperature

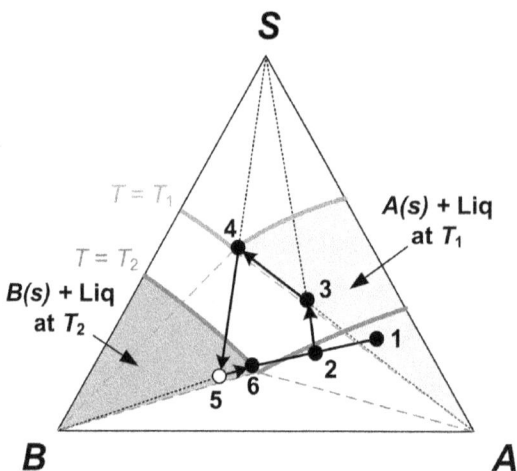

Figure 4.2: Isothermal viewpoint of crystallization-based separation process.

T_2. The mother liquor (point 6) is recycled to the feed stream, completing the process cycle. In this isothermal viewpoint, the emphasis is on the identification of saturation regions at specific temperatures. Again, the key concept is to use mixing or another driving force to move from one region to the other.

Once the separation process has been synthesized, mass balance equations can be written to calculate all the streams in the process. For example, referring to the process flowsheet in Figure 4.3, mass balance of each component over the entire process gives

$$F_1 x_{A,1} = F_A \tag{4.1}$$

$$F_1 x_{B,1} = F_B \tag{4.2}$$

$$F_1 x_{S,1} + F_7 = F_8 \tag{4.3}$$

where F_i denotes the mass flow rate of stream i, and $x_{j,i}$ is the mass fraction of component j in stream i. Note that the flow rate and composition of the feed (stream 1) should be known. Therefore, F_A and F_B can be directly obtained from eqs. (4.1) and (4.2).

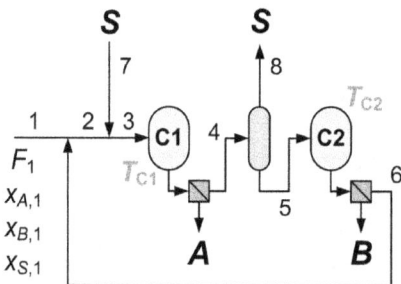

Figure 4.3: Process flowsheet for separating A and B from a solution containing both components.

In order to calculate the flow rate of the internal streams, it is necessary to consider the mass balance equations for individual units. The overall mass balance around crystallizer C1, which operates at T_{C1}, gives:

$$F_3 = F_A + F_4 \tag{4.4}$$

Similarly, overall mass balance around crystallizer C2, operating at T_{C2}, gives

$$F_5 = F_B + F_6 \tag{4.5}$$

Mass balance equations can also be written separately for components A and B around crystallizer C2, as follows:

$$F_5 x_{A,5} = F_6 x_{A,6} \tag{4.6}$$

$$F_5 x_{B,5} = F_B + F_6 x_{B,6} \tag{4.7}$$

Finally, the mass balance equations for the evaporator are

$$F_4 = F_5 + F_8 \tag{4.8}$$

$$F_4 x_{A,4} = F_5 x_{A,5} \tag{4.9}$$

$$F_4 x_{B,4} = F_5 x_{B,5} \tag{4.10}$$

By substituting eq. (4.9) into eq. (4.6) and eq. (4.10) into eq. (4.7), a system of two equations with two unknowns (F_4 and F_6) is obtained as shown here:

$$F_4 x_{A,4} = F_6 x_{A,6} \tag{4.11}$$

$$F_4 x_{B,4} = F_B + F_6 x_{B,6} \tag{4.12}$$

Equations (4.11) and (4.12) can be solved simultaneously to give

$$F_4 = \frac{F_B x_{A,6}}{x_{A,6} x_{B,4} - x_{A,4} x_{B,6}} \tag{4.13}$$

$$F_6 = \frac{F_B x_{A,4}}{x_{A,6} x_{B,4} - x_{A,4} x_{B,6}} \tag{4.14}$$

The values of F_3 and F_5 can then be obtained from eqs. (4.4) and (4.5), respectively. Equations (4.8) and (4.3) allow the calculation of F_8 and F_7, respectively. Having calculated all the streams in the process, it is then possible to estimate the required equipment size as well as the corresponding capital cost. Also, by performing energy balance calculations over the crystallizers and evaporator, the cooling and heating duties of each unit can be determined, thus allowing for the estimation of the operating cost of the process.

Note that $x_{A,4}$ and $x_{B,4}$ represent the solubility of components A and B at temperature T_{C1}, which are related to each other, since they define a single point along the saturation curve of A at T_{C1}. Also, since the location of the saturation curve depends on temperature, both $x_{A,4}$ and $x_{B,4}$ are functions of T_{C1}. Similarly, $x_{A,6}$ and $x_{B,6}$ are functions of T_{C2}. In other words, the choice of the two temperatures (T_{C1} and T_{C2}) determines all flow rates in the process via eqs. (4.13) and (4.14).

Example 4.1(I) – Separation of inorganic salts

A catalyst plant byproduct stream contains two inorganic salts, PX and QX. Since this stream is a combination of effluents from several plants, the feed salt ratio (ratio of PX to QX in the feed) varies significantly over time. While it is possible to use this mixture directly as a fertilizer solution, isolating the two salts as pure products would improve the economic value of the byproducts. An initial investigation has indicated that it is possible to selectively crystallize PX and QX with reasonably high purity from this byproduct stream. Therefore, it was decided to synthesize crystallization-based process alternatives for the separation process and study their economic feasibility.

The SLE phase behavior of the system is considered. As is the case with many other inorganic salt systems, solubility data at different temperatures can be readily found in the literature. Some of the literature data are plotted as isothermal cuts at different temperatures in Figure 4.4. Note that since these two salts share a common ion, (X^-), the PX/QX/water system can be treated as a molecular system with regard to its phase diagram. As seen from the figure, the location of the double saturation point shifts quite significantly with temperature change. At −17 °C (the topmost curve), the double saturation point contains nearly the same amount of PX and QX, whereas at 98 ° C (the bottommost curve), the ratio of PX: QX is about 1: 4. Such a shift makes it possible to move from the crystallization region of one salt at one temperature to the crystallization region of the other salt at another temperature. Based on utility considerations, 95 °C and 0 °C were chosen to be the two crystallizer temperatures.

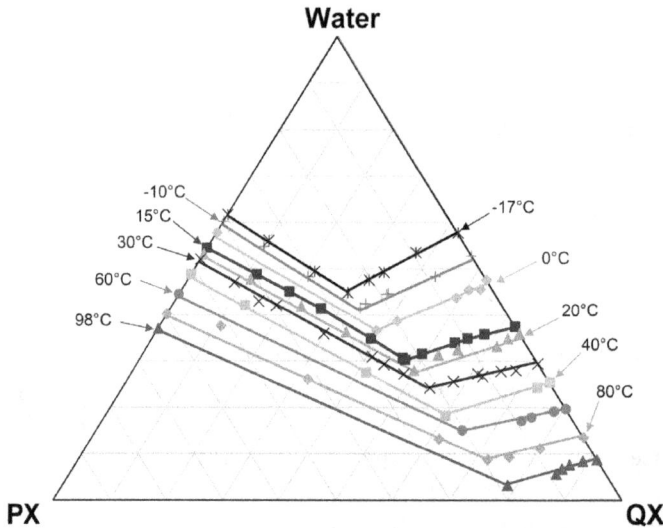

Figure 4.4: Solubility data of salts PX and QX in water at various temperatures.

Two process alternatives can be generated, as shown in Figure 4.5. In alternative 1 (Figure 4.5a), PX is crystallized first at 95 °C, then water is added to mother liquor and the solution is cooled to 0 °C to crystallize QX. In alternative 2 (Figure 4.5b), QX is crystallized first at 0 °C, then mother liquor is heated up and concentrated to crystallize PX at 95 °C. Although the processes are shown to separate both PX and QX with 100% recovery, in reality, the recovery would be lower because part of the recycle stream in both alternatives has to be purged to prevent accumulation of other components that may be present in the feed.

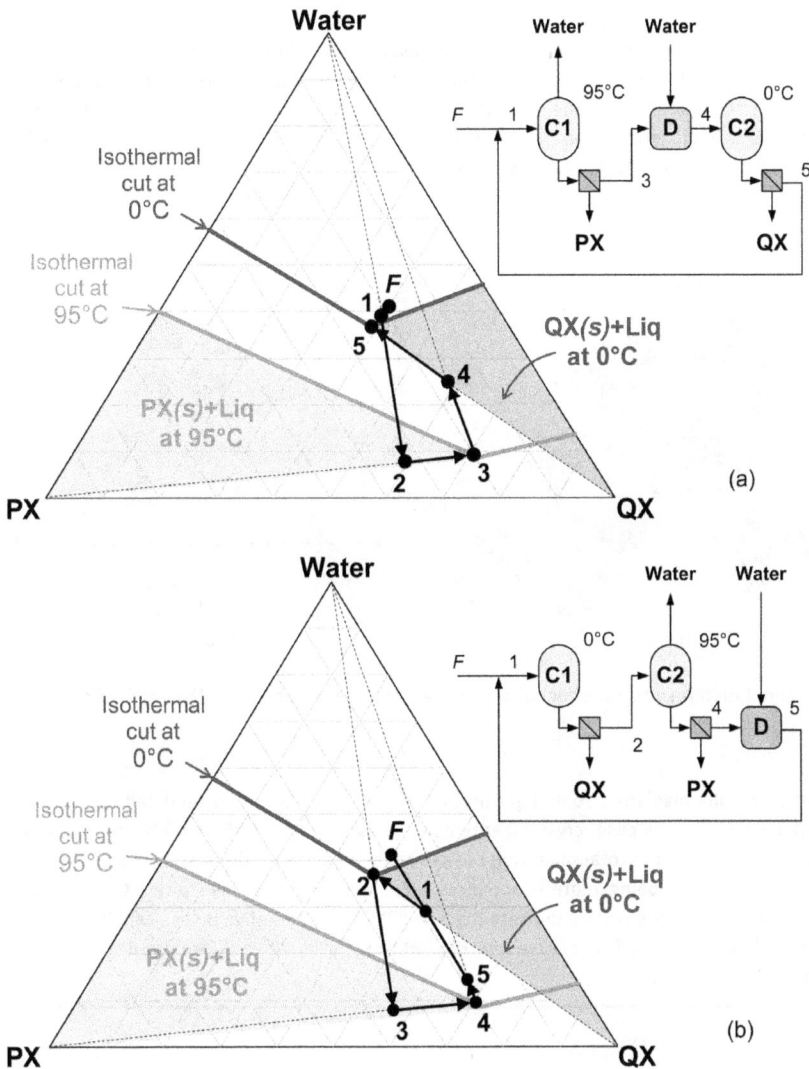

Figure 4.5: Process alternatives for separating PX and QX: (a) crystallizing PX first and (b) crystallizing QX first.

To allow convenient evaluation of the economic feasibility of each process alternative, an SLE model for the ternary system PX/QX/water was developed, based on the literature data. The model is embedded into an Excel spreadsheet, which performs mass and energy balance calculations to find the flow rate and enthalpy of all streams in a steady-state continuous process. The SLE model is used to obtain the crystallizer outputs for the feed composition calculated by the Excel spreadsheet. With the help of the spreadsheet, the effect of changing design parameters such as feed salt ratio, crystallization temperatures, and the fraction of recycle stream being purged on the mass and energy balance can be simulated. The operating cost of the plant, which includes the cost of utilities such as steam for heating and electricity for refrigeration, can also be readily calculated.

Figure 4.6a shows an example of the simulation results, which indicate that the total operating cost for the "PX first" alternative reaches a minimum when the PX and QX crystallizers are operated at 95 °C and −2 °C, respectively. Basically, the total operating cost is lower when the difference between the two crystallization temperatures is larger. This is because a larger shift in the double saturation temperature leads to smaller recycle streams.

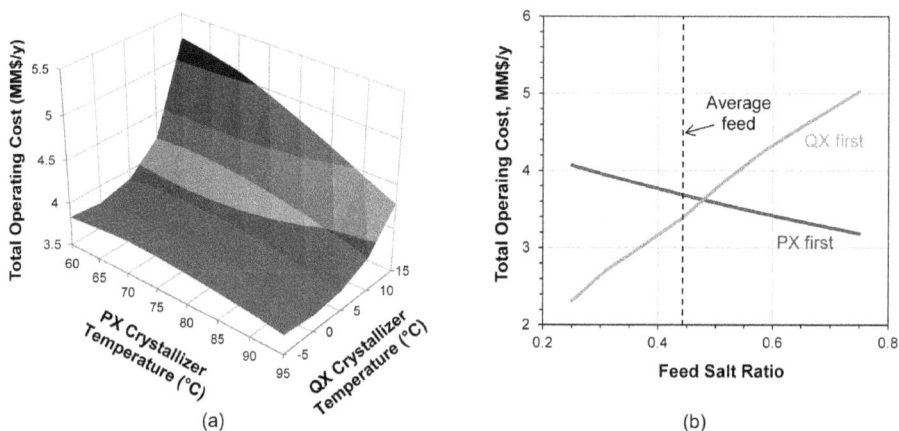

(a)   (b)

Figure 4.6: Simulation results: (a) effect of crystallizer temperatures and (b) effect of feed salt ratio.

The simulation results also shed some light on the effect of the fluctuating feed salt ratio, as depicted in Figure 4.6b (in this case, crystallizer temperatures are fixed at 95 and 0 °C for both alternatives). It is evident that the total operating cost strongly depends on the feed composition, with the "PX first" alternative being more economical when the feed contains more PX (higher salt ratio). More interestingly, the results indicate that the "PX first" alternative is less sensitive to fluctuations in feed composition. For this reason, the "PX first" alternative is preferred, from operational stability point of view.

Example 4.2 – Production of potash from sylvinite
Instead of using two crystallizers to obtain pure components from a solution, a solid mixture can be separated by selectively dissolving a component in a dissolver, followed by crystallizing it out in a subsequent crystallizer. This selective dissolution process has been used commercially in the production of potash (KCl) from sylvinite (KCl+NaCl mixture) [4]. The phase diagram of KCl/NaCl/water system is shown in Figure 4.7, with two isothermal cuts at T_c and T_d ($T_c<T_d$). Note that the two sets of saturation curves at the different temperatures cross each other. This is due to the opposite dependence of KCl and NaCl solubility on temperature – KCl solubility increases with temperature, while NaCl solubility is lower at higher temperature.

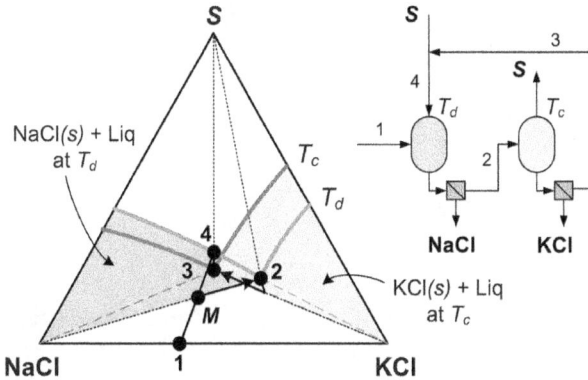

Figure 4.7: Production of potash from sylvinite by selective dissolution process.

The feed, represented by point 1 on the phase diagram, is a solid mixture of KCl and NaCl. This mixture goes into a dissolver operating at temperature T_d, where it is mixed with a solution represented by point 4 to give point M. Focusing on the saturation curves at T_d, point M falls in the NaCl saturation region. This means that all KCl in the solids would dissolve in the liquid phase, which is represented by point 2, and the solid would contain only NaCl. Note that in order to minimize the amount of solution that must be added to the dissolver, point M is located near the boundary of the NaCl saturation region at T_d, and point 2 is located close to the double saturation point at T_d.

Switching the focus to the saturation curves at T_c, point 2 is located inside the KCl saturation region at this temperature. Therefore, pure KCl can be crystallized out from this solution at temperature T_c. Some water is evaporated to maximize the KCl recovery, and the mother liquor composition (point 3) is located close to the double saturation point at T_c. By adding back water to point 3, the solution introduced to the dissolver (point 4) is obtained, and the cycle repeats itself.

4.2.2 Quaternary system

Consider a quaternary system consisting of three solutes, A, B, and C, and a solvent S. The three-dimensional polythermal phase diagram as well as its Jänecke projection is shown in Figure 4.8. The feed composition represented by point F is located in compartment A, where A can be recovered by crystallization. The fresh feed is mixed with a recycle stream to give an overall composition given by point 1 (Figure 4.8a). At this time, it is known that there is such a recycle stream, but its exact composition is not yet known. (For a sleek preview, look at Figure 4.8d.)

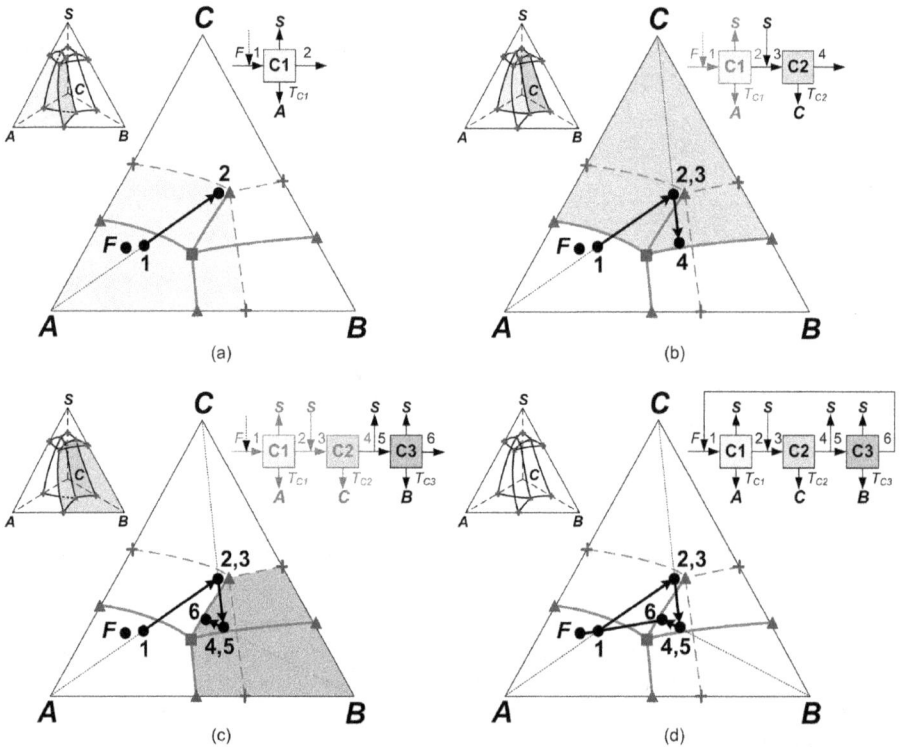

Figure 4.8: Crystallization-based separation process of three products from a quaternary mixture: (a) crystallizing pure A, (b) crystallizing pure C, (c) crystallizing pure B, and (d) closing the recycle loop.

Point 1 is also located in compartment A, which is a three-dimensional space enclosed by six "walls": AC-ABC-$ABCS$-ACS, AB-ABC-$ABCS$-ABS, and ACS-$ABCS$-BS-AS, and three walls on the ABS, ACS, and ABC faces of the pyramid. Eutectic points AC, AB, and ABC are located on the base of the tetrahedron, while AS, ACS, ABS, and $ABCS$ are elevated. The eutectic points are not labeled to avoid cluttering. By cooling point 1 to temperature T_{C1}, pure A is crystallized out, and the composition moves to point 2. Depending on how much solvent is in the fresh feed, some may have to be removed. Focusing next on compartment C, which is highlighted in Figure 4.8b, there appears to be an overlap with compartment A, but this is just because of the projection. Note that AC-ABC-$ABCS$-ACS is a leaning "wall", with compartment A below it and compartment C above it. Therefore, compartment C can be reached by adding solvent to point 2, so that it would move vertically upward to point 3 in compartment C (which appears to be the same point in the Jänecke projection). It is now possible to crystallize pure C, and the liquid composition moves to point 4. Next, a similar strategy can be applied to move into compartment B, which is highlighted in Figure 4.8c. Since the BC-ABC-$ABCS$-BCS surface also leans toward B, removing solvent from point 4 leads to point 5 in compartment B. Pure B can be crystallized out, and the liquid composition moves to point 6, which is recycled by mixing with F. The result is point 1, which is back in compartment A, as shown in Figure 4.8d.

Despite the use of projections to help visualize the high-dimensional phase diagram, the same strategy as the one for ternary system has been followed – visiting one compartment after another and bypassing the boundary between compartments by solvent addition or removal – to achieve complete separation of the mixture. Process alternatives can also be generated by considering different sets of movements to visit the compartments in different sequences. They can then be evaluated via mass balance calculations and cost estimations to determine the best alternative.

Example 4.3 – Separation of three amino acids

A process for separating three amino acids using crystallization has been reported by Takano et al. [101]. This process can be rationalized by understanding the SLE phase behavior of the quaternary system involving the three amino acids and water as the solvent, as depicted in Figure 4.9. Shown on the left-hand side is the Jänecke projection of three isothermal cuts at different temperatures. A sketch of the original three-dimensional phase diagram, featuring three sets of saturation surfaces at the different temperatures, is also shown in the figure.

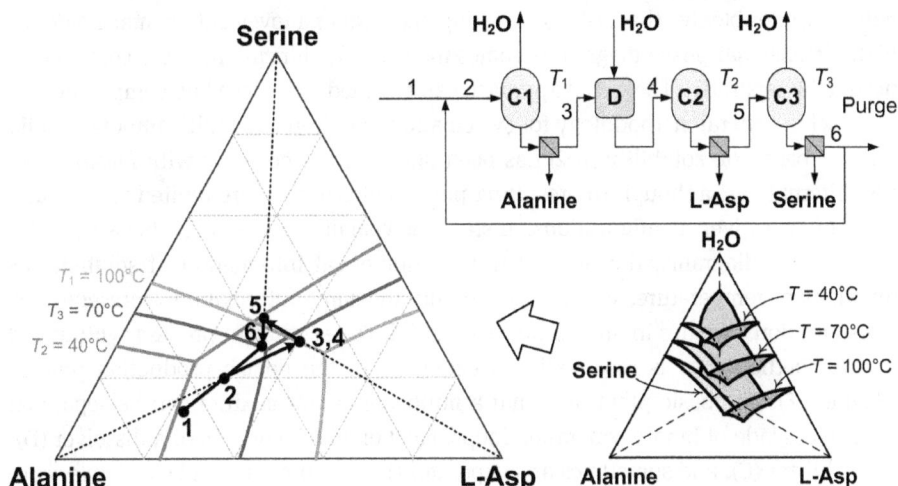

Figure 4.9: Process for separating three amino acids by crystallization (data from [101]).

The feed mixture, indicated by point 1, is first mixed with the recycle stream to give point 2, which is located in the crystallization region of (DL)-alanine at 100 °C. Removal of water from this stream while keeping a constant temperature at 100 °C causes (DL)-alanine to crystallize out, leaving behind mother liquor with the composition of point 3. By stopping just before the liquid composition hits the double saturation curve, co-crystallization of (DL)-serine is prevented. Water is then added to move the composition to point 4, which is located inside the crystallization region of L-asparagine at 40 °C. If just enough water is added to avoid getting into the region of multiple solids, cooling to 40 °C results in crystallization of pure L-asparagine, and the mother liquor (point 5) is located close to the double saturation curve between L-asparagine and (DL)-serine at 40 °C. Finally, this mother liquor is heated to 70 °C, at which the composition falls within the crystallization region of (DL)-serine. Water is removed again to crystallize (DL)-serine, and the mother liquor (stream 6) is recycled to the feed. Partially purging this stream avoids accumulation of any minor impurities present in the feed stream.

In this example, isothermal cuts, which are easier to determine experimentally, have been used instead of polythermal projection. By observing the location of these double saturation curves at different isothermal cuts, it is easy to determine whether moving from one compartment to another by switching to another temperature is possible. However, selecting the suitable movement (solvent addition or removal) requires information about the solvent content at the double saturation curves at different temperatures. As the solvent content is not shown in the Jänecke projection, this information can only be obtained from the three-dimensional isothermal cut.

4.2.3 Multicomponent system

In principle, the same strategy of moving around in composition space and visiting various compartments to crystallize pure components can be applied to systems involving any number of components. Projections and cuts can be used to help visualize the SLE phase diagrams for multicomponent systems, which are high-dimensional in nature. Undoubtedly, the task of selecting the proper movement to maneuver the high-dimensional phase diagram becomes increasingly difficult, as more components need to be separated. The visual approach can be aided or replaced by computations.

Such a general methodology for systematic separation of a multicomponent solid mixture based on solubility [102] has been developed to come up with feasible process alternatives, although the resulting process alternatives are limited to permutations of two given configurations. Instead of visualizing the SLE phase behavior using phase diagrams, the method utilizes numerical information of solubility as functions of temperature, which is essentially the same information represented by the saturation varieties in an SLE phase diagram. The separation process synthesized by this methodology is essentially an extension of the potash production process (Figure 4.7). The basic premise is that a multicomponent mixture can be separated using a cascade of binary separation loops, each of which consists of a dissolver (D), a crystallizer (C), and sometimes an evaporator (E). The process can be a *simple separation*, in which a single pure component is recovered from each loop, or a *complex*

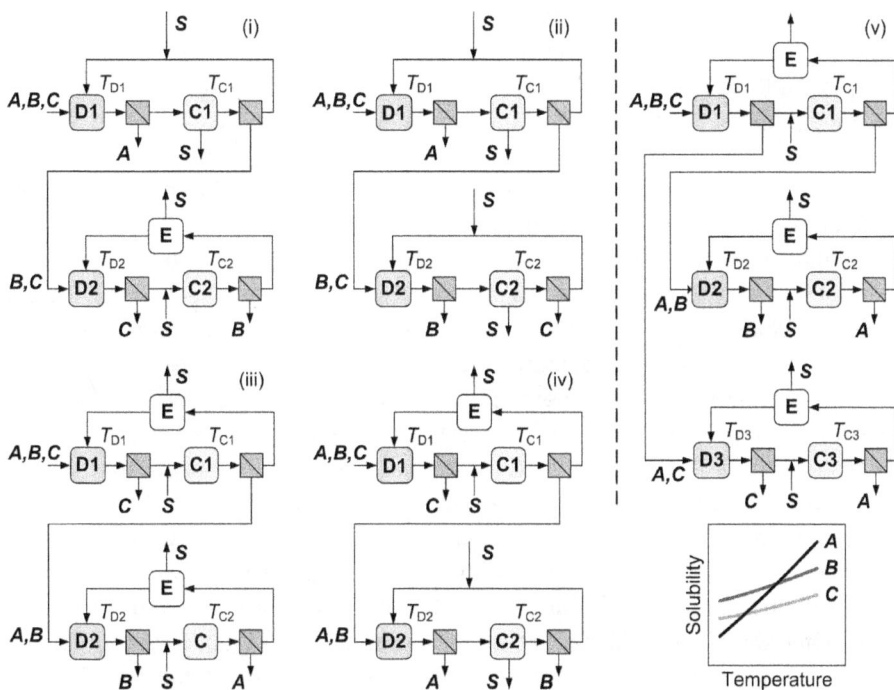

Figure 4.10: Cascade process alternatives for simple separation (i–iv) and complex separation (v) of a three-component mixture having the given joint solubility behavior.

separation, in which a fraction of a certain component leaves a loop after the dissolver and the remainder comes out after the crystallizer. Figure 4.10 shows examples of cascade separation processes to separate a solid mixture of three components A, B, and C using a solvent S. The dependence of the joint solubility of these components on temperature is shown at the bottom of the figure. All these process alternatives are made up of only two types of loops: the *conventional scheme*, in which solvent is added to the dissolver and removed at the crystallizer as in the potash process, and the *alternate scheme*, in which solvent is added just before the crystallizer and removed from the recycle stream to the dissolver. The difference between one process alternative and the other lies in the separation sequence and the number of components separated in the filters.

Process alternatives (i) to (iv) are examples of simple separation. The conventional scheme is used in the first loop in (i) and (ii), as well as in the second loop in (iv), to obtain pure A in the dissolver. Since A has the lowest joint solubility at low temperature, the dissolver is operated at a lower temperature than the crystallizer, allowing A to be the only component in the solid phase from the dissolver. B and C, which have lower joint solubility at the higher crystallizer temperature, are recovered in the crystallizer. On the other hand, the alternate scheme is used in the first loop in (iii) and (iv) to obtain pure C in the dissolver. The dissolver temperature is higher than the crystallizer temperature, leading to only C (which has the lowest joint solubility at high temperature) in the solid phase from the dissolver. Basically, by comparing the joint solubility of all components entering each loop at two specified temperatures, either the conventional or the alternate scheme is selected, and the higher temperature is assigned to either the dissolver or the crystallizer. Different process alternatives are obtained by repeating this procedure for different separation sequences. Furthermore, a set of design equations can be developed to calculate the recycle and solvent make-up flow rates, based on the solubility values and the various process alternatives can then be compared by evaluating the corresponding dissolver and crystallizer costs.

Process alternative (v) is an example of complex separation, which can be more practical when there is a large amount of component A in the feed, such that not all of it can be dissolved in the dissolver. Thus, some A leaves the first loop together with C, while the remainder comes out together with B in the crystallizer. Consequently, two more loops are required to obtain three pure component streams. The complex separation alternative may also be preferable in other situations, such as when separation between some components is unnecessary. For example, if several impurities end up in the same waste stream, there is no need to separately recover them as pure components. A complex separation process can also be synthesized by comparing the joint solubility of all components entering each loop, and similar design equations can be used for evaluation of process alternatives.

4.2.4 Conjugate salt system

The same principles can also be used to synthesize the separation process for electrolyte mixtures such as conjugate salt systems. As an example, consider the conjugate salt system of AX, AY, BX, and BY in water. The feed contains the same molar amount of A^+, B^+, X^-, and Y^-, and it is desirable to crystallize out the compatible pair, AY and BX. Two isothermal cuts at T_1 and T_2 are shown in Figure 4.11, along with the Jänecke projection and the AY–BX diagonal cut of the phase diagram.

The feed (point 1) is first mixed with a recycle stream to obtain point 2, followed by removing water to crystallize AY at temperature T_1. Afterward, water is added back to the mother liquor (point 3) to obtain point 4, which is cooled down to temperature T_2, to crystallize out BX. The final mother liquor (point 5) is recycled. Since all the process points lie along the AY-BX diagonal, the process path can be shown on the diagonal cut for a clearer view of water addition and removal. This diagonal cut is essentially the same as the phase diagram for a ternary molecular system.

Because of the phase behavior, sometimes operation along the diagonal does not work. As an example, consider the conjugate salt system consisting of HA, H_2B, and their sodium salts in Figure 4.12. The Jänecke projection of isothermal cuts at T_1 and

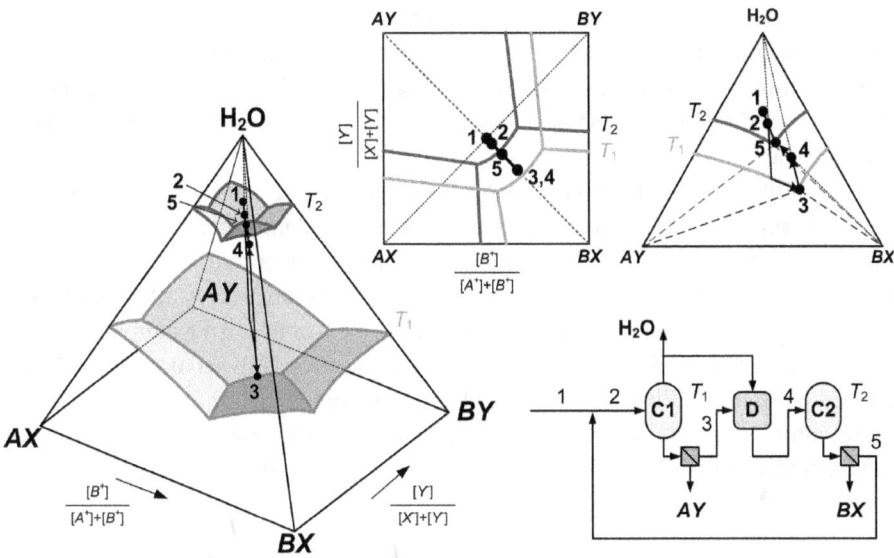

Figure 4.11: Separation of a conjugate salt pair by crystallization.

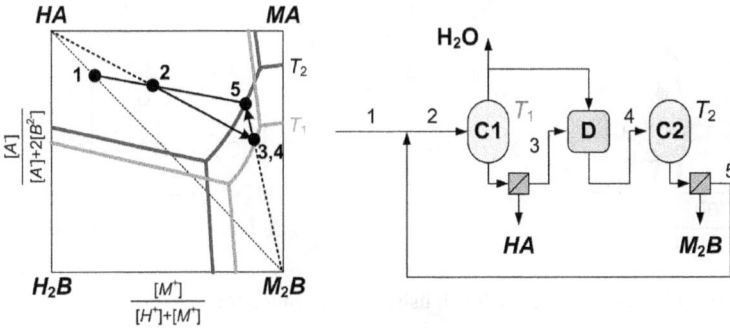

Figure 4.12: Separation of a compatible salt pair with an off-diagonal process path.

T_2 are shown on the diagram. The objective is to obtain HA and M_2B from the mixture indicated by point 1 in the diagram. While point 1 lies on the diagonal (HA-M_2B), it is not possible to go from the HA region to M_2B region if movements were limited along the diagonal, because there is an H_2B region lying between those two regions.

However, it is possible to use a little improvisation to overcome this problem. Mixing the feed (point 1) with recycle stream 5 gives point 2, which is located off-diagonal. Evaporating water at T_1 leads to crystallization of HA, because point 2 is inside the HA region at that temperature. The liquid composition moves to point 3, which is close to the double saturation curve. Adding water moves the composition to point 4, which is located at the same point as 3 on the Jänecke projection. Since point 4 is inside the M_2B region at T_2, M_2B can be crystallized out at T_2 and the liquid composition moves to point 5, which is then recycled. Although the resulting separation process is similar to the one in Figure 4.11, the process path cannot be plotted on a diagonal cut, so a Jänecke projection is more suitable for visualizing the movements.

It is also possible to recover the incompatible salt pairs instead of the compatible ones, but the process becomes considerably more complicated because the saturation regions of the incompatible pair are not adjacent to each other. After crystallizing one of the desired salts, it is usually necessary to crystallize other salts first, before the region of the second product can be reached. An example of such a process is shown in Figure 4.13. Although the objective is to separate BY and AX, AY and BX are crystallized out as part of the strategy to reach the saturation region of AX, after separating BY. These salts are recycled to the dissolver and mixed with the fresh feed, so that the overall process only gives BY and AX as products.

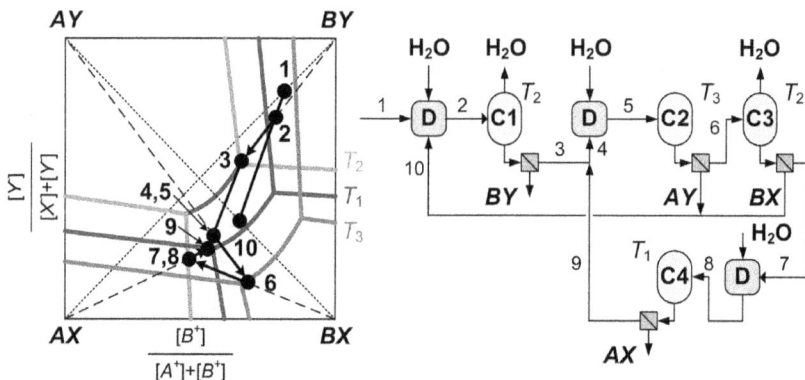

Figure 4.13: Separation of an incompatible salt pair using crystallization [73].

Several generic separation schemes have been proposed for various cases with different separation objectives (separating the compatible salts, separating the incompatible salts, or separating two compatible salts and one incompatible salt) as well as different characteristics of the SLE phase behavior (location of ternary saturation points at different temperatures) [73]. All these schemes can be obtained using the same approach of moving from one region to the other to crystallize the desired products, keeping in mind that crystallization of an undesired product and recycling it may be a useful means to eventually getting to the desired region. Furthermore, this is, essentially, the most practical approach for systems with complex SLE phase behaviors, such as those involving compounds and hydrates, which have to be considered on a case-by-case basis.

Example 4.4 – Isolation of sodium chloride from a salt lake

Salt lakes contain rich resources of inorganic salts – mostly sodium chloride, but sometimes also comparatively more valuable commodities such as lithium salts. The brine from the lake is normally pumped to a series of solar ponds where crystals of different salts are obtained successively through evaporative crystallization. A common problem is the determination of which particular salt (or mixture of salts) would crystallize out under a specific set of operating conditions, from a feed with a given composition. Consider a simplified scenario in which the salt lake brine comprises of only four major ions: Na^+, K^+, Cl^-, and SO_4^{2-}. The composition and the SLE phase diagram for this system at 25 °C, calculated based on literature data, are shown in Figure 4.14. Ionic coordinates are used due to the fact that the salts dissociate into ions in solution. Regions of various salts – potassium chloride, sodium chloride, sodium sulfate, Glauber's salt (sodium sulfate decahydrate), double salt sodium potassium sulfate, and potassium sulfate – can be seen on the diagram.

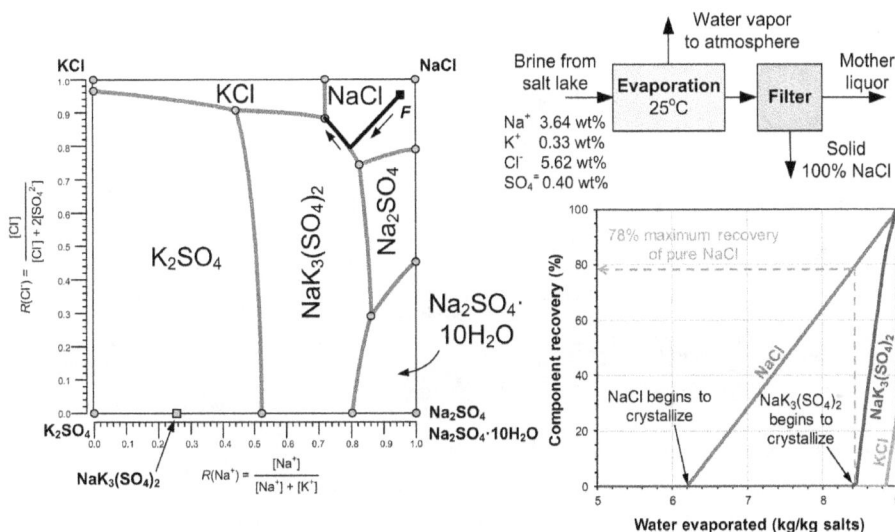

Figure 4.14: Crystallization of sodium chloride at 25 °C.

Since the feed composition (point F) is located inside the NaCl region, NaCl would crystallize out first upon evaporation of water. The liquid composition would move diagonally (as indicated by the arrow) as it loses Na^+ and Cl^- to the precipitating solid, until it eventually reaches the boundary with the neighboring region, which is the region of double salt $NaK_3(SO_4)_2$. Further evaporation of water causes the two salts to crystallize together, and the liquid composition to move along the boundary, in the direction of decreasing water content. Finally, as the composition arrives at the intersection with the KCl region, the solution becomes saturated with KCl as well, and further water removal causes all three salts (NaCl, $NaK_3(SO_4)_2$, and KCl) to precipitate out at the same time. The component recoveries (defined as the ratio of the amount of a component in the solid product to its amount in the brine) as a function of the amount of water evaporated are also shown in Figure 4.14. About 78% of the NaCl in the brine can be recovered as pure component.

The case where the mother liquor from the solar pond is sent to a crystallizer operating at 100 °C is interesting. Figure 4.15 shows the SLE phase diagram for the salt system at 100 °C. After crystallizing pure NaCl, the mother liquor is now leaner in both Na^+ and Cl^- compared to the original feed, as indicated by point F'. Note that the region boundaries have shifted and at this temperature, there is no decahydrate region. Since point F' is located inside the Na_2SO_4 region at 100 °C, Na_2SO_4 comes out first upon removal of water, followed by NaCl, $NaK_3(SO_4)_2$, and finally KCl. It is therefore possible to evaporate just enough water to produce pure Na_2SO_4 in the crystallizer. From the plot of calculated component recoveries as a function of the amount of water evaporated, the maximum recovery of pure Na_2SO_4 is about 65%, which occurs after about 1.7 kg water/kg salts has been evaporated. Na_2SO_4 eventually redissolves and precipitates as $NaK_3(SO_4)_2$ upon evaporation of more water.

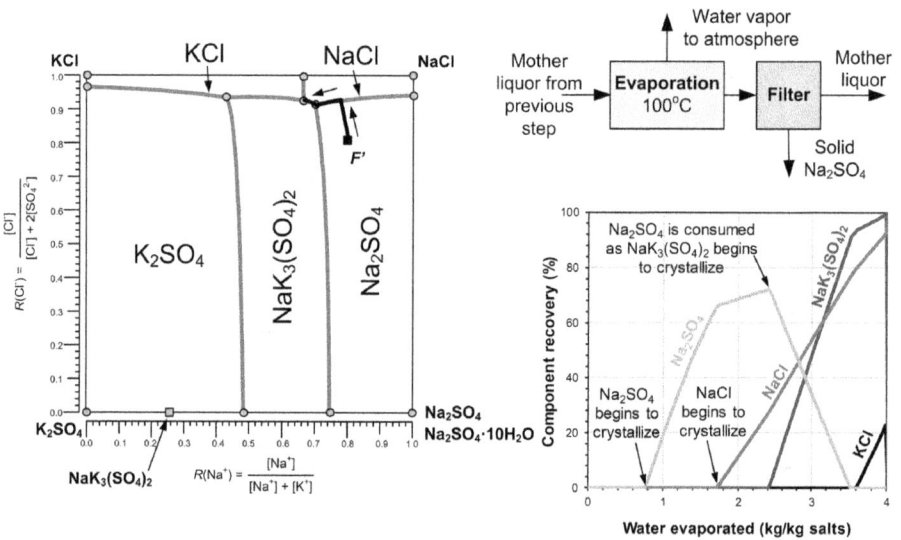

Figure 4.15: Crystallization of sodium sulfate at 100 °C.

Example 4.5(I) – Purification of a chlorination product

A company was developing a multistep process for producing a monomer D from an acid raw material (H_2A), as illustrated in Figure 4.16. The first step is chlorination of H_2A to form an intermediate H_2B. The conversion in the chlorination reaction is about 85%. H_2B, along with unreacted H_2A and some impurities, is recovered by crystallization and undergoes coupling to form the sodium salt, Na-C, of another intermediate, C. The sodium salt is subsequently neutralized with HCl to recover C. The monomer product, D, is finally obtained by the dehydration of C. The presence of these impurities, particularly the unreacted H_2A, is believed to cause side reactions during the coupling step, which reduce the overall yield. Therefore, it was decided to separate the unreacted H_2A prior to the coupling step, rather than letting it crystallize together with H_2B.

While it is possible to separate H_2A and H_2B by crystallization alone, it was proposed that NaOH be added to the crystallizer to recover NaHB, which is used in the subsequent coupling reaction. Therefore, a sketch of the SLE phase diagram of the H^+, Na^+/A^{2-}, B^{2-} system was developed, based on the results of an initial investigation of the solubility of the relevant components as well as their sodium salts. Figure 4.17a shows the Jänecke projection with respect to the solvent (water) of two isothermal cuts at 30 and 60 °C, which correspond to the lower and upper

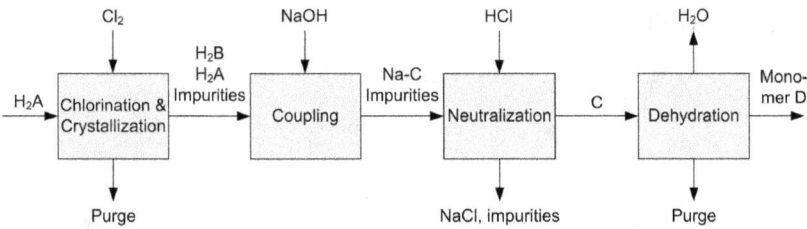

Figure 4.16: Simplified block diagram of a monomer production process.

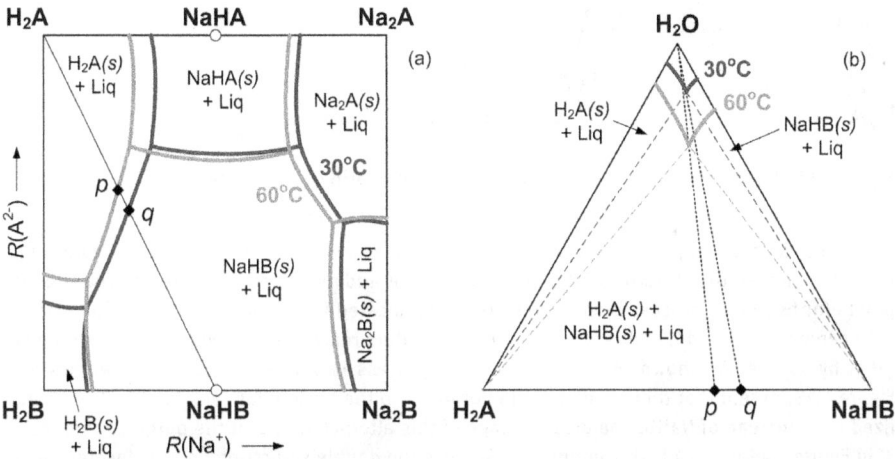

Figure 4.17: SLE phase diagram of the H^+, Na^+/A^{2-}, B^{2-} system showing isothermal cuts at 30 and 60 °C: (a) Jänecke projection and (b) cut along H_2A–NaHB diagonal.

limits of the desirable crystallization temperatures. Note that there are regions for the mono-salts NaHA and NaHB, which can also precipitate out of the mixture, depending on how much sodium is added to the system.

One alternative to achieve the separation objective is to recover H_2A and NaHB, which is feasible as their compartments share a common boundary. To do so, just enough sodium is added to the system to place the feed composition along the line connecting H_2A and NaHB. Figure 4.17b shows a sketch of the cut along this line, which takes the shape of a triangle featuring two sets of saturation curves at 30 and 60 °C. Projection of the double saturation points from the H_2O vertex onto the H_2A–NaHB line gives points p and q, which are also indicated in Figure 4.17a.

Figure 4.18 shows the SLE phase diagram obtained from experimental results, along with a block diagram of a process for separating H_2A and NaHB. The experimental data points at 30 and 60 °C are identified on the diagram by the squares and triangles, respectively. The process begins with the addition of NaOH, a recycle stream, and the product from the chlorination reactor (stream F) in a dissolver, D, to give the feed to the first crystallizer (stream 1), which is located in the NaHB region at 60 °C. After crystallizing NaHB, water is added to the mother liquor (stream 2) to produce stream 3 inside the H_2A region at 30 °C. By cooling down this mixture to 30 °C, H_2A is crystallized out and purged. The mother liquor (stream 4) is concentrated and recycled. In reality, some of the impurities that have very low solubility are co-crystallized with NaHB and H_2A, but this is not expected to be a problem. As the majority of the impurities stay in the liquid phase, a partial purge is necessary to prevent them from accumulating.

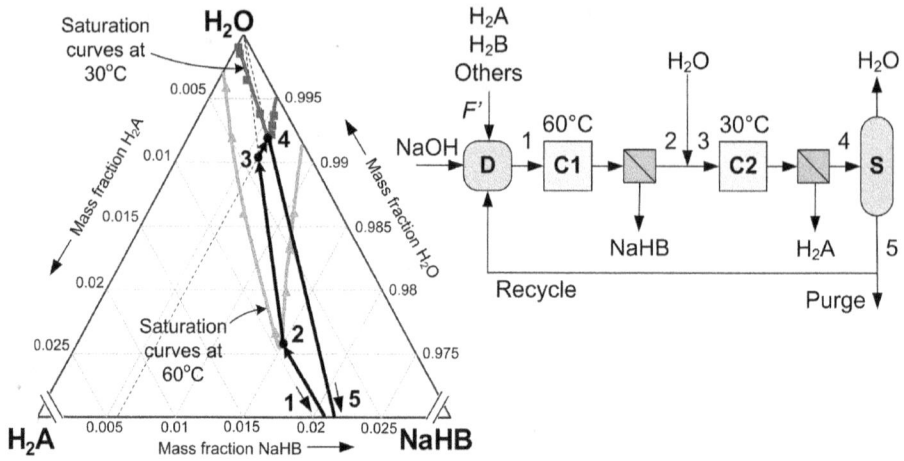

Figure 4.18: A process alternative for separating H_2A and NaHB by crystallization.

The locations of the process points (1–5) on the Jänecke projection are depicted in Figure 4.19a. Point F indicates the mixture of F' and NaOH, before adding the recycle stream. The process points for two other process alternatives are also indicated in the same figure. One alternative is to separate H_2A and H_2B directly from the chlorination reactor outlet, without first adding sodium hydroxide. As shown in Figure 4.19b, the process is very similar to the one shown in Figure 4.18, except that there is no addition of NaOH to the dissolution tank, and H_2B is crystallized in C1 instead of NaHB. The disadvantage of this alternative is that the purge stream (point 5' in Figure 4.19a) has a high content of H_2B, since the double saturation point is located closer to H_2B than to H_2A.

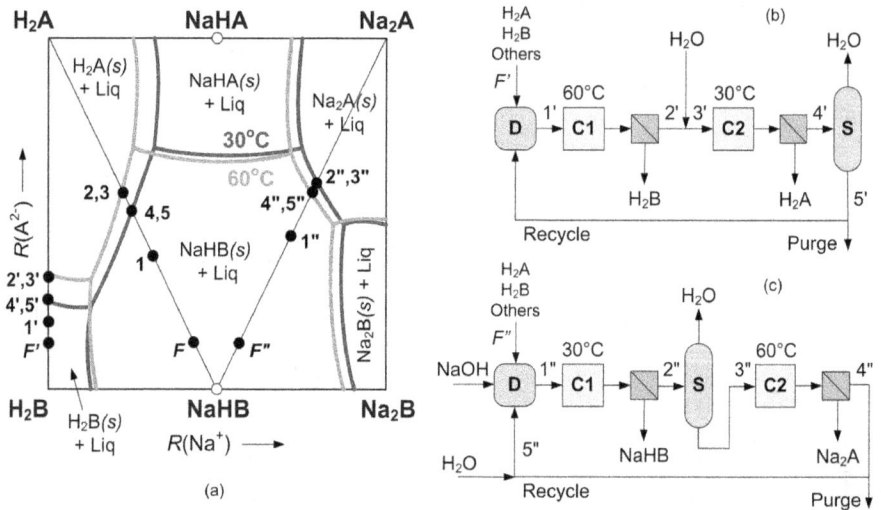

Figure 4.19: Other process alternatives: (a) location of process points on the Jänecke projection, (b) separation of H_2A and H_2B, and (c) separation of Na_2A and NaHB.

Yet another alternative is to add more NaOH to the dissolver, so that the mixture composition is located farther away from the H_2A-H_2B edge compared to the first alternative, such as at point F'' in Figure 4.19a. As shown in Figure 4.19c, NaHB and Na_2A can be crystallized out at 30 and 60 °C, respectively. However, this alternative is not desirable because the recovered Na_2A cannot be directly recycled to the chlorination step, without first converting it to H_2A.

4.3 Effect of Solvent

One key idea in the crystallization-based separation processes discussed so far is crossing a thermodynamic boundary by removing or adding solvent. The effectiveness of such movements depends on the SLE phase behavior of the overall system, which includes the solvent as well. Therefore, the choice of solvent or a mixture of solvents is very important in the conceptual design of a crystallization process.

4.3.1 Solvent selection

Consider two systems depicted in Figure 4.20a, consisting of the same two solutes A and B but with two different solvents S_1 and S_2. The phase behavior on the AB side is the same for both systems. The location of the AB binary eutectic is therefore fixed, because it is a characteristic of the binary system. However, the eutectic troughs are different depending on the solvent. If the eutectic trough between compartments A and B leans more toward B, as in the left figure, removing solvent from point 4 gives point 5, deep inside compartment B. This corresponds to a higher maximum per-pass recovery of pure B, which depends on the distance between point 5 and point 6 at the boundary, for solvent S_1. Such a solvent can be said to be a "good" solvent for separation purposes. On the other hand, if the situation is as

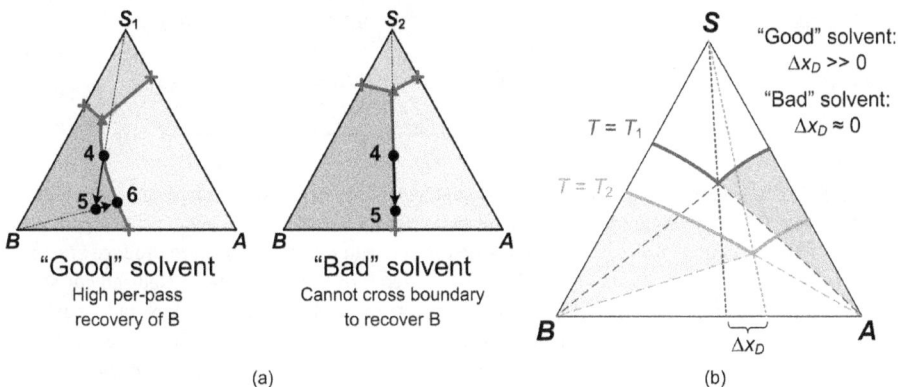

Figure 4.20: Phase diagram for solvent selection: (a) polythermal view and (b) isothermal view.

shown on the right figure, in which the eutectic trough lines up with the solvent apex, it is not possible to cross into compartment B by removing solvent from 4. Therefore, S_2 is considered a "bad" solvent for separation.

An isothermal view of the same situation is shown in Figure 4.20b. It is often more practical to consider two isothermal cuts at given temperatures, corresponding to the likely operation temperatures based on available utilities such as steam, cooling water, or chilled water. The larger the difference in the relative amounts of the two solutes in the double saturation point at the two different temperatures (Δx_D), the better the solvent is for separation. Consequently, either the inclination of the eutectic trough or the magnitude of Δx_D can be used as a criterion for selecting a suitable solvent for a crystallization-based separation process.

Example 4.6 – Separation of isoflavones
Isoflavones are plant-derived phenolic compounds with estrogenic activity, which can be found in leguminous plants such as peas, beans, and clovers. A crystallization-based process to separate isoflavones – daidzein (D) and genistein (G) – from a mixture isolated from soybean flour has been developed by Harjo et al. [103]. A key step in the development is finding a suitable solvent for the separation. For this purpose, the SLE phase behavior of a ternary system containing D, G, and solvent is considered. Table 4.2 summarizes the results of double saturation point measurements at 15 and 60 °C for four common organic solvents in pharmaceutical processing, namely methanol, ethanol, ethyl acetate, and acetone.

Table 4.2: Double saturation points of the ternary systems of isoflavone D, G, and solvents at two different temperatures [103].

Solvent	Temperature ° C	Solvent-free composition of the double-saturation point (x_D)	Difference (Δx_D)
Methanol	15	0.222	0.030
	60	0.252	
Ethanol	15	0.341	0.050
	60	0.291	
Ethyl acetate	15	0.093	0.027
	60	0.120	
Acetone	15	0.266	0.015
	60	0.251	

Based on the results, ethanol is selected as solvent as it is found to have the largest value of Δx_D. A sketch of the SLE phase diagram of the ternary system of D, G, and ethanol is shown in Figure 4.21, along with the separation process. The feed is first mixed with a recycle stream to produce stream 1, which is located inside the region of pure D at 60 °C. Therefore, pure D can be crystallized out at this temperature. The mother liquor (stream 2) is located close to the double-saturation point to maximize the per-pass recovery of D. Next, ethanol is added to give a composition inside the region of pure G at 15 °C (point 3), so that pure G can be obtained by crystallization at 15 °C. The remaining mother liquor (stream 4) is sent to a distillation unit to remove ethanol prior to recycling. The bottoms product (stream 5) is partially purged to avoid accumulation of minor impurities present in the feed.

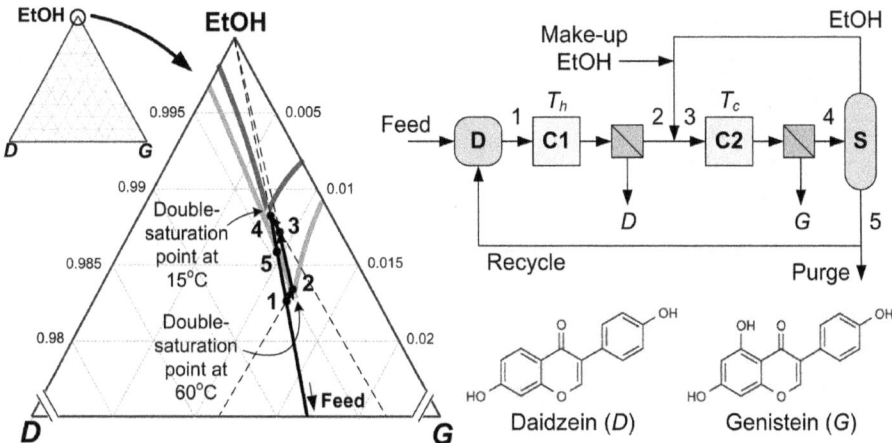

Figure 4.21: Process for separating isoflavones by crystallization.

4.3.2 Solvent switching

The SLE phase behavior of most systems that have been discussed so far features a significant variation of the multiple saturation point composition with temperature, which allows for the use of movements to cross from one compartment to another (or from the saturation region of one component to the saturation region of another component) with ease. In other words, these belong to the case of the "good" solvent in Figure 4.20a. However, this is not always the case, as some systems exhibit only a small shift in the multiple saturation point composition, even when two significantly different temperatures are selected. Or even worse, as in the case of a "bad" solvent, there is barely any shift at all. When the shift is small, the process points would be located very closely together, the per-pass recovery of the pure component would be low, and the recycle flow rate may be prohibitively large.

One alternative to overcome this problem is to use the solvent switching technique, in which different solvents are employed to crystallize different components. This is illustrated in Figure 4.22 for the separation of A and B from their mixture, using two different solvents S_1 and S_2. The polythermal SLE phase diagrams indicate very different behaviors between the ternary systems involving S_1 and S_2.

The separation process begins by mixing feed F with a recycle stream 7 (both are solid mixtures), to obtain stream 1. This stream is then dissolved in solvent S_1 to give a solution of point 2 located in compartment B, so that pure B can be crystallized out in the first crystallizer. The solvent is then completely removed from the mother liquor to give a mixture of solids of an overall composition represented by

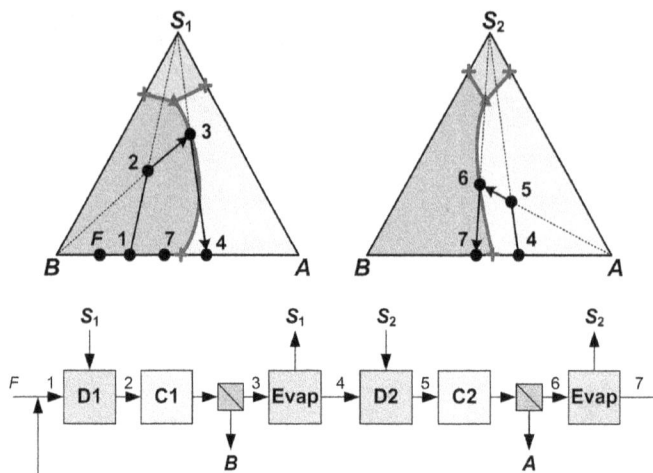

Figure 4.22: Separation of two solutes by crystallization in different solvents.

point 4, which is dissolved in solvent S_2. With this solvent, the resulting solution (point 5) is located well inside compartment A, because the compartment boundary leans away from A instead of toward A as with solvent S_1. Consequently, pure A can be crystallized with a relatively high per-pass recovery. The solvent is again completely removed from the mother liquor (point 6), to give a mixture of solids of an overall composition represented by point 7, which is recycled.

Example 4.7 – Separation of 1β-methylcarbapenem key intermediates
A practical method for separation of the methyl esters of two diastereomeric, α-methylated, 2-azetidinon-4-yl acetic acid derivatives, with distinctly different relative solubility in two solvents, is described by Bender et al. [104]. They reported that the solubility of the more soluble diastereomer (A) in isopropanol is 1.65 times that of the less soluble diastereomer (B), whereas the solubility of A in toluene is only 1.15 times that of B. The proposed separation process is illustrated in Figure 4.23. The feed is a solid mixture of A and B, which is enriched in B or slightly enriched in A. By mixing with recycle stream and adding isopropanol, A is selectively dissolved, allowing the isolation of crystalline B with over 95% purity. The remaining solution (stream 2) becomes highly enriched in A. The solvent is then switched to toluene, and just the enough amount is added, such that all of B dissolves into solution, leaving behind pure crystalline A. The solution, slightly enriched in B, is recycled back to the feed.

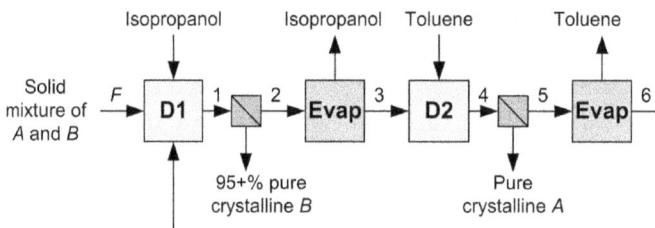

Figure 4.23: Process for separating 1β-methylcarbapenem key intermediates by solvent switching.

This process can be understood by examining the process path on the SLE phase diagram, as shown in Figure 4.24. For convenience, isothermal cuts at the desired dissolution temperature are used. The difference in relative solubility of A and B in the two solvents is apparent from the location of the solubility curves at the given temperature. A is more soluble in isopropanol, resulting in a much larger saturation region of B compared to that of A in the phase diagram on the left-hand side. On the other hand, the solubility difference in toluene is relatively small, resulting in regions of similar size in the phase diagram on the right-hand side.

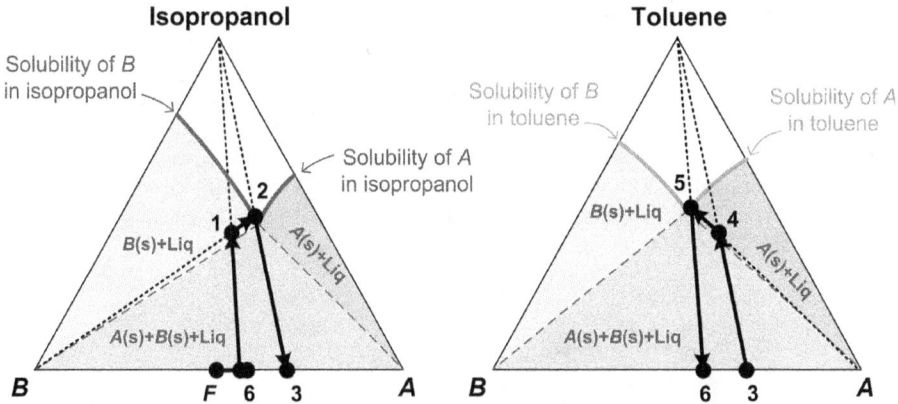

Figure 4.24: Representation of process paths on isothermal phase diagrams.

Addition of isopropanol to the mixture of feed and recycle streams gives point 1 in the B(s)+Liq region, so that only pure B is left in the solid phase. The remaining solution is point 2, which gives a solid mixture of point 3 upon removal of isopropanol. Addition of toluene to point 3 gives point 4 in the A(s)+Liq region, so that pure solid A can be obtained. Removal of toluene from the remaining solution (point 5) gives point 6, which is recycled.

With knowledge of the SLE behavior of the system, optimization of the process is relatively straightforward. To maximize the per-pass yield of pure B and pure A in the two dissolution tanks, points 2 and 5, respectively, should be as close as possible to the respective double saturation point. Furthermore, if the relative solubility of A and B in these solvents change with temperature, it would also be interesting to maximize the difference in relative solubility by selecting different operation temperatures for the two dissolution tanks.

4.4 Hybrid Separation Process

Sometimes, it is not practical to use only crystallization to completely separate all the components. It is often advantageous to include the use of another driving force in a hybrid separation process. For example, vapor–liquid equilibrium (VLE) driving force can be used to get a desired product by distillation, while, at the same time, generating a mixture that is suitable for separation by crystallization. Hybrid processes with adsorption, extraction, membrane separation, and chromatography have also been implemented in industry.

Example 4.8 – Phenol/1-heptanol separation

Phenol is a major pollutant present in wastewaters from several industrial activities (coal mining and gasification, petroleum refining, pharmaceutical production, steel and iron manufacture, tanning and finishing of leather, and sebacic acid manufacture). Liquid-liquid extraction is one possible alternative for removing phenol from wastewater, and a suitable extraction solvent is 1-heptanol [105]. This leads to another separation problem, which is the separation of phenol and 1-heptanol.

Separation of these two components using distillation is limited by the presence of a maximum boiling azeotrope at 76 mol% phenol. Furthermore, there is only a small difference between the boiling point of pure 1-heptanol (176 °C) and pure phenol (182 °C). Combining distillation with crystallization is an interesting option, since the melting point of phenol is 40 °C, which is high enough to obtain a solid product at room temperature, and has a very large difference with the melting point of 1-heptanol (−36 °C). Figure 4.25 shows the VLE phase behavior of phenol/1-heptanol system at atmospheric pressure, along with the SLE phase behavior.

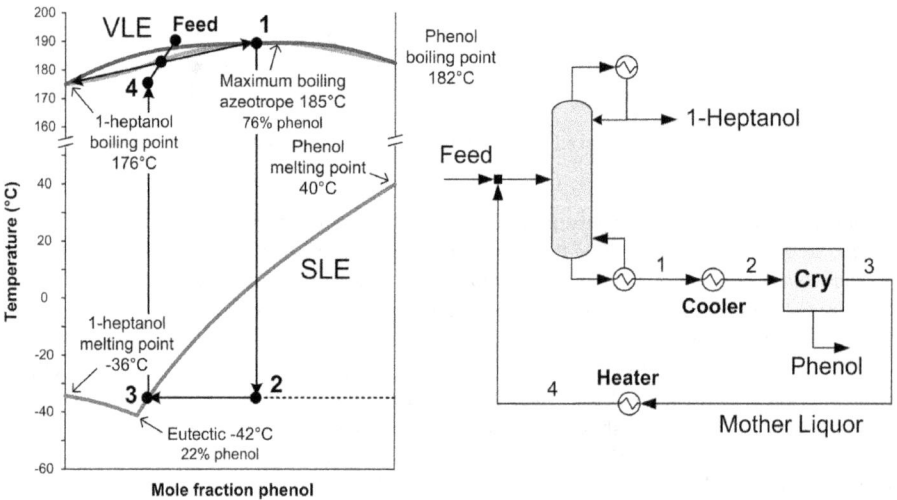

Figure 4.25: Hybrid process for separation of 1-heptanol and phenol [106].

The SLE behavior features a eutectic point at 22 mol% phenol and −42 °C. By combining the VLE and SLE driving forces, a hybrid separation process can be synthesized. The feed is first combined with recycle stream 4 and sent to a distillation column to obtain highly pure 1-heptanol as distillate. The composition of the bottoms product (stream 1) is limited by the azeotrope. This product is cooled down to point 2, which is located inside the crystallization region of pure phenol. A crystallizer can be used to recover pure phenol and mother liquor (stream 3), with a composition close to the eutectic point. This mother liquor is then heated and recycled.

Example 4.9(I) – Ice crystallization for concentrating an aqueous alcohol solution
Ethanol, as well as other alcohols, is routinely separated from water by distillation. However, when starting with a very low alcohol concentration, such as from a waste stream, the energy consumption can be very high because a very high reflux ratio would be required. Therefore, a company was interested in using a crystallization/distillation hybrid process where the feed solution is concentrated by crystallizing ice, before being subjected to distillation. Figure 4.26 shows the SLE phase diagram of ethanol-water system, with a special focus on the temperature range from −10 to 0 °C. The presence of various hydrates have been reported in the literature [107, 108], but these regions are not very relevant, since the hydrates can only be crystallized at very low temperatures. Solubility data from literature [109], which make no distinction of the identity of the solid phase, are indicated by the diamonds.

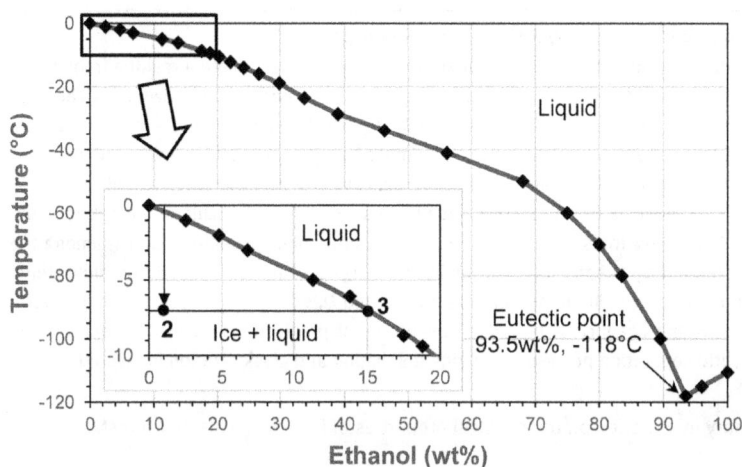

Figure 4.26: SLE phase diagram of ethanol-water system (data from [109]).

The block diagram of the hybrid process is shown in Figure 4.27. As an example, consider a feed containing 1 wt% ethanol (stream 1), with a flow rate of 1 metric ton per hour. Simulation results indicate that if this feed were sent directly to a distillation column with 40 equilibrium stages, the required reboiler duty for 99.9% ethanol recovery (95 wt% ethanol as distillate) would be 0.0787 Gcal/h. In the hybrid process, this stream is cooled and sent to a crystallizer instead. Cooling to −7 °C would cause ice to crystallize because the mixture (point 2 in Figure 4.26) is inside the solid−liquid region. The mother liquor (point 3) contains 15 wt% ethanol. By mass balance, 933 kg of ice is crystallized, leaving just 67 kg of concentrated ethanol solution. With the same number of equilibrium stages and ethanol recovery in the distillation column, the required reboiler duty is only 0.0133 Gcal/h, or almost 6 times smaller compared to a distillation-only process. However, the fact that refrigeration is required to operate the crystallizer at −7 °C must be taken into account in evaluating the economics of this process. Depending on the unit costs of refrigeration and steam for the reboiler, the hybrid process may be an attractive alternative.

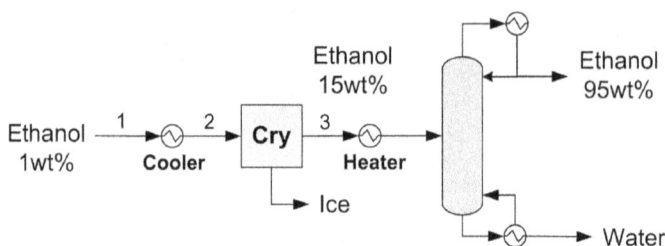

Figure 4.27: Flowsheet for a hybrid process to separate ethanol and water.

Example 4.10(I) – Hybrid process for hydrocarbon mixture separation

A petrochemical company wanted to develop a separation process to fractionate a residue from an existing process, which contains five chemicals of interest: A, B1, B2, C, and D. The known physical properties of these compounds are summarized in Table 4.3. High-purity samples of the chemicals have been successfully obtained by batch distillation, which must be conducted under a very low pressure and high reflux ratios. The results from batch distillation also suggest that the order of boiling point is A<B1<B2<C<D, and that chemicals B1 and B2 are close boilers, which are particularly difficult to separate by distillation. Preliminary trials indicate that B1, which is the most valuable chemical among the five, can be crystallized out by keeping the residue in the refrigerator for an extended period of time. Crystallization would also be a suitable option to recover C and D, which are solids at room temperature. In contrast, recovering chemical A by crystallization is not practical, as its melting point is too low. Therefore, a hybrid separation process combining distillation and crystallization is favored.

Table 4.3: Preliminary information on the physical properties of five chemicals of interest.

Chemical	Physical appearance (at room temperature)	Boiling point (°C)	Melting point (°C)
A	Colorless liquid	234	<−45
B1	White solid	–	–
B2	Colorless liquid	–	–
C	White solid	261	>25
D	White solid	272	>40

To come up with the necessary information to synthesize the process, an experimental study was conducted to determine the VLE and SLE behaviors of relevant systems. Figure 4.28 shows the SLE phase diagrams for two binary systems of interest, namely B1/B2 and C/D, both of which exhibit simple eutectic behavior. The location of the eutectic point is estimated based on experimental data, which are indicated by the open circles.

A hybrid separation process that utilizes crystallization to recover B1, C, and D and distillation to recover A and B2 is depicted in Figure 4.29. The residue (stream 1) is first sent to distillation column D1 to separate A as distillate (stream 2). Next, a mixture of B1 and B2 containing about 32% B1 (stream 4) is obtained as distillate from column D2. This mixture is further separated by a combination of crystallization and distillation. The bottoms from column D2 (stream 10) is mainly a mixture of C and D, which are then separated using distillation and crystallization.

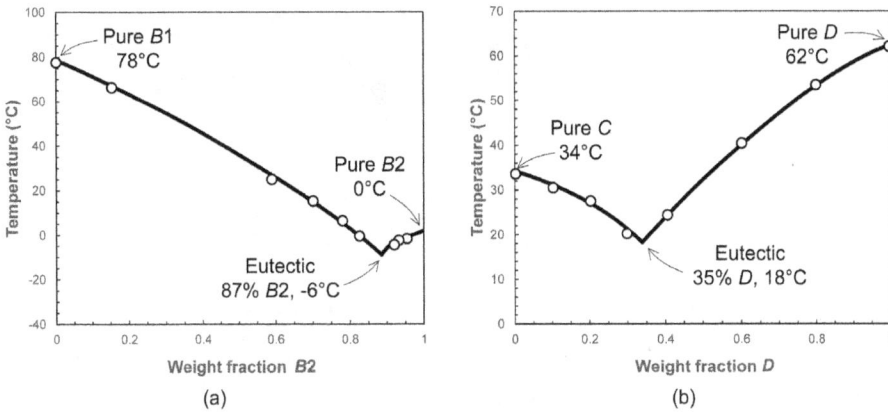

Figure 4.28: SLE phase diagram based on experimental data: (a) B1/B2 binary system and (b) C/D binary system.

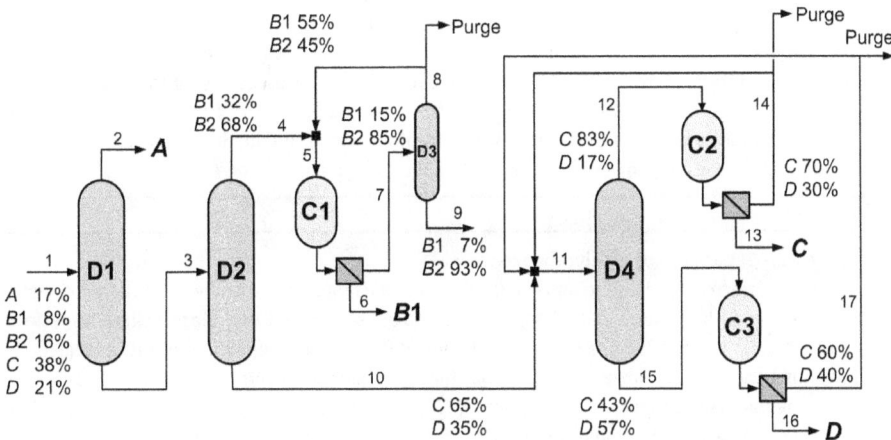

Figure 4.29: Process for hydrocarbon mixture separation.

Figure 4.30a shows the process paths for the B1/B2 separation, plotted on the phase diagram. The distillate from column D2 (stream 4) is combined with a recycle stream to produce the feed to crystallizer C1 (stream 5). Cooling of this stream causes crystallization of B1. The crystallization temperature is selected such that the mother liquor composition (point 7) is located close to the eutectic point, so as to maximize the recovery of pure B1. The mother liquor, which is enriched in B2 (stream 7) is sent to distillation column D3 to obtain a B1-rich distillate (stream 8) and a B2-rich bottoms product (stream 9). Note that stream 7' is stream 7 at a higher temperature. A fraction of stream 8 is purged to avoid accumulation of minor components, and the rest is recycled to crystallizer C1. Due to the low economical value of B2, stream 9 is simply purged, without conducting further purification. Therefore, the main role of distillation in this hybrid process is to produce the B1-rich recycle stream, thereby minimizing the loss of B1 to the purge stream.

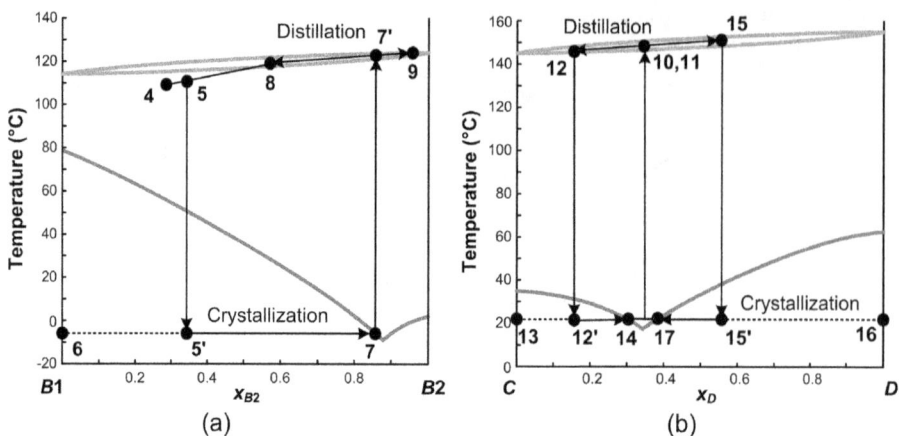

Figure 4.30: Process paths on the phase diagram: (a) $B1/B2$ separation and (b) C/D separation.

Figure 4.30b shows a plot of the process paths for the C/D separation on the phase diagram. Since the composition of column D2 bottoms (point 10) is close to the eutectic composition, it is difficult to use crystallization to separate this mixture. Therefore, stream 10 is mixed with recycle streams and sent to distillation column D4, which gives C-rich and D-rich streams as distillate (stream 12) and bottoms (stream 15), respectively. Each of these streams is sent to a crystallizer to obtain pure C from the C-rich stream and pure D from the D-rich stream. The mother liquors from the crystallizers (streams 14 and 17) are recycled. In other words, distillation is used for initial enrichment to obtain compositions inside the crystallization regions of C and D.

Example 4.11 – Separation of dichlorobenzenes
The separation process for dichlorobenzene (DCB) isomers using a hybrid of crystallization and distillation was patented by Bayer in 1981 [110]. Since the DCB isomers are very close boilers, especially the para and meta isomers (whose boiling points only differ by 1 °C), separation of these isomers by distillation is very difficult. However, it turns out that their melting points are significantly different, especially the para isomer with a melting point of 53 °C. The polythermal projection of the SLE phase diagram involving para, meta, and ortho isomers of DCB is shown in Figure 4.31a, and the separation process according to the Bayer patent is shown in Figure 4.31b. The process paths are plotted on two figures in order to maintain clarity, as there are several overlapping paths.

The process starts with feeding a mixture of fresh feed (F) and a recycle stream (9) to a distillation column. Since the ortho isomer has a slightly higher boiling point at 180.4 °C, it is possible to recover a reasonably pure o-DCB stream by distillation, while at the same time producing a distillate (stream 2) which is located inside the compartment of p-DCB (Figure 4.31a). Pure p-DCB is then crystallized out, and the mother liquor moves to point 3. After filtering out the solids, cooling is continued to crystallize a mixture of p-DCB and m-DCB (stream 8), causing the liquid composition to move along the eutectic trough to point 4. This liquid is fed into a distillation column together with a recycle stream, effectively giving an overall feed stream marked as "4+7" in Figure 4.31a. Distillation produces a bottoms stream enriched in o-DCB (stream 5) and an o-DCB-free distillate (stream 6), which is located in the m-DCB compartment. Therefore, pure m-DCB can be crystallized out. The mother liquor from crystallizer C3 (stream 7), located at approximately the same location as stream 8, is recycled to the second distillation column. Meanwhile, stream 5 is mixed with the solid mixture (stream 8) to produce stream 9, which is the recycle stream to the first distillation column.

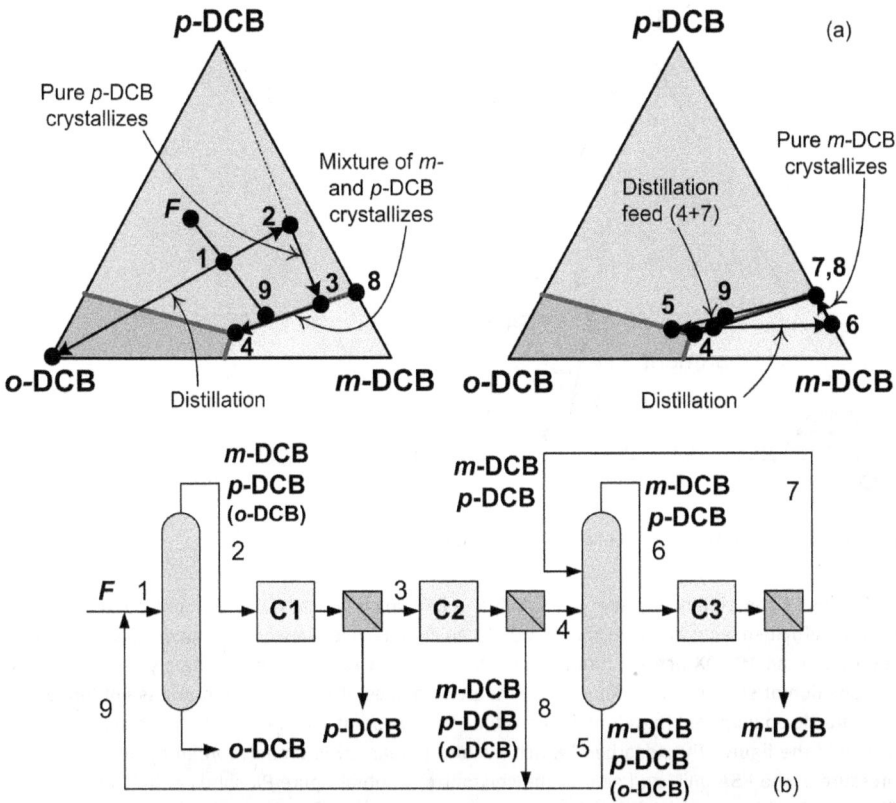

Figure 4.31: Separation of dichlorobenzenes by crystallization-distillation hybrid process: (a) process paths and (b) process flowsheet.

As illustrated in this example, complete separation of the three components in the feed mixture is achieved by a combination of crystallization and distillation. In particular, crystallization is used to obtain pure p-DCB and m-DCB, which are difficult to obtain by distillation. On the other hand, distillation preceded by co-crystallization of m-DCB and p-DCB has been used as a means to move from p-DCB compartment into m-DCB compartment. This example demonstrates a clever exploitation of both the SLE and VLE behavior of the system in devising the overall strategy to achieve the desired separation.

Example 4.12 – Adsorption/crystallization hybrid process for p-xylene separation
As discussed in Example 3.1, the maximum yield in the crystallization of p-xylene is limited by the presence of eutectic boundaries. Therefore, processes that combine crystallization with another driving force, such as adsorption, are widely employed in commercial production of p-xylene. An example is pressure swing adsorption (PSA)/crystallization hybrid process described by Doyle et al. [111]. The process is characterized by the use of PSA to produce a medium-purity p-xylene feedstock stream that is then processed by crystallization to produce high purity p-xylene. Figure 4.32 illustrates the concept of this hybrid separation process on the SLE phase diagram.

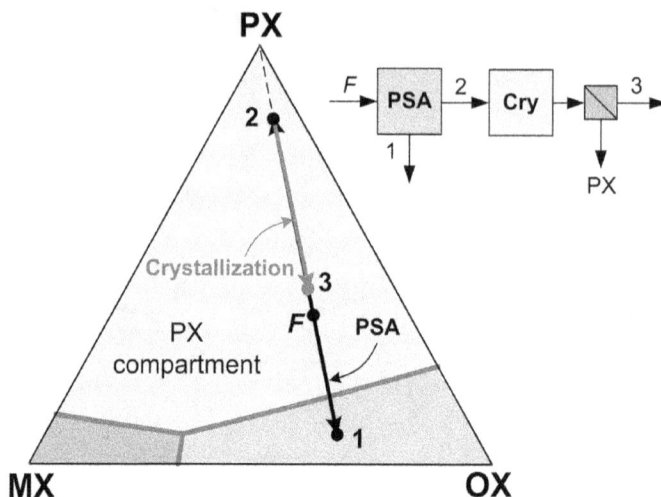

Figure 4.32: Separation of *p*-xylene by PSA/crystallization hybrid process.

The feed F containing a mixture of PX, MX, and OX (which typically also contains other aromatics such as ethylbenzene) is fed to the PSA unit, which selectively adsorbs PX under high pressure, resulting in an MX/OX-rich mixture (stream 1). In contrast to separation using crystallization, the composition of stream 1 only depends on the performance of the adsorbent, and is not limited by the eutectic boundaries. Therefore, it is possible to have point 1 located in the OX compartment, as shown in the figure. The adsorbed PX-rich mixture (stream 2) is then desorbed by decreasing the pressure of the PSA unit, and sent to the crystallizer to obtain pure PX solids at low temperature. Since the feed is already rich in PX, a high per-pass recovery of PX is obtained even if the mother liquor composition (point 3) is not located close to the compartment boundary. In practice, both streams 1 and 3 are typically sent to an isomerization reactor to convert OX and MX to PX, which is then recycled back to the feed stream. If desired, pure OX can be recovered from stream 1 by crystallization, before recycling the remaining mother liquor.

In the process using crystallization only (described in Example 3.1), the mother liquor is also typically isomerized to OX, which means that in both processes, nearly all of the xylenes in the feedstock can be recovered as PX. However, the combination of PSA and crystallization increases the per-pass recovery of PX in the hybrid process, resulting in a smaller recycle stream flow rate compared to the crystallization-only process.

4.5 Summary

This chapter has discussed how crystallization-based separation processes can be synthesized by putting together basic movements such as cooling, heating, component removal, and component addition. The basic idea is to move the composition of a stream from one compartment or saturation region to another and crystallize the corresponding components. Sometimes, a combination of crystallization and other driving forces is preferable in achieving the separation objective, resulting in a hybrid process. Understanding the SLE phase behavior and, if applicable, the VLE phase behavior, which can be visualized using relevant phase diagrams, is crucial for process synthesis. The effect of distillation, adsorption, chromatography, and so on can be similarly captured on a phase diagram. A polythermal projection or a series of isothermal cuts is normally used in the visualization. The important feature to look for is the location of compartment boundaries or double saturation points at different temperatures.

Exercises

4.1. A company is developing a process for producing a dietary supplement pill containing active ingredient A, obtained from its source by solvent extraction. Unfortunately, an impurity B is also extracted in a considerable amount, so that the extract is a mixture of A, B, and extraction solvent $S1$, as indicated by point F in Figure 4.33.

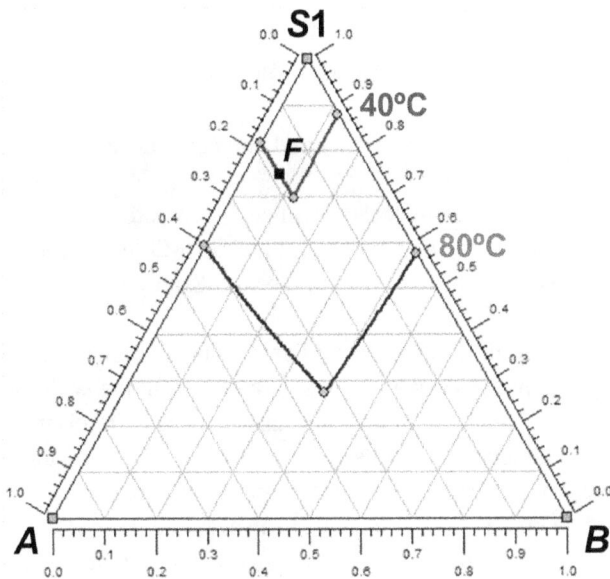

Figure 4.33: SLE phase diagram of a ternary system (problem 4.1).

a. If the solvent is evaporated from this mixture at 80 °C, which component would crystallize out first?

b. Beginning with this solvent removal process, synthesize a process to completely separate A and B, and draw the process path on the diagram.

4.2. The process for producing potash from sylvinite discussed in Example 4.2 begins with selective dissolution at temperature T_d to obtain pure NaCl crystals, followed by crystallization at temperature T_c to obtain pure KCl ("NaCl first" process). It is possible to construct a process alternative that begins with selective dissolution to obtain pure KCl crystals, followed by crystallization to obtain pure NaCl ("KCl first" process), as shown in Figure 4.34.

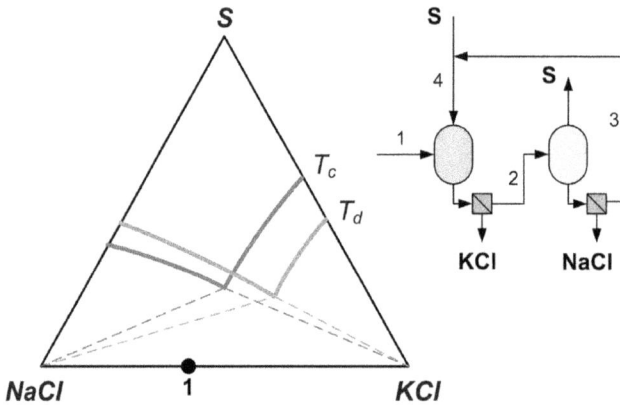

Figure 4.34: SLE phase diagram and flowsheet for separating KCl and NaCl (problem 4.2).

a. Draw the process paths on the phase diagram, and indicate the location of points 2, 3, and 4.
b. At what temperature should the dissolver and crystallizer be operated?
c. What factor is likely to determine whether the "NaCl first" or "KCl first" process alternative would be more economical? (Hint: compare with Example 4.1.)

4.3. A solid mixture of two salts, A and B, is to be separated using a crystallization-based process. The flowsheet for this process is given in Figure 4.35 along with the relevant SLE phase diagram. The solid feed (F) is totally dissolved in a dissolver by adding solvent S. The resulting solution (stream 1) is mixed with the recycle (stream 4) and sent to a cooling crystallizer (C1) operating at T_c. After separating B from the effluent of this crystallizer, the mother liquor (stream 3) is sent to an evaporative crystallizer (C2) operating at $T_h > T_c$. Pure A is crystallized out and the final mother liquor (stream 4) is recycled. The feed contains 80% B and 20% A, and has a flow rate of 1,000 kg/h. The locations of points 3 and 4 are given on the phase diagram.
a. Based on mass balances, determine the flow rate of the solvent added to the dissolver (stream 5) as well as all other streams in the process.
b. Locate points F, 1, and 2 on the phase diagram. Draw the process paths.
c. If stream 1 is an unsaturated solution, would the dissolver temperature be below or above T_h? Briefly explain your reasoning.

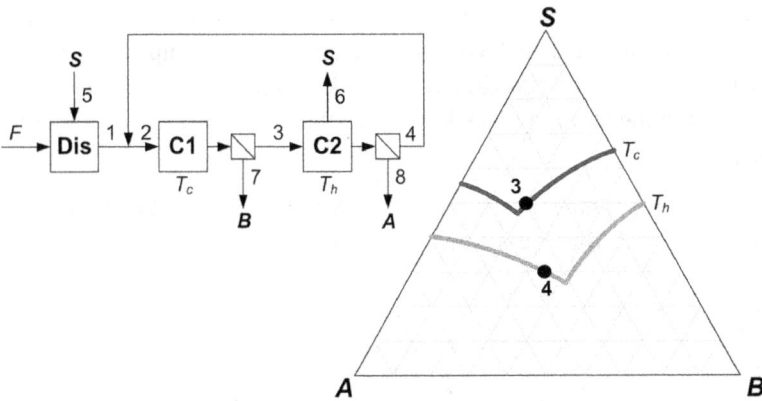

Figure 4.35: SLE phase diagram and flowsheet for a crystallization-based separation process (problem 4.3).

4.4. A fine chemical product is produced via a coupling reaction under basic conditions. The reactor outlet contains a mixture of reactant HA, product H_2B, and their sodium salts. Two isothermal cuts of the phase diagram for the $H^+,Na^+/A^-,B^{2-}$ system are shown on the diagram in Figure 4.36. The feed composition is indicated by point F.

a. Synthesize a crystallization-based separation process to obtain HA and Na_2B from the mixture and draw the process paths on the diagram.

b. What is the molar ratio of the HA product to Na_2B product that would be obtained?

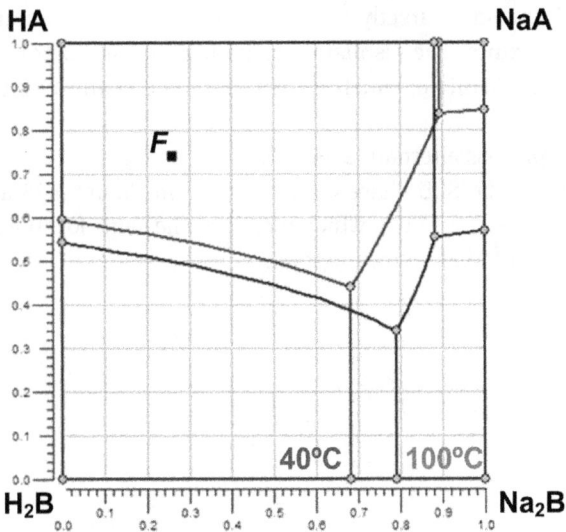

Figure 4.36: Isothermal cuts of the phase diagram for the $H^+,Na^+/A^-,B^{2-}$ system (problem 4.4).

4.5. Consider the process for isolating NaCl from a salt lake discussed in Example 4.4. Using the information provided in Figure 4.14, determine the composition of the mother liquor (liquid to further processing) if 800 kg of water is evaporated from 1,000 kg of brine. Assume that the brine does not contain any other ion besides Na^+, K^+, Cl^-, and SO_4^{2-}.

4.6. Which of the phase behaviors shown in Figure 4.37 is most suitable for separating a mixture of A and B using a third component as a solvent? Explain your reasoning.

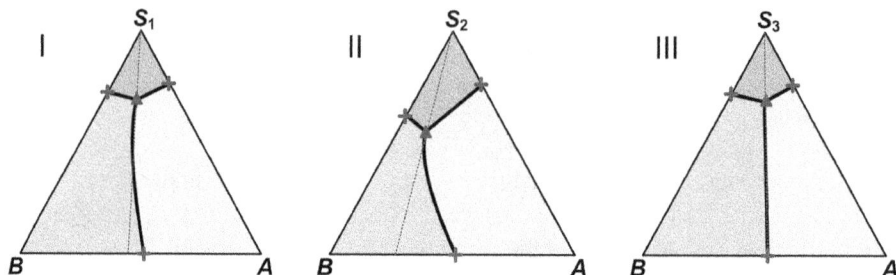

Figure 4.37: SLE phase diagram of ternary systems involving different solvents (problem 4.6).

4.7. The Bayer process uses three crystallizers for the separation of dichlorobenzenes (Example 4.11), with the second one (C2) producing a solid mixture of m-DCB and p-DCB. Despite the potentially higher energy consumption for distillation, it is conceptually possible to send stream 3 directly to a distillation column, without crystallizing the m-DCB/p-DCB mixture. The distillation column would still produce a mixture of m-DCB and p-DCB as distillate, which is then sent to crystallizer C3 to obtain pure m-DCB.
a. Draw the flowsheet for this process alternative based on the description.
b. Identify the process points on the SLE phase diagram given in Figure 4.38 as well as the process paths. Points F, 2, and 3, which remain at the same locations as in Figure 4.31a, are already plotted on the diagram.

p-DCB

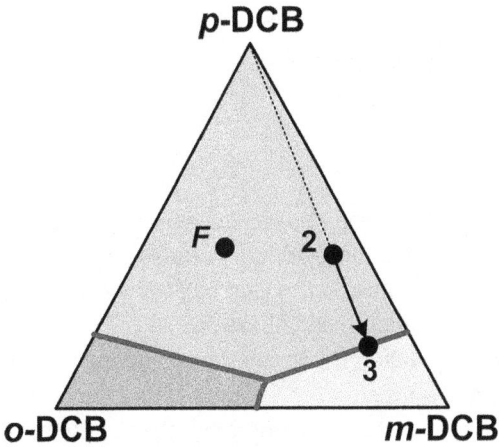

Figure 4.38: SLE phase diagram for dichlorobenzene system (problem 4.7).

4.8. Bauer, Jr. et al. [112] describes a process for purifying unsaturated carboxylic acids, such as methacrylic acid (MAA) or acrylic acid, using a combination of crystallization and distillation. One example in the patent illustrates the application of the claimed process for separating a feed mixture containing 60 wt% MAA and 40 wt% isobutyric acid (iBuA). This feed is mixed with a recycle stream and fed to a distillation column, producing distillate and bottoms streams containing 10 wt% MAA and 95 wt% MAA, respectively. The bottoms stream is sent to a crystallizer, which produces substantially pure MAA crystals. The mother liquor containing 93.6 wt% MAA is recycled to the distillation column.

a. Draw a block diagram of the process based on this description.
b. Sketch a phase diagram showing both the VLE and SLE parts, complete with the process paths. MAA (melting point 15 °C, boiling point 161 °C) and iBuA (melting point −47 °C, boiling point 155 °C) exhibit simple eutectic behavior, and they do not form an azeotrope.

5 Crystallization Processes Involving More Complex Phase Behaviors

5.1 Adductive Crystallization

An adduct is basically a compound formed by direct addition or combination of two or more distinct molecules into a single product. Since adduct can also be seen as a result of a chemical reaction, adductive crystallization, which has been applied in numerous industrial processes, is a type of reactive crystallization.

When a chemical component forms an adduct with another component, it is sometimes desirable to crystallize the adduct instead of the pure component. This may be because it is easier to get to the adduct compartment rather than to the pure component compartment, the adduct crystals are easier to handle, or sometimes the adduct itself is the favorable product; for example, pharmaceutical co-crystals. By forming the co-crystal, the physicochemical properties such as solubility and stability may be enhanced, while the biological properties of the drug molecule remain unchanged [113]. This technique has been widely used in drug development. An example is carbamazepine, a mood stabilizing drug used primarily in the treatment of epilepsy and bipolar disorder, which can form a co-crystal with a variety of compounds. Carbamazepine is practically insoluble in water, but has a high dose requirement for therapeutic effect. To meet this requirement, the aqueous solubility can be improved by forming a co-crystal. As an additional advantage, the co-crystal has reduced polymorphic tendency relative to pure carbamazepine, which makes it easier to consistently produce the desirable form [114]. In any case, solid–liquid equilibrium (SLE) phase diagram is useful in identifying the suitable region for the operation.

5.1.1 SLE phase behavior involving adduct

The SLE phase diagram of binary systems involving an adduct, which is equivalent to the compound formed between the two components, is shown in Figure 2.5. When a third component, such as a solvent, is added to the system, the phase diagram takes the shape of a triangular prism, as shown in Figure 5.1a. In this ternary system, one pair of components (A and B) form a congruently melting adduct (C), while other pairs exhibit simple eutectic behavior. The polythermal projection (Figure 5.1b) features a compartment for the adduct, in addition to the three compartments corresponding to A, B, and S. The isothermal cut at T_1 (Figure 5.1c) also has an additional region corresponding to the adduct.

https://doi.org/10.1515/9781501519901-005

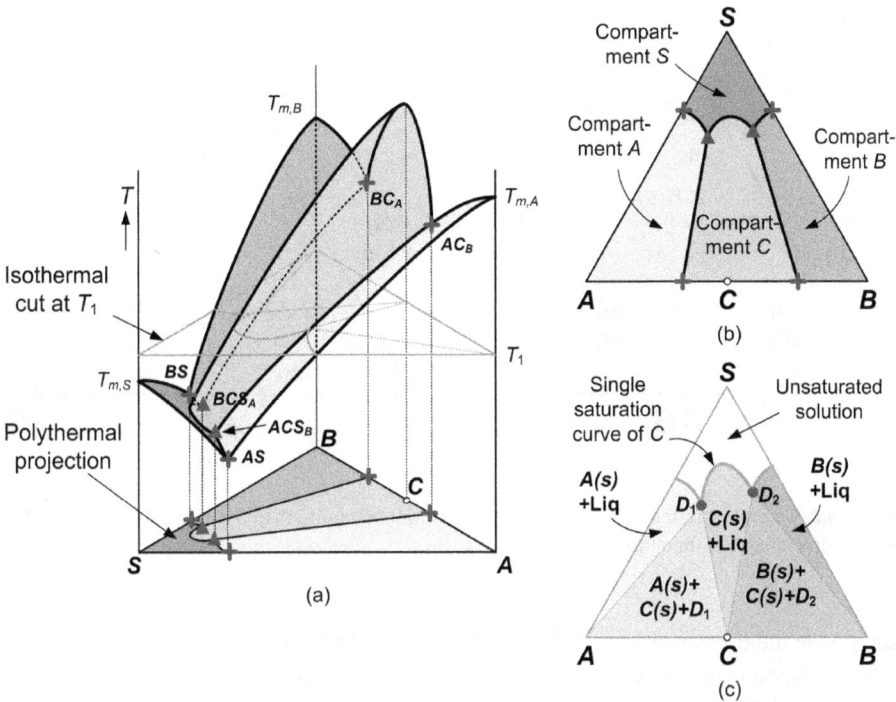

Figure 5.1: Ternary SLE phase diagram with compound formation behavior: (a) T–x diagram, (b) polythermal projection and (c) isothermal cut at T_1.

5.1.2 Process synthesis

The same strategy of process synthesis can be applied to systems involving adducts. Berry and Ng [68] identified the flowsheet alternatives for reactive crystallization processes, most of which involve adducts, and pointed out the use of basic movements such as temperature, pressure, and composition swings in those flowsheet alternatives. Therefore, the idea of moving around in composition space and visiting the compartment of the adduct provides the logic for constructing the flowsheet alternatives.

As an example of how the SLE phase behavior affects the design of an adductive crystallization process, consider the crystallization of a pharmaceutical co-crystal. Figure 5.2 shows the typical SLE phase behavior of an active pharmaceutical ingredient, API (A), a co-former (B) (a compound that can form a co-crystal AB with the API), and a solvent (S) [115, 116]. In Figure 5.2a, the co-former has a similar solubility as the API, and the co-crystal has a lower solubility than the two. Consequently, the co-crystal can be precipitated from solution by simply mixing solutions containing the API and the co-former as shown by point M inside the AB crystallization region in

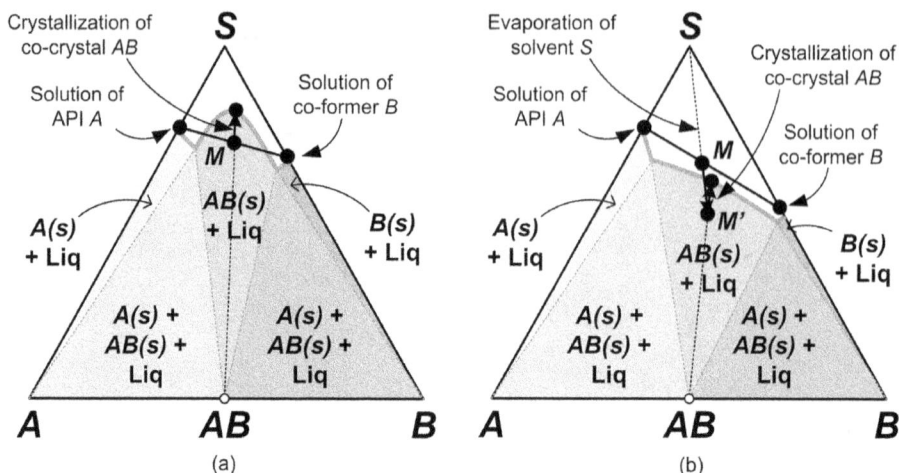

Figure 5.2: Co-crystal formation: (a) co-crystal with lower solubility than the API and (b) co-crystal with higher solubility than the API.

the phase diagram. However, it is also possible that the co-crystal has a higher solubility compared to the API, as illustrated in Figure 5.2b, so that point *M* lies in the unsaturated liquid region [117]. The process path for crystallizing the co-crystal *AB* involves evaporation of the solvent to produce point *M'* inside the *AB* crystallization region. In any case, a proper ratio of the co-former to API has to be selected, in order to get into the crystallization region of the co-crystal.

Example 5.1(I) – Crystallization of bisphenol A

Bisphenol A (BPA) is an important chemical for producing polycarbonate. The conventional route to make BPA is by condensation reaction from acetone and phenol. A process using resin catalyst [9], shown in Figure 5.3, is the most widely used and becomes the basis for many commercial processes of companies such as Badger, Mitsubishi Chemical (BISA-MAX), and Sinopec/Lummus. Phenol is used in excess amount to push acetone conversion to nearly complete. Any remaining acetone, along with water produced as a byproduct in the reaction, is removed in a concentrator before being sent to a crystallizer, where the 1:1 adduct of phenol and BPA is crystallized out. The mother liquor, which contains more than 90% phenol, is recycled. The adduct is melted and sent to a stripper in which phenol is completely evaporated, leaving behind the BPA in melt form. The melt is then sent to a finishing unit such as a prilling tower to obtain the final product in solid form.

The SLE phase diagram of phenol/BPA system is shown in Figure 5.4, with the diamonds indicating solubility data from the literature [43]. The feed to the crystallizer is essentially a phenol-rich mixture of phenol and BPA, as indicated by point *F*. When this mixture is cooled down to 60 °C or so, the composition enters the adduct region, and the adduct is crystallized. Crystallizing the adduct instead of pure BPA has the advantage of lower crystallization temperature, because the location of the adduct crystallization region is below the BPA region.

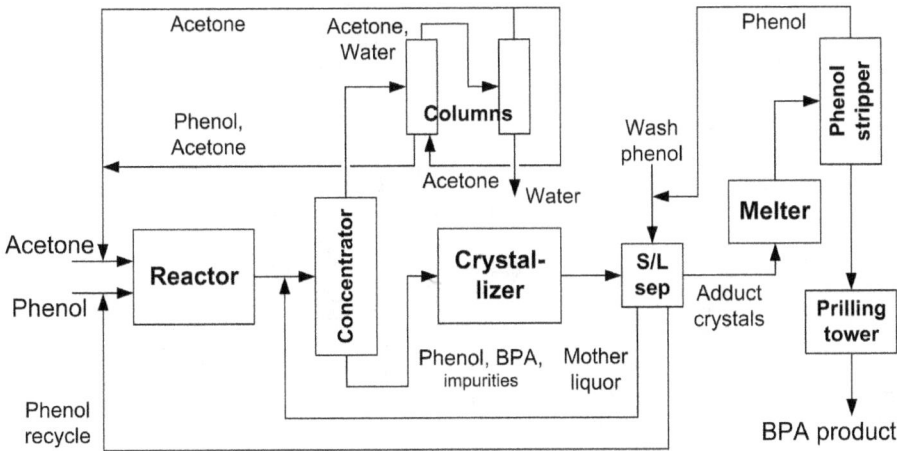

Figure 5.3: Conventional process for producing BPA with resin catalyst [9].

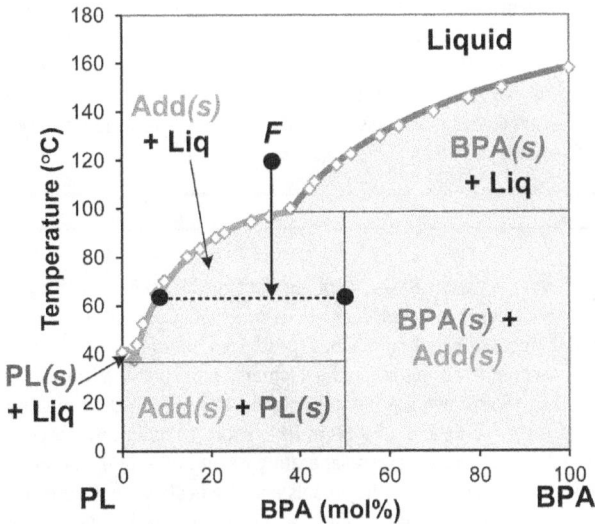

Figure 5.4: SLE phase diagram of phenol/BPA binary system (data from [43, 118]).

A new process based on direct crystallization of BPA has been proposed [119]. Considering acetone and water as solvent, the polythermal phase diagram of the ternary system (in mass fraction coordinates) featuring compartments of phenol, adduct, and BPA is shown in Figure 5.5. Note that the compositions shown on the phase diagram are mass-based, and the 1:1 adduct contains roughly 70 wt% BPA. The mixture after reaction (stream 1) is mixed with a recycle stream 2, and more solvent is added to create a crystallizer feed (stream 3) that is located within the BPA compartment. BPA is directly crystallized at a relatively low temperature due to the presence of the

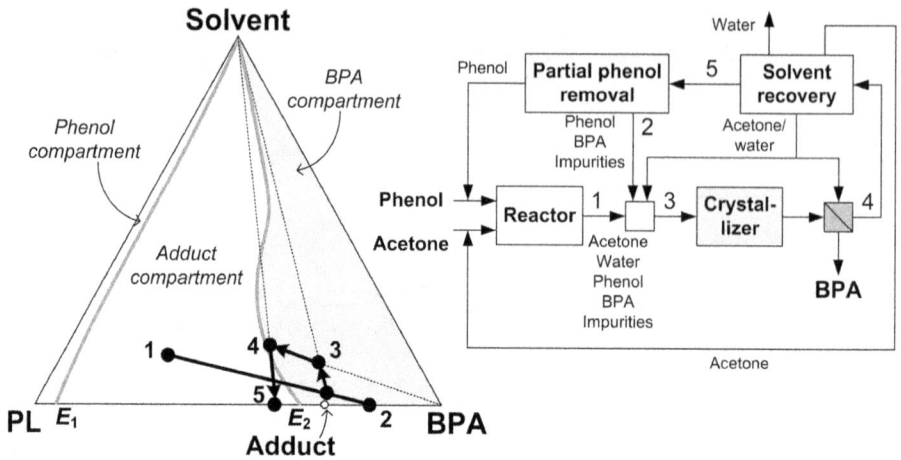

Figure 5.5: New process for producing BPA by direct crystallization of BPA.

solvent. The mother liquor (stream 4) is sent to a solvent recovery system to separate acetone and water. Some of the acetone and water is sent to the dissolution tank, and also to the filter for washing the BPA crystals. The phenol/BPA mixture (stream 5) is subjected to partial phenol removal (which can consist of distillation as in the conventional process, but phenol does not have to be completely removed) to generate the recycle stream 2. This new process is potentially attractive because it is simpler compared to the conventional process. In particular, complete evaporation of phenol and prilling are not needed to obtain solid BPA.

Example 5.2 – Separation of cresols

Cresols are widely used as household cleaners and disinfectants, perhaps most famously under the trade name Lysol. Its three isomers (*para*, *meta*, and *ortho*) serves as precursors or synthetic intermediates for various compounds and materials, including plastics, pesticides, pharmaceuticals, and dyes. Crystallization is an attractive technique to separate the isomers, particularly p-cresol (A) and m-cresol (B), which are close boilers (both having a boiling point of 202 °C) but their melting points differ quite significantly (35 °C for p-cresol and 10.9 °C for m-cresol). To overcome eutectic limitations, t-butyl alcohol (TBA) (S) is introduced as a mass separating agent (MSA) [45]. The polythermal projection of the ternary phase diagram involving the cresols and TBA is shown in Figure 5.6. One problem is that p- and m-cresol form an adduct with a molar ratio of 1:2, marked as AB_2 in the phase diagram, which makes it difficult to simply rely on MSA addition and removal to isolate A and B in pure form. Interestingly, both isomers also form 1:1 adducts with TBA, marked as AS and BS. As a result, there are six compartments, one each for A, B, AB_2, AS, BS, and S.

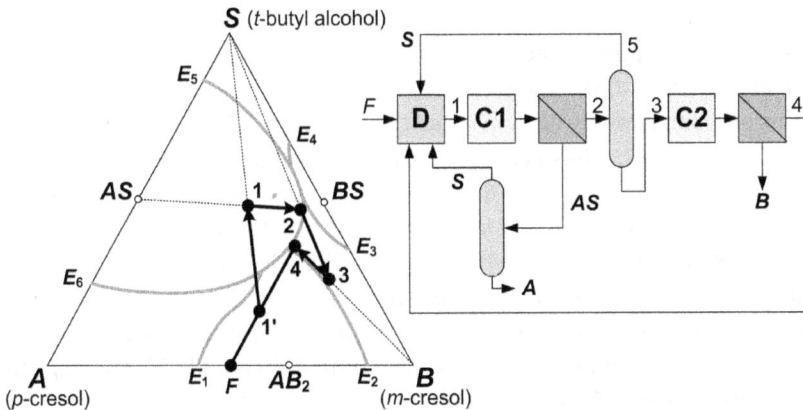

Figure 5.6: Process for separating p-cresol and m-cresol using t-butyl alcohol as solvent.

A feasible process alternative for separating an equimolar mixture of the cresols is also shown in Figure 5.6. First, the feed is mixed with a recycle stream (stream 4) and additional TBA to give a composition given by point 1 on the diagram, which is located in the compartment of adduct AS. Therefore, the adduct between p-cresol and TBA can be recovered by cooling, stopping at point 2 just before reaching the compartment boundary. Removing TBA from this mixture gives point 3 in compartment B, so that m-cresol can be recovered in a second crystallizer. The mother liquor (point 4) is recycled, while the adduct is decomposed to obtain p-cresol and TBA.

5.2 Chiral Resolution by Crystallization

Crystallization is one of the most effective methods for chiral resolution. The term "chiral" is derived from the Greek word for hand, χειρ (cheir), and is used, in general, to describe an object that is not superimposable on its mirror image, just like the left and right hands. A molecule can be chiral when four different groups are attached to an atom, usually a carbon atom, known as the *chiral center*. Molecules with the same formula and atomic connectivity but different three-dimensional (3D) arrangement around the chiral center are called *stereoisomers*. The systematic names of chiral stereoisomers follow the Cahn-Ingold-Prelog (CIP) nomenclature, which involves an R or S descriptor to each chiral center to uniquely specify the configuration of the entire molecule [120]. However, descriptors based on other nomenclatures have also been used in their common names, such as d and l or + and − (based on light polarization), and D and L (based on the location of -OH group in the Fischer projection; commonly used for sugars and carbohydrates) [121].

Chirality plays an important role in biological activities, since the pharmacological effects of the individual stereoisomers of a drug can be strikingly different. An example is naproxen, with the (S)-isomer, sold commercially as ALEVE, having 28 times the anti-inflammatory drug activity as the (R)-isomer [122]. Another example is propoxyphene. The (2R, 3S)-isomer is an analgesic used for pain relief, sold commercially until 2010 as DARVON [123], while the (2S, 3R)-isomer is an antitussive used as cough suppressant, trademarked as NOVRAD [124]. Perhaps the most infamous example of an extreme case is the drug Thalidomide, which was introduced to the market in 1957 in Germany and prescribed to pregnant women to combat symptoms associated with morning sickness. While the (R)-isomer is a safe sedative, (S)-thalidomide is a teratogen, which causes birth defects and deformities such as missing arms or legs [125]. The drug was withdrawn in 1961 after thousands of babies were born with defects. Clearly, separating chiral stereoisomers is an important problem in pharmaceutical processing.

5.2.1 SLE phase behavior of chiral systems

There are two important types of stereoisomers. A pair of molecules that are nonsuperimposable mirror images of each other are called *enantiomers* or optical isomers, which are the true chiral molecules by definition. The second type is *diastereomers*, which have different configurations at one or more chiral centers, but are not mirror images of each other. As an example, consider the different stereoisomers of tartaric acid shown in Figure 5.7. Since the two carbon atoms in the middle of the tartaric acid molecule are chiral centers, there are four stereoisomers. The (2R,3R) and (2S,3S)

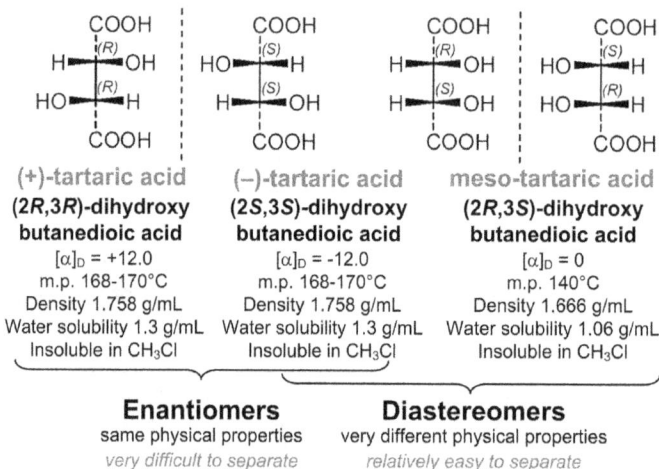

Figure 5.7: Different types of stereoisomers.

isomers are enantiomers, since they are mirror images of each other. Due to the very similar physical properties, including melting point, density, and solubility, they are very difficult to separate. Their optical properties are opposite of each other, as indicated by the specific angle of optical rotation [α], thus the name (+) and (–) tartaric acid. Meanwhile, (2R,3S) and (2S,3R) isomers are identical, so they are basically just a single isomer known as meso-tartaric acid. This isomer is said to be a diastereomer of the other two isomers. Meso-tartaric acid does not rotate the polarization plane, and has very different melting point, density, and solubility compared to (+) and (–) tartaric acid. The differences in physical properties of diastereomers make it easier to separate them.

The SLE phase behavior of a pair of enantiomers is usually characterized by symmetry, such as those shown in Figure 5.8, because both enantiomers have the same melting point and solubility in a third component. Simple eutectic behavior, which is usually referred to as *conglomerate* (Figure 5.8a), only makes up 5–7% of all enantiomeric systems [62]. The existence of a racemic compound (50:50 composition, Figure 5.8b) is by far the most common, about 90% of the time. The rest exhibit solid solution behavior, referred to as *pseudoracemates*. Different types exist, including uniform melting point (Figure 5.8c), minimum melting (Figure 5.8d), or maximum melting (Figure 5.8e). Diastereomers, on the other hand, are not symmetric, and most of them exhibit simple eutectic behavior.

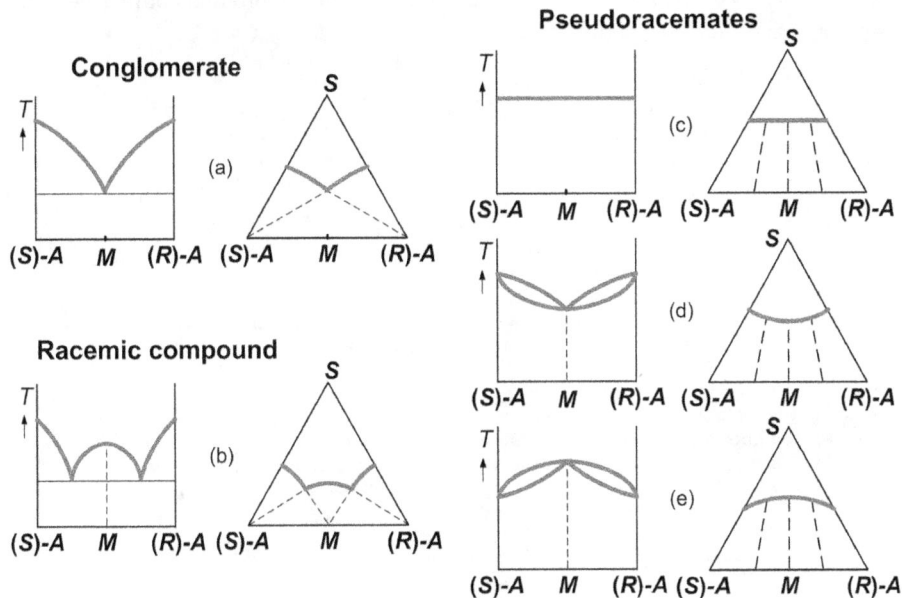

Figure 5.8: Binary and ternary SLE phase diagrams of systems involving chiral enantiomers [126].

5.2.2 Process synthesis

Thermodynamic-based resolution process of chiral enantiomers using crystallization is difficult due to the symmetry in the SLE phase diagram. Successful application of the techniques described in Chapters 3 and 4 often requires a high initial enantiomeric purity of the feed material [127], which has to be achieved via some other separation methods. If a racemic compound is present, enantiomeric purification can be achieved via a two-step crystallization process, in which the racemic compound is crystallized first, followed by crystallization of a pure enantiomer in the second step [128, 129, 130]. However, its applicability requires a significant shift in the composition of double saturation points between the pure enantiomers and the racemic compound, and the initial feed also has to be sufficiently enriched in the enantiomer that needs to be purified.

To overcome the limitations imposed by the phase behavior of the enantiomeric system, *resolving agents* are widely employed to industrially produce pure enantiomers. A resolving agent converts the enantiomers into diastereomers, which exhibit a more favorable phase behavior. The double saturation point is shifted away from the 50:50 racemate composition, and solid compound formation is less likely. A generic flowsheet that describes the basic steps for this type of separation is shown in Figure 5.9. To bypass the eutectic (or 1:1 compound), dissociable compounds are formed between the racemate and the resolving agent. After separation, the dissociable compounds are decomposed to recover the products of interest. The resolving agent *B* is recycled.

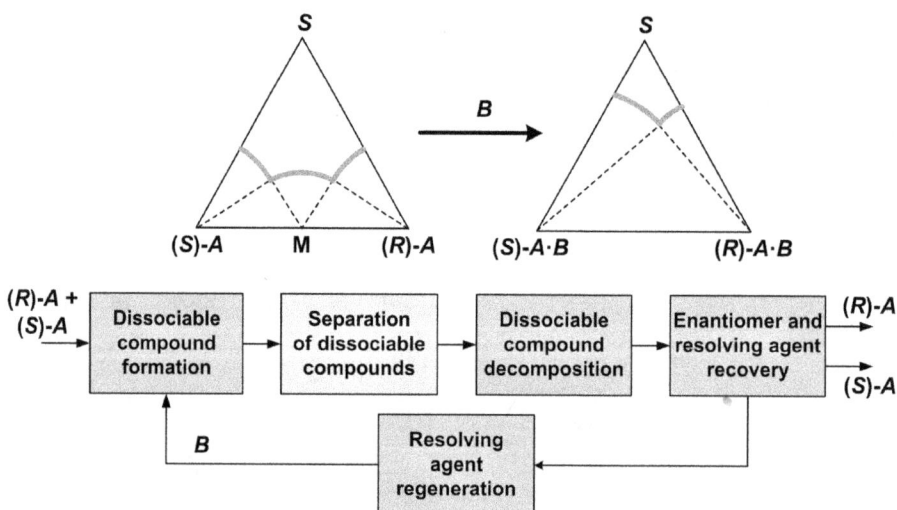

Figure 5.9: Generic scheme of chiral resolution using a resolving agent [126].

Figure 5.10 is the generic scheme for resolution by converting the enantiomers into diastereomeric salts [126]. If a chiral resolving agent, which is a racemic mixture of (R)-B and (S)-B, is added in a stoichiometric amount to a racemic mixture of enantiomers (R)-AH and (S)-AH, four diastereomeric salts would be produced: (R)-$AH \cdot (R)$-B, (R)-$AH \cdot (S)$-B, (S)-$AH \cdot (R)$-B, and (S)-$AH \cdot (S)$-B. The SLE phase behavior of the system involving the solution of these four salts (the solvent is usually water) resembles that of a conjugate salt system, with no compound formation. Therefore, the fractional crystallization technique, discussed in section 4.2.4, can be applied to separate the compatible salt pair, which is (S)-$AH \cdot (R)$-B and (R)-$AH \cdot (S)$-B, in this case.

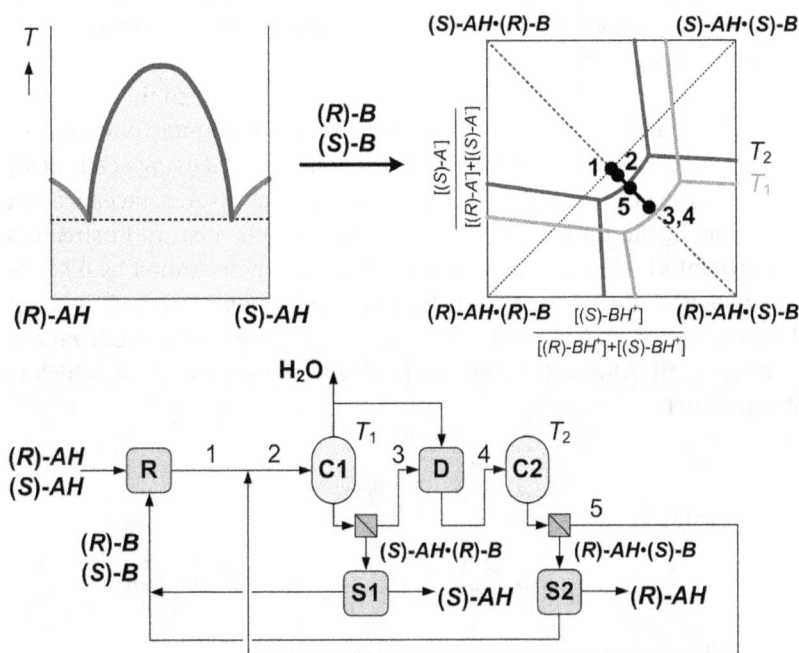

Figure 5.10: Chiral resolution by fractional crystallization of diastereomeric salts.

The flowsheet in Figure 5.10 shows the separation process, with stream numbers corresponding to the process points on the Jänecke projection of the SLE phase diagram. The overall composition, after mixing the feed and resolving agent in the reactor R, lies at the center of the projection (point 1), as both are equimolar mixtures of (R) and (S) enantiomers. Addition of a recycle stream produces a composition of point 2 inside the region of (S)-$AH \cdot (R)$-B, which is crystallized out in crystallizer C1 at temperature T_1. Water is added to the mother liquor (point 3) to give a composition inside the region of (S)-AH·(R)-B (point 4). Note that point 4 coincides with point 3 on the Jänecke projection. This salt is crystallized out in crystallizer C2 at temperature T_2, and the mother liquor (point 5) is recycled. Both solid products are decomposed in S1 and S2 to obtain pure enantiomers (R)-AH and (S)-AH, respectively, and the recovered resolving agent is recycled to reactor R.

Often, only one of the enantiomers is the desired product while both enantiomers are produced by reaction. In this case, *racemization*, a process in which an unequal mixture of enantiomers is converted into a 50:50 mixture, can be used to improve the yield in the crystallization process. Figure 5.11 shows a process in which only one enantiomer, the (S) form, is the desirable product [126]. An enantiopure resolving agent, (S)-B, is used to convert the enantiomeric mixture into a mixture of diastereomers, (R)-A·(S)-B and (S)-A·(S)-B. The composition in solution (point 1) lies inside the (S)-A·(S)-B saturation region, so that (S)-A·(S)-B can be crystallized out in the first crystallizer. The mother liquor (point 2), enriched in (R)-A·(S)-B, is sent to extraction unit 1 to remove (S)-B using acetic acid, resulting in a mixture of (R)-A and (S)-A that is enriched in (R)-A. A racemization reactor (block Rac in the flowsheet) is used to convert this mixture in stream 3 back to 50:50 (point 4), which is then recycled. Stream 9 is separated by distillation into acetic acid and (S)-B in stream 10. The solid product (S)-A·(S)-B is decomposed in extraction unit 2, and pure (S)-A is obtained by crystallization. Using this strategy, all (R)-A in the feed is effectively converted to (S)-A, which is the desirable product.

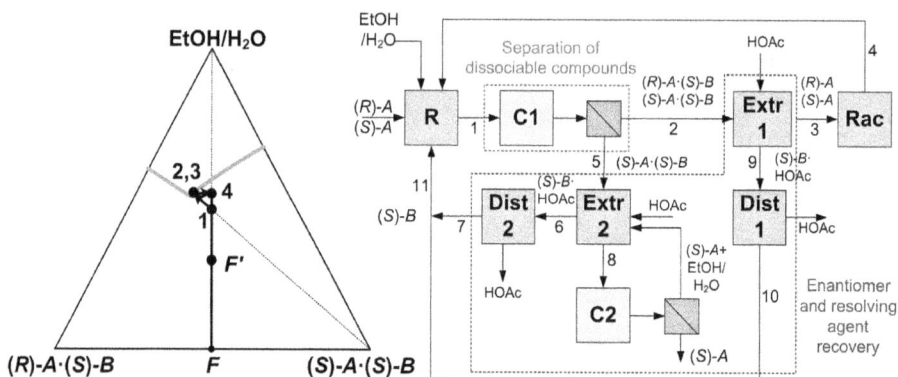

Figure 5.11: Combination of crystallization and racemization for chiral resolution.

Example 5.3 – Resolution of ibuprofen

Ibuprofen is a common analgesic, and has two enantiomers, (R) and (S), that can form a racemic compound. As shown in the phase diagram in Figure 5.12a, the compound melts at a much higher temperature (over 70 °C) compared to the pure enantiomers (about 50 °C). Ibuprofen is commonly sold as a racemic mixture, which is essentially the compound. In fact, the values of physical properties such as melting point and heat of fusion reported in common databases (NIST, DIPPR, etc.), are those of the compound instead of the pure enantiomers. However, research findings show that (S)-ibuprofen exhibits superior performance compared to the racemate [131], which have led to extensive research efforts in developing a resolution process for ibuprofen. The use of resolving agents such as (S)-lysine and N-methyl-D-glucamine (NMDG) has been reported [132, 133].

The phase diagram for the ibuprofen/NMDG–ibuprofen system with 95 vol% acetone in water as the solvent has been experimentally measured [134]. The Jänecke projection is shown in Figure 5.12b, with the rectangles indicating experimental data. NMDG forms a complex with each ibuprofen enantiomer, but there is no compound between NMDG-(R)-ibuprofen and NMDG-(S)-ibuprofen. Also, the double saturation point between NMDG-(R)-ibuprofen and NMDG-(S)-ibuprofen is located closer to the NMDG-(S)-ibuprofen side, instead of being in the middle of the NMDG-(R)-Ibu-NMDG-(S)-Ibu edge.

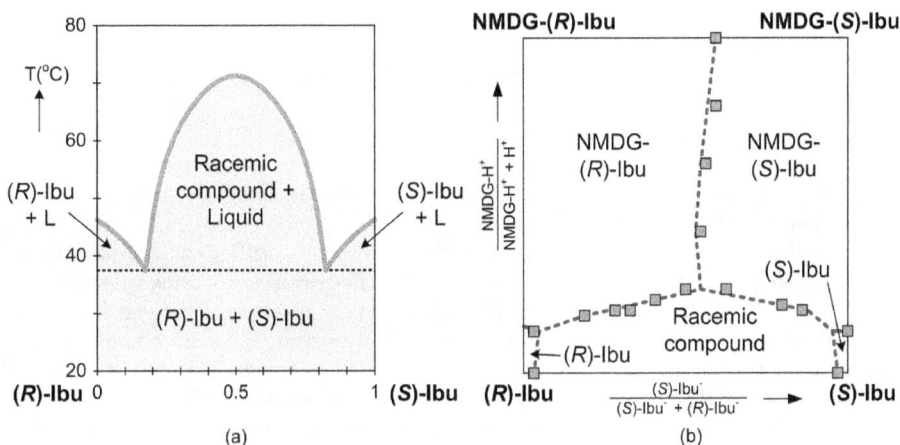

Figure 5.12: SLE phase diagrams for ibuprofen resolution: (a) binary system of ibuprofen enantiomers and (b) conjugate salt system of ibuprofen-NMDG (Jänecke projection of isothermal cut) [134].

Based on the phase diagram, process alternatives to obtain the desirable product (S)-ibuprofen using NMDG as resolving agent can be synthesized. Figure 5.13 shows a process flowsheet of a crystallization process to recover pure NMDG-(S)-ibuprofen under equilibrium conditions. The composition change of the process streams is shown on the Jänecke projection of the phase diagram.

Racemic ibuprofen enters the process as the feed (F), and is mixed with the recycled racemic ibuprofen, solvent (S), and resolving agent NMDG in a reactor (R1) where diastereomeric salts are formed. The solution of diastereomeric salts (stream 2) is mixed with a recycle stream (stream 9) to give stream 3, which is within the saturation compartment of NMDG-(R)-ibuprofen, so that pure NMDG-(R)-ibuprofen is recovered in the first crystallizer (C1). The crystallization is stopped when the

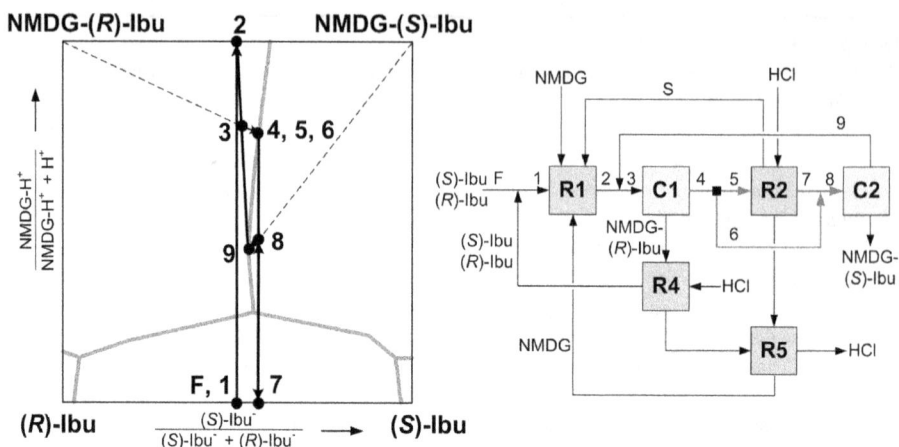

Figure 5.13: Process alternative for resolution of ibuprofen using NMDG as resolving agent [134].

composition of the mother liquor (stream 4) nearly reaches the double saturation curve. NMDG-(*R*)-ibuprofen can be racemized (R4) and recycled back to the feed.

In order to pass from the saturation compartment of NMDG-(*R*)-ibuprofen to that of NMDG-(*S*)-ibuprofen, a portion of the mother liquor is sent to reactor R2 where ibuprofen is liberated by switching the solvent to water and reacting the diastereomeric salts with HCl. The reaction forms the complex NMDGH$^+$Cl$^-$, which is highly soluble in water, while the liberated ibuprofen precipitates spontaneously due to its very low solubility in water. NMDG can then be recovered from the salt using an ion exchange resin (R5).

Meanwhile, the liberated ibuprofen (stream 7) is mixed with the remaining mother liquor from C1 (stream 6) to form a diastereomeric salt solution (stream 8). This stream is within the saturation compartment of NMDG-(*S*)-ibuprofen, so that NMDG-(*S*)-ibuprofen can be recovered in the second crystallizer (C2). The solid NMDG-(*S*)-ibuprofen is subjected to downstream processing to liberate (*S*)-ibuprofen, and the mother liquor (stream 9) is recycled back to C1. Therefore, in principle, the undesired product NMDG-(*R*)-ibuprofen is crystallized first in order to move the composition into the region of the desired product. This process has been verified experimentally in the laboratory to demonstrate that high purity (*S*)-ibuprofen can indeed be obtained.

5.2.3 Hybrid process for chiral resolution

Often, a combination of crystallization and chromatography is used for chiral resolution [135, 136]. Consider the phase diagram in Figure 5.14a, featuring the compartments of three solutes – enantiomer *R*, enantiomer *S*, and a racemic compound *RS* – as well as that of solvent *D*. The feed to be separated is a racemic mixture, represented by point F. Since this point is located in the *RS* compartment, crystallization would produce the racemic compound *RS*, which means there is no resolution. Meanwhile, it is possible to produce highly pure enantiomers from a racemic mixture using chiral

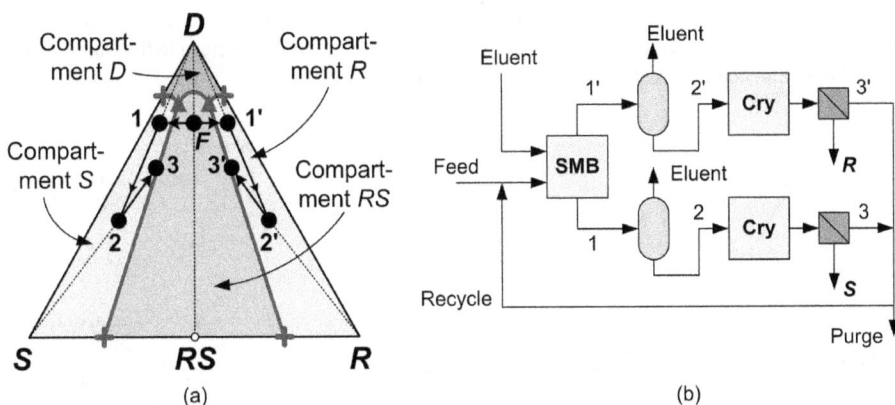

Figure 5.14: Resolution of enantiomers using a crystallization-chromatography hybrid process [136]: (a) process path and (b) process flowsheet.

chromatography, which is often performed in a simulated moving bed (SMB) in a continuous process, but the throughput tends to be low. A better option is to use SMB to obtain a solution that is sufficiently enriched in one enantiomer, followed by using crystallization to get the pure product, as shown in Figure 5.14b.

The initial separation is performed by SMB to give an extract having the composition of point 1, which is inside compartment S. After removing some solvent to get point 2, pure S can be obtained by crystallization. Similarly, the raffinate from the SMB is given by point 1', which is inside compartment R. After solvent removal, pure R can be crystallized out. The mother liquor from the crystallizers, streams 3 and 3', is mixed with the feed stream F to become the feed stream to the SMB. This combination improves the overall throughput of the process.

Example 5.4 – Resolution of mandelic acid
Lorenz et al. [137] used mandelic acid (MA), a useful precursor to various drugs, as a model system to illustrate efficient enantioseparation, using a coupling between SMB and crystallization. Their experimental results indicate that the chromatographic separation between the (+) and (–) enantiomers of MA is a difficult one. Simulation results of the SMB performance based on the determined adsorption isotherms lead to a plot of possible productivity of the SMB unit as a function of purity requirement, as shown in Figure 5.15a. While it is possible to achieve a high purity using SMB alone, the productivity decreases significantly with increasing purity requirement.

The polythermal SLE phase diagram of the MA/water system is shown in Figure 5.15b, along with three isothermal cuts. Symbols on the figure indicate experimental data points [137, 138]. The eutectic points are located at 31% and 69% (+)-MA, with a eutectic temperature of 115 °C as determined by differential scanning calorimetry. Emanating from these points are two eutectic troughs, which constitute the boundary between the compartments of the two pure enantiomers and that of the racemate. The double saturation points at different temperatures are all located on these eutectic troughs, which

happen to be relatively linear for this system. The ternary eutectics involving water, as well as the eutectic points between water and the pure enantiomers, are located extremely close to the water vertex,
so that the water compartment is too small to be displayed in the figure.

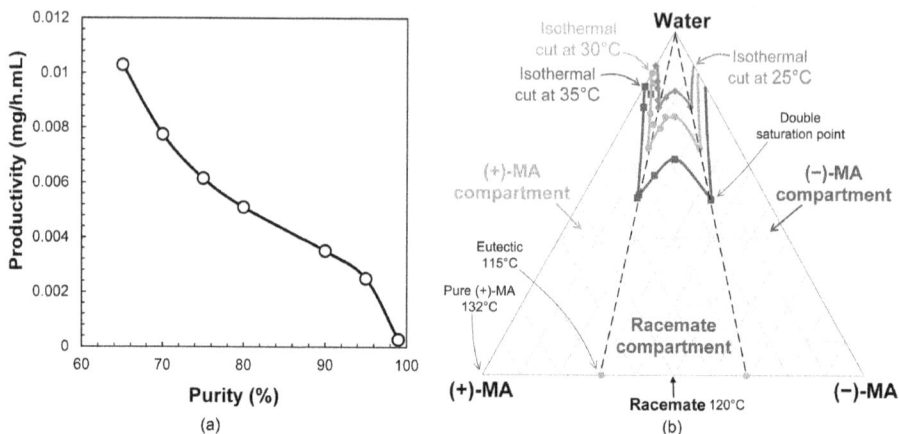

Figure 5.15: Basic information for mandelic acid separation by chromatography and crystallization
(data from [137]): (a) productivity as a function of product purity at feed concentration = 0.001 g/L
and (b) SLE phase diagram of mandelic acid/water system.

Assuming that the SMB unit is operated at low MA concentration, the coupled SMB/crystallization
process is the same as the one shown in Figure 5.14b, with water as the eluent and the two MA
enantiomers (+ and −) as the products, instead of R and S. Figure 5.16 shows the corresponding
process path when crystallization is performed at 35 °C. The purity of the SMB product (consider
stream 1 for (+)-MA) should be sufficiently high that upon concentration the crystallizer feed
(stream 2) falls inside the pure enantiomer region (instead of the solid mixture region), at this temperature. A close-up view of half the upper part of the triangle is shown on the right-hand side,
focusing on three scenarios that illustrate the effect of the product purity coming out of the SMB
unit, as well as the tradeoff between the productivities of the SMB and crystallization units. It is
assumed that the (+)-rich stream is concentrated to the same water content prior to crystallization.
If the chromatographic separation gives a (+)-rich stream with a composition of point 1a, corresponding to just over 70% purity, the crystallizer feed would be at point 2a, which is inside the
solid mixture region. Therefore, a mixture of (+)-MA and the racemate would crystallize out, and
the liquid composition (point 3a) is at the double saturation point. In other words, the crystallizer
feed purity is not sufficiently high to obtain pure (+)-MA crystals. A higher purity of the (+)-rich
stream, as given by point 1b, would give point 2b inside the pure (+) solid region as the crystallizer
feed, leading to crystallization of pure (+)-MA. The mother liquor composition is given by point 3b,
which is located on the saturation curve of (+)-MA. An even higher purity of the SMB product stream
(point 1c) would give point 2c, which is deeper inside the pure (+) solid region, as the crystallizer
feed. The corresponding mother liquor composition is given by point 3c, which is closer to the
water vertex compared to point 3b. Therefore, the solubility of (+)-MA is lower, and the feed at
point 2c would give a higher per-pass recovery of (+)-MA compared to that at point 2b. On the other
hand, based on Figure 5.15a, the higher SMB product purity in point 1c leads to a lower productivity
compared to point 1b.

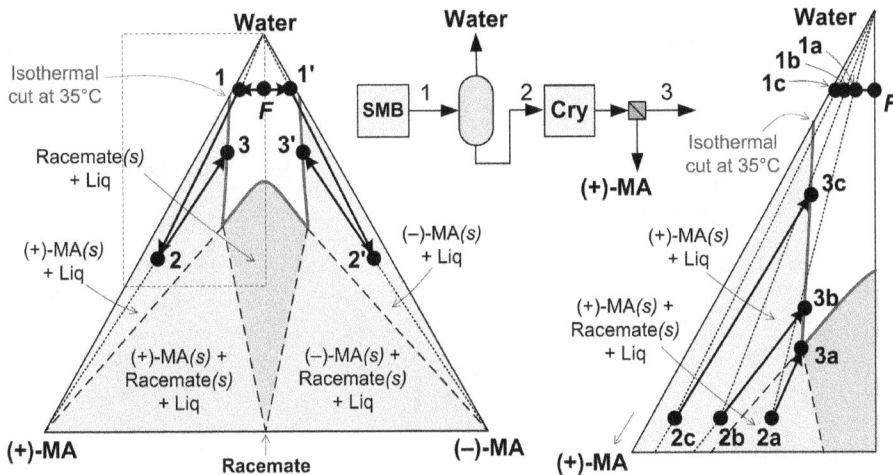

Figure 5.16: Process path of the chromatography/crystallization hybrid process for mandelic acid resolution.

5.3 Solid Solution Crystallization

Solid solution can form when molecules of one component are incorporated into the crystal lattice of another component. This can happen either by substitution of molecules in the crystal lattice with another component having a similar structure, or by insertion of a smaller molecule into the interstitial space between larger molecules. From process point of view, solid solution behavior leads to a more complex crystallization process because the solid phase is thermodynamically not pure. A rather complex purification process is often needed to produce a final product that meets the given purity specification.

5.3.1 SLE phase behavior of solid solution systems

Figure 5.17 shows how the SLE phase diagram for the ternary system looks like, if one binary pair (namely A–B) exhibits solid solution behavior. The 3D figure (a) features only two solubility surfaces: one for the solid solution (referred to, here, as AB, for simplicity) and another for solvent S. Two isothermal cuts at different temperatures are shown in Figure 5.17b and 5.17c. Similar to the simple eutectic system, the cuts reveal regions of single phase (unsaturated liquid) and multiple phases (solid and liquid) in equilibrium at the given temperature. However, the solid phase in the two-phase

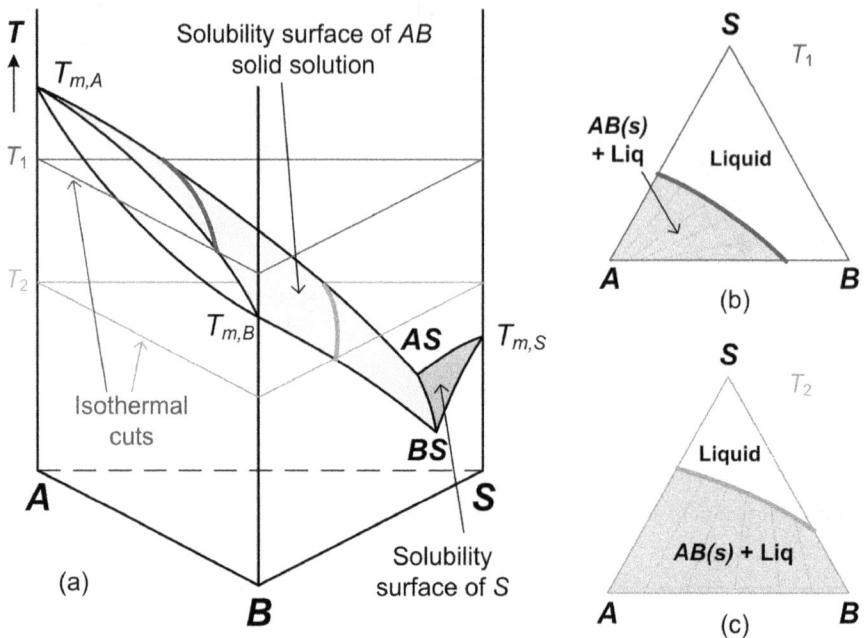

Figure 5.17: Ternary SLE phase diagram with type I solid solution behavior: (a) T–x diagram, (b) isothermal cut at T_1, and (c) isothermal cut at T_2.

region is a solid solution instead of a pure component. As indicated by the tie-lines, the solid phase composition depends on the liquid phase composition.

If the binary pair A-B exhibits a type V solid solution behavior (Figure 2.6e) instead, the phase diagram of the ternary system would look like Figure 5.18. There are separate solubility surfaces for the A-rich and B-rich solid solutions, intersecting along a eutectic trough. The isothermal cuts feature regions in which a solid solution exists in equilibrium with a liquid phase. At temperature T_2 (Figure 5.18c), there is also a region where two solid solutions (two different solid phases) are in equilibrium with a liquid phase. Points m and n are located on the boundaries of the miscibility gap between A and B.

5.3.2 Process synthesis

The basic strategy to purify a solid solution is to perform crystallization in multiple stages until the desired purity is achieved. As an illustration, the process path for a continuous, 3-stage countercurrent process to purify a solid solution containing A and B using solvent S is shown in Figure 5.19. Limited miscibility behavior is assumed

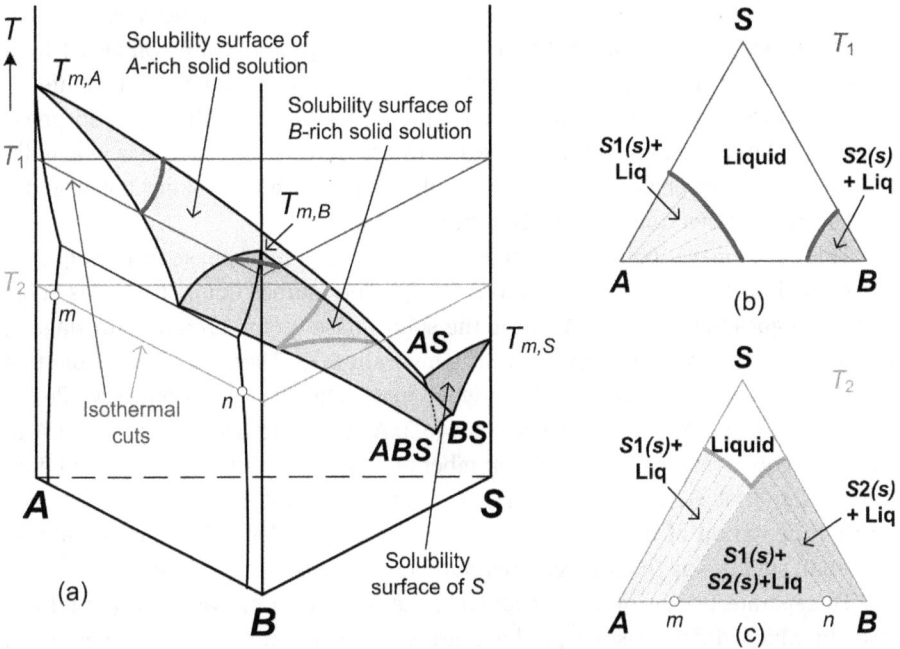

Figure 5.18: Ternary SLE phase diagram with Type V solid solution behavior: (a) T–x diagram, (b) isothermal cut at T_1, and (c) isothermal cut at T_2.

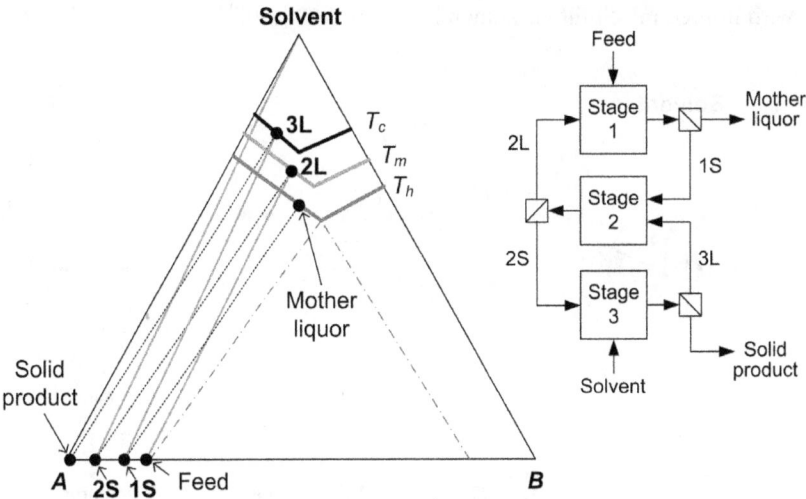

Figure 5.19: Purification of solid solution by multistage crystallization.

between A and B. The dotted dash lines indicate the boundaries between different regions at temperature T_h. The solid and liquid compositions are connected by a tie-line, the orientation of which depends on composition as well as temperature. In the first stage crystallizer operating at T_h, the feed is mixed with the liquid phase from stage 2 to form a mixture in the region where A-rich solid solution is in equilibrium with a liquid phase at T_h, which would split to give the mother liquor and a solid solution with a composition given by point 1S.

This solid phase (point 1S) is then mixed with the liquid phase from stage 3 in the second stage crystallizer, operating at T_m. The mother liquor from this stage (point 2L) goes back to stage 1, while the solid phase (point 2S) moves to stage 3, where it is mixed with fresh solvent at temperature T_c. The solid phase from this stage, which has been significantly enriched in A, is taken as the product, while the mother liquor (point 3L) goes back to stage 2. As the solid becomes richer in A in each subsequent stage, the required number of stages depends on the targeted final purity. Note that T_h, T_m, and T_c can be the same temperature; the choice depends on the orientation of tie-lines at various temperatures. Generally, they should be chosen in such a way that the required number of stages is minimized.

To separate two or more components, the basic idea of fractional crystallization, in which different solid products are crystallized at different temperatures, can be applied. With solid solution behavior, the solid products are no longer pure components, and further purification by multistage crystallization may be necessary to meet the desired purity specifications. As an illustration, a separation process for separating a pair of components, A and B, which exhibits solid solution behavior with limited miscibility is shown in Figure 5.20 [139].

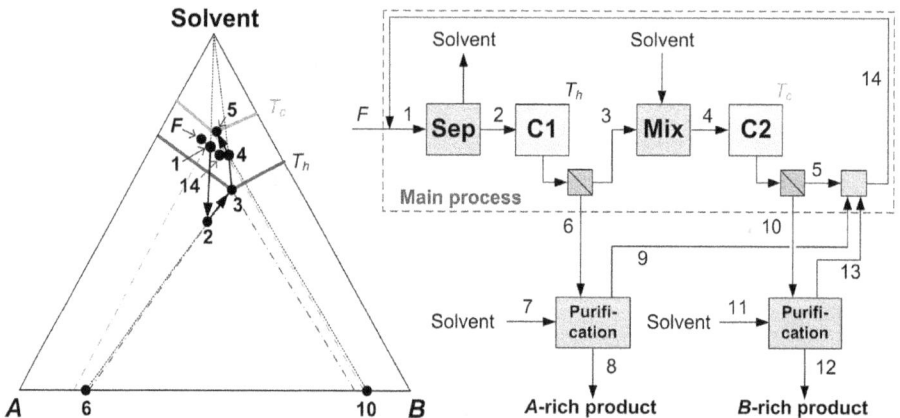

Figure 5.20: Crystallization process for separating a mixture of two components exhibiting solid solution behavior.

Using the concept of movement in composition space, the first step is to generate a composition inside the A-rich saturation region at T_h (point 2), so that an A-rich solid product can be crystallized out. Then, solvent is added to the mother liquor to move the composition into the B-rich saturation region at T_c (point 4) so as to obtain a B-rich solid product. The two solid products are then purified using a multistage crystallization scheme as in Figure 5.19, with the final mother liquor recycled to the main process (streams 9 and 13).

Example 5.5 – Isolation and purification of Schisandrin B

Schisandrin B (Sch B) is an active ingredient in *Fructus schisandrae*, a Chinese herb that is said to be good for liver protection and detoxification [140]. Like with other herbs in traditional Chinese medicine, extraction is the common way to recover the active ingredient [141, 142]. Besides schisandrin B, a typical extract also contains its stereoisomers, (±)γ-Schisandrin, which normally exist in a racemic mixture and can be considered as one component. Since (–)Sch B has been found to be more potent, it is desirable to obtain pure (–)Sch B from the extract. These schisandrins form a solid solution, and the phase diagrams with ethanol at 40 °C and 10 °C are shown in Figure 5.21.

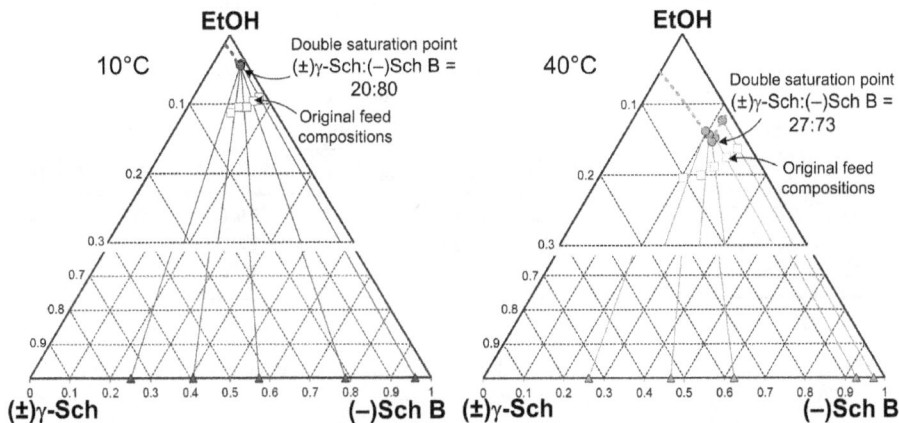

Figure 5.21: Phase diagram of (±)γ-Sch/(–)Sch B/ethanol at 10 and 40 °C [143].

Based on experimental determination of solid and liquid compositions from various samples, the composition of the double saturation point at the two temperatures is known. The ratio of (±)γ-Sch to (–)Sch B at the double saturation point is 27:73 at 40 °C and 20:80 at 10 °C. With such a shift in the double saturation point composition, it is possible to obtain both (–)Sch B and (±)γ-Sch as separate products, using the process shown in Figure 5.22. A (–)Sch B-rich product is obtained at 40 °C in the first crystallizer, and a (±)γ-Sch-rich product is crystallized at 10 °C in another crystallizer.

Because of the tie-line orientation, the mother liquor from the first crystallizer is not enriched enough in (±)γ-Sch to cross over to the (±)γ-Sch region. Therefore, a second crystallization step at 40 °C is added to remove another (–)Sch B-rich product, which is recycled to the feed stream because it is not sufficiently pure. To demonstrate the feasibility of this process, three crystallization experiments were performed, with the results summarized in Table 5.1. All compositions are shown in weight percent on a solvent-free basis.

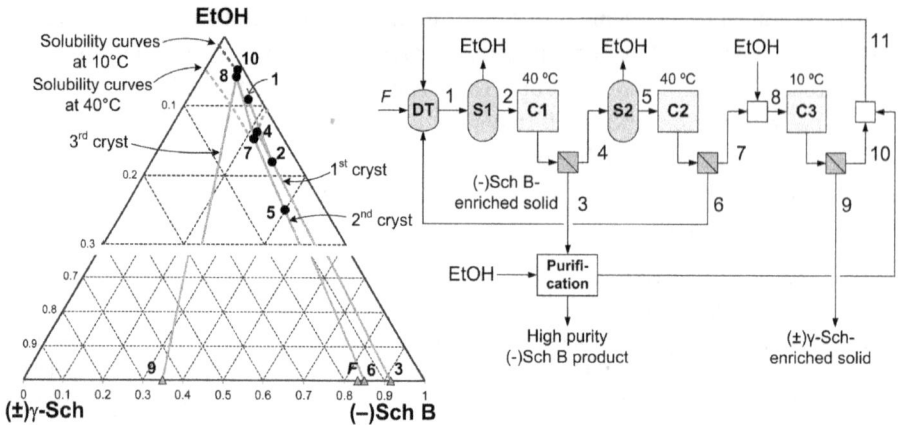

Figure 5.22: Process for separating and purifying (–)Sch B from its mixture with (±)γ-Sch [143].

Table 5.1: Results of experimental verification of crystallization-based Sch B separation process (data from [143]).

Crystallization experiments	Initial composition (wt%)		Liquid composition (wt%)		Solid composition (wt%)	
	(±)γ-Sch	(–)Sch B	(±)γ-Sch	(–)Sch B	(±)γ-Sch	(–)Sch B
First cryst. at 40 °C	16.7	83.3	23.6	76.4	8.4	91.6
Second cryst. at 40 °C	23.6	76.4	28.0	72.0	15.2	84.8
Third cryst. at 10 °C	28.0	72.0	27.9	71.9	65.0	35.0

The feed to the first crystallizer (stream 1) contains approximately 83% (–)Sch B. Crystallization at 40 °C gives a solid product containing almost 92% (–)Sch B. The mother liquor (stream 4), which contains about 76% (–)Sch B and 24% (±)γ-Sch is sent to a second crystallization step, also operating at 40 °C . The mother liquor from this second crystallizer (stream 7) is sufficiently enriched in (±)γ-Sch, so that a third crystallization step operating at 10 °C produces a solid product containing 65% (±)γ-Sch (stream 9). This (±)γ-Sch-enriched product was not further purified because it was not the focus of the study.

Instead of a countercurrent process as discussed previously, a simpler, crosscurrent process (in which the solid was mixed with pure solvent in each stage) was used to verify the feasibility of (–) Sch B purification by recrystallization. The results are summarized in Table 5.2. It can be seen that the (–)Sch B content in the solid product increases to over 96% after the first stage, and to 98.5% after the second stage. Therefore, it was demonstrated that highly pure (–)Sch B can be isolated from its mixture with (±)γ-Sch by crystallization. Although the process is relatively complex due to the presence of solid solution, it can be synthesized using the previously discussed techniques based on the understanding of SLE phase behavior.

Table 5.2: Results of experimental verification of Sch B purification by recrystallization (data from [143]).

Recrystallization experiments	Initial composition (%)		Liquid composition (wt%)		Solid composition (wt%)	
	(±)γ-sch	(−)Sch B	(±)γ-sch	(−)Sch B	(±)γ-sch	(−)Sch B
First cryst. at 40 °C	8.4	91.6	15.3	84.7	3.8	96.2
Second cryst. at 40 °C	3.8	96.2	6.2	93.8	1.5	98.5

Example 5.6 – Separation of fullerenes by crystallization

Fullerenes are allotropes of carbon characterized by their hollow cage structures. They have wide applications including solar or photovoltaic cells and lubricants used in bowling balls, tennis and badminton rackets. In their pure form, C_{60} and C_{70}, the two most abundant fullerene species, are the best available electron acceptors in organic solar cells and photovoltaic devices [144]. Common synthesis methods of fullerenes, such as combustion of hydrocarbons, produce a mixture of fullerenes. Fullerenes of high purity are very expensive because they are usually separated by liquid chromatography, which has low product throughput, involves expensive stationary and mobile phases, and is difficult to scale up.

The solubility of C_{60} in o-xylene in the temperature range of −20 to 80 °C features a maximum at around 30 °C, with decreasing solubility beyond this point, while the solubility of C_{70} in o-xylene increases monotonically with temperature in the same temperature range [145]. This peculiar behavior opens up the opportunity for using crystallization as an attractive alternative to separate pure C_{60} and C_{70} solids from their mixture [146, 147, 148]. Figure 5.23 shows the SLE phase diagram of the C_{60}/C_{70}/o-xylene system, highlighting the regions of C_{60}-rich solid solution at 110 °C and C_{70}-rich solution at −16 °C. A batch process for recovering highly pure C_{60} and C_{70} solids is depicted in Figure 5.24, with the process points plotted on the phase diagram of Figure 5.23.

The feed is a typical mixture of fullerenes containing C_{60}, C_{70}, and higher fullerenes such as C_{76} and C_{84}. The mixture is first dissolved in o-xylene and filtered to remove any undissolved solids.

The fullerenes solution (stream 1) is fed to a crystallizer operating at 110 °C, in which a C_{60}-rich solid solution (stream 1S) is crystallized out. These solids are purified in a series of stages, each of which involves mixing with o-xylene and equilibration at 110 °C to produce C_{60} solids with progressively higher content of C_{60}. The final solids (stream 4S) are sent to a vacuum dryer to give highly pure C_{60} solids as product. The mother liquor from the first crystallizer (stream 1L) is sent to vacuum evaporator to remove some o-xylene, such that the remaining liquid (stream 5) falls inside the crystallization region of C_{70}-rich solid solution at −15 °C. Following crystallization at −15 °C, the C_{70}-rich solids is purified in a series of recrystallization and finally vacuum dried to produce highly pure C_{70} solids. The mother liquors from various crystallization stages (streams 2L, 3L, 4L, 5L, 6L, 7L, and 8L) can be combined and processed in the next batch.

Crystallization experiments were conducted to verify the feasibility of the process. Table 5.3 summarizes the results. Starting with a feed containing 60 wt% C_{60}, 25 wt% C_{70}, and 15 wt% higher fullerenes, the first crystallization at 110 °C produces solids containing 94 wt% C_{60}, which are further purified to 97 wt% in three recrystallization stages. The mother liquor from the first crystallization is processed at −15 °C and produces solids containing 78 wt% C_{70}. After purification in three recrystallization stages, a C_{70}-rich product containing 98 wt% C_{70} is obtained.

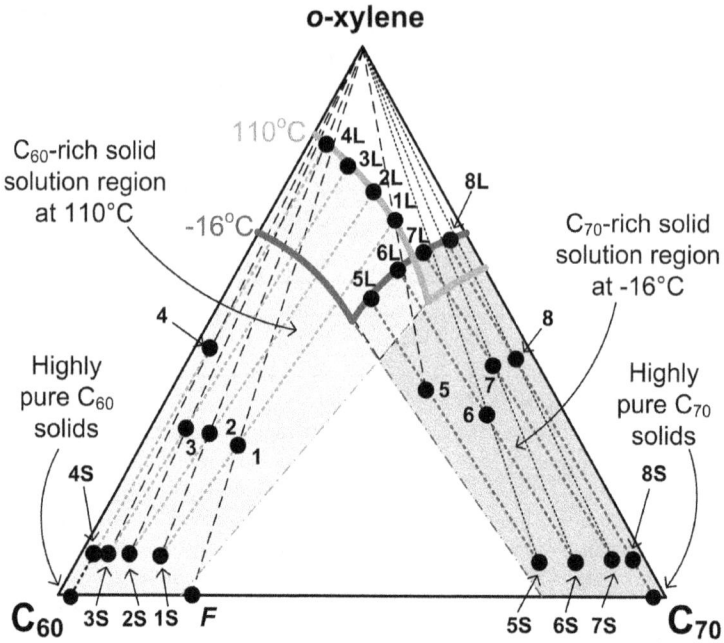

Figure 5.23: SLE phase diagram of $C_{60}/C_{70}/o$-xylene ternary system [146].

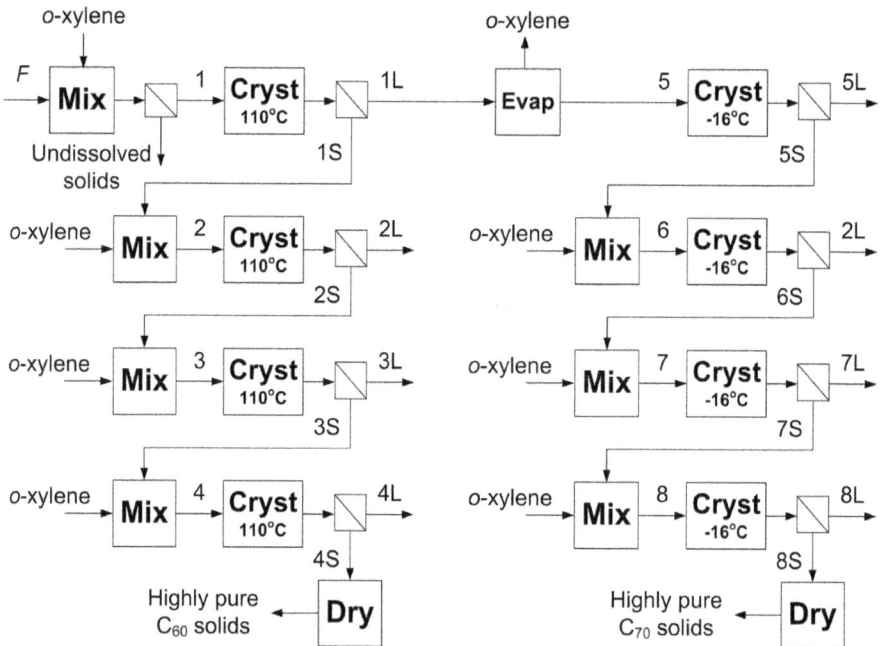

Figure 5.24: Crystallization-based separation process to obtain highly pure C_{60} and C_{70} solids [146].

Table 5.3: Results of fullerenes crystallization experiments (data from [146]).

	C_{60}-rich side		C_{70}-rich side	
	Stream	wt% C_{60} (solvent-free basis)	Stream	wt% C_{70} (solvent-free basis)
After first crystallization	1S	93.91	5S	77.70
After second crystallization	2S	96.03	6S	83.11
After third crystallization	3S	96.84	7S	94.99
After fourth crystallization	4S	96.97	8S	98.07

5.4 Amino Acid and Protein Crystallization

Amino acids, which are widely used in food, chemical, agricultural, cosmetic, and pharmaceutical products, are generally produced by fermentation process, protein hydrolysis, or chemical synthesis, although the major method in the recent years has been fermentation process. Regardless of the production method, the presence of impurities is inevitable, and crystallization is usually involved in the purification step to obtain amino acid crystals with high purity [14, 149].

Amino acids are ampholytes, meaning that it can accept or donate proton depending on pH. This is possible because an amino acid molecule contains both an amine (-NH$_2$) group and a carboxylic (-COOH) group. Consider phenylalanine ((S)-2-amino-3-phenylpropanoic acid) as an example, which for simplicity is referred to as HA. At low pH or acidic condition, the amine group accepts proton and a complex cation HA.H$^+$ is formed. With Cl$^-$ as the counterion, it forms HA.HCl, which can be considered as an adduct between HA and HCl. At high pH (basic environment), the carboxylic group loses a proton and produces an anion A$^-$. In the presence of sodium counterion, it forms salt NaA. At intermediate pH, both of these happen at the same time, and a zwitterion is formed.

5.4.1 SLE phase behavior of amino acid systems

It is well known that an amino acid has the lowest solubility at its isoelectric point (the pH at which the molecule carries no net electrical charge). This is because the electrostatic repulsion force among the amino acid molecules is at the minimum, so that they tend to aggregate together. The solubility generally increases at low and high pH values, but only up to a certain concentration of acid or base, beyond which the solubility decreases again, as shown in the top left of Figure 5.25.

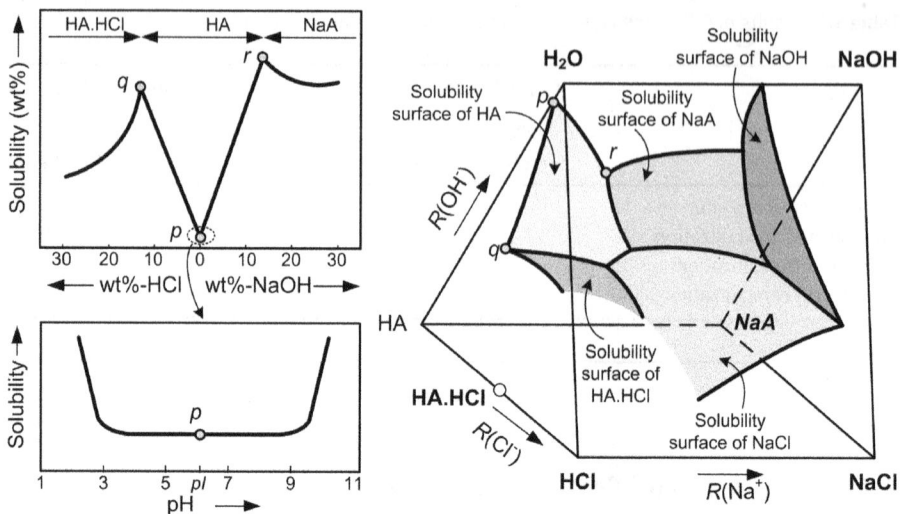

Figure 5.25: SLE phase behavior of an amino acid system [80].

This behavior is easier to understand by looking at the 3D phase diagram shown on the right-hand side. This is an isothermal cut, and the system involves two cations (H^+ and Na^+) and three anions (OH^-, Cl^-, and A^-), leading to the triangular prism shape of the composition space. Various solubility surfaces are shown; but note that there is no solubility surface for HCl since it exists as a liquid at this temperature. The solubility of HA in H_2O is at point p, which is usually very close to the H_2O vertex because of the low solubility of the amino acid in water.

The curves pq and pr correspond to the ones in the figure on the left-hand side, except that the scales and coordinates are different. From the phase diagram, it can be seen that when acid concentration is higher than point q, HA.HCl is crystallized instead of HA. Similarly, when the base concentration is higher than point r, NaA crystallizes out. Points q and r are the double saturation points, at which two components are both saturated. It should be noted that as the phase behavior depends on the counterions, solubility is not a unique function of pH. At the same value of pH, the solubility can be different if the counterion is different – for example, if H_2SO_4 is used instead of HCl.

The SLE phase behavior of protein systems can also be represented in a similar fashion. Protein has many acid and basic groups, which can acquire positive or negative charges by donating or receiving protons, and the net charge of the protein strongly depends on pH [150]. The phase diagram would look similar to the one shown in Figure 5.25. However, since protein can form a large number of different protein salts, it is difficult to clearly demarcate the boundaries between the solubility surfaces of individual salts and different solid compositions. Therefore, it is convenient to consider that the protein salts exhibit solid solution behavior, in which the solid composition is a continuous function of the liquid composition.

5.4.2 Process synthesis

Despite the complexity of the phase diagram, the same strategy based on movements on phase diagram can be applied to synthesize crystallization-based processes for amino acid and protein separation. A key part of the strategy is the addition of acid and base to reach the region where an amino acid, protein, or their salt can be crystallized out. It is also important to avoid the accumulation of any inorganic salt that may be formed as the byproduct of the acid and base addition, either by crystallizing it out or removing it by another means such as ultrafiltration.

Example 5.7 – Crystallization of L-glutamic acid

L-Glutamic acid (H_2Glu), a nonessential amino acid produced by fermentation of corn glucose, and its salts are flavor enhancers that are widely used as food additives. NH_4OH and HCl are used as pH-controlling agents during and after fermentation. Two ammonium salts of glutamic acid can be formed: the monoammonium acid, which normally crystallizes in monohydrate form, and the diammonium acid. Therefore, the phase diagram would have these two solubility surfaces beside those of glutamic acid, glutamic acid hydrochloride, and ammonium chloride, as shown conceptually in Figure 5.26. The coordinates are

$$R(NH_4^+) = \frac{[NH_4^+]}{[H^+] + [NH_4^+]} \tag{5.1}$$

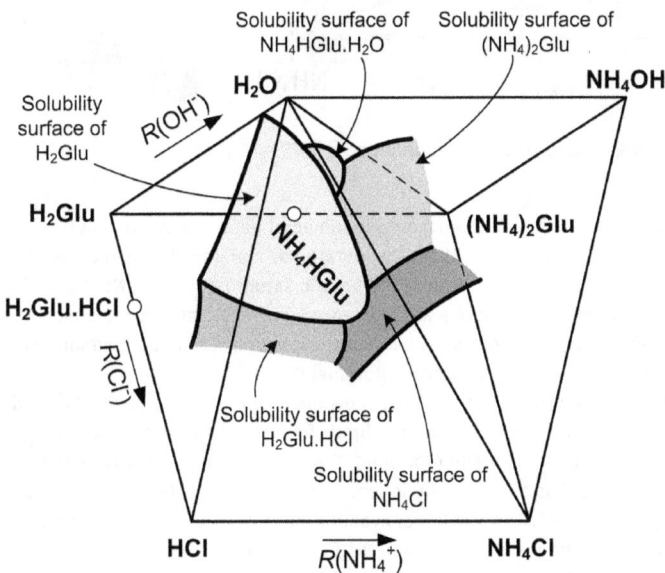

Figure 5.26: Image of the SLE phase diagram of glutamic acid/HCl/NH_4OH system based on experimental data.

$$R(Cl^-) = \frac{[Cl^-]}{2[Glu^{2-}] + [Cl^-] + [OH^-]} \tag{5.2}$$

$$R(OH^-) = \frac{[OH^-]}{2[Glu^{2-}] + [Cl^-] + [OH^-]} \tag{5.3}$$

The prism has been tilted to provide a clearer view of the solubility surfaces, the relative positions of which are deduced based on experimental data [80]. A Jänecke projection toward the base is taken of the left half of the prism (H$_2$O-H$_2$Glu-(NH$_4$)$_2$Glu-NH$_4$Cl-HCl), to obtain a two-dimensional projection shown in Figure 5.27.

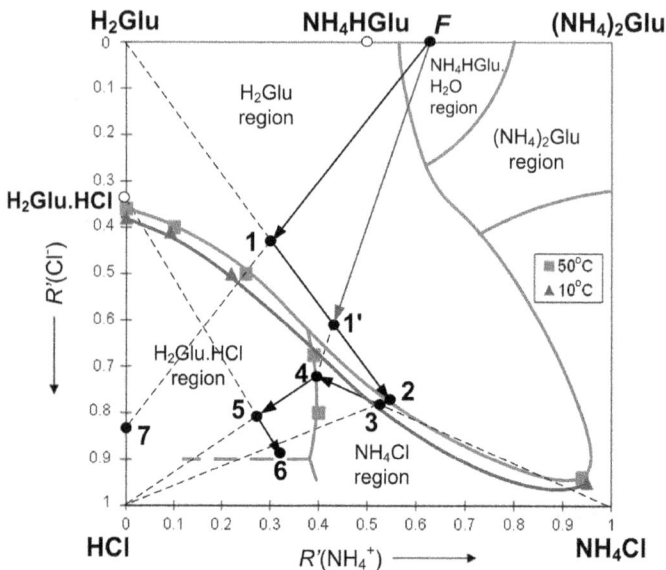

Figure 5.27: Process paths for glutamic acid separation process [80].

Double saturation points at 10 and 50 °C, obtained from experimental data, are marked on the diagram. Two sets of double saturation curves at these temperatures, marking the boundaries between different regions, are also drawn by connecting the double saturation points. Note that the boundaries of NH$_4$Glu.H$_2$O and (NH$_4$)$_2$Glu regions are only approximate. No attempt was made to find the exact location of these boundaries because, without the crystallization of ammonium glutamate salts, they are not expected to affect the process flowsheet.

Based on the phase diagram, two process alternatives shown in Figure 5.28 can be synthesized. The first alternative (Figure 5.28a) starts with mixing the broth from fermentation (point *F*) with a mixture of HCl and H$_2$Glu.HCl (mixture composition of point 7) to give a mixture that is inside the H$_2$Glu region (point 1). In practice, HCl is added as an aqueous solution. Due to its very low solubility in water, glutamic acid is crystallized. The amount of evaporation in the crystallizer is controlled such that the mother liquor composition is at point 2, and NH$_4$Cl does not co-crystallize. To improve the overall yield, it is also desirable to crystallize H$_2$Glu.HCl, which can then be recycled to the feed stream. Before doing so, NH$_4$Cl is first crystallized by mixing stream 2 with HCl to create a mixture of point 3, which is inside the NH$_4$Cl region, followed by water evaporation in the second crystallizer (C2). The purpose is to remove the NH_4^+ from the feed and Cl^- from the added HCl so that they would not accumulate. Since the solubility of NH_4Cl at 10 °C is much lower than at 50 °C, this crystallization is performed at 10 °C so that less water evaporation is necessary. In order to get into the $H_2Glu.HCl$ compartment, HCl is added to the mother liquor (stream 4), resulting in a mixture of point 5. Crystallization of $H_2Glu.HCl$ is then performed from point 5 to point 6 in the third crystallizer (C3). The mother liquor of point 6, which has much less glutamic acid compared to the previous mother liquor (point 2), is purged, to avoid accumulation of impurities from the broth. The $H_2Glu.HCl$ is recycled back to the feed stream. This process alternative agrees with the process described in the literature. Both NH_4Cl and the purged mother liquor can actually be used in fertilizer application.

The second alternative (Figure 5.28b) is to mix the feed with the recycled mother liquor stream (point 4) in order to get into the H_2Glu region. This results in a mixture of point 1', and crystallization of H_2Glu is performed in crystallizer C1 until point 2 is reached. HCl is then added to stream 2 to create stream 3 inside the NH_4Cl compartment, just like in the first alternative. After NH_4Cl crystallization in C2, the mother liquor (point 4), still rich in H_2Glu, is partially recycled to the feed stream. Of course, a portion of stream 4 should be purged to avoid the accumulation of impurities. This second alternative consumes less HCl than the first one, but the glutamic acid recovery is slightly lower due to higher H_2Glu concentration in the purge stream. Nonetheless, with the SLE phase diagram at hand, it is possible to come up with new alternatives that have not been considered before, thus opening up possibilities for improvement of the existing process.

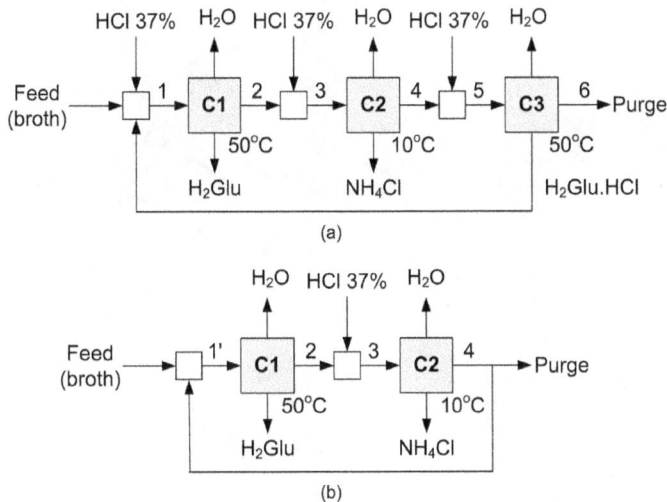

(a)

(b)

Figure 5.28: Process for separating glutamic acid by crystallization: (a) conventional process (b) alternative process.

Example 5.8 – Crystallization of ovalbumin

Lysozyme, an enzyme in chicken egg white, can be used as a natural preservative in food industry as it can effectively inhibit bacteria growth. For lysozyme production, a concentrated salt solution is commonly used as precipitating agent in the crystallization process. The concentration of counterions in lysozyme should be reduced to a proper level before it can be sold as a product. Tam et al. [150] synthesized a process for isolating lysozyme from egg white in aqueous NaCl solution using SLE phase diagram as guidance.

Figure 5.29 shows the conceptual solventless projection for the lysozyme system, along with the original 3D isobaric-isothermal diagram. The diagrams are not drawn to scale. The net charge of the protein is −13 as pH approaches 14, as determined from titration curve [151]. Therefore, pure lysozyme is denoted by $H_{13}Lys$, and various lysozyme salts with Na^+ and Cl^- as counterions are denoted by $Na_i(H_{13-i}Lys)(HCl)_j$, with i=1, 2, . . ., 13. The projection is obtained by normalization using Na^+ and Cl^- as the reference components, and the coordinate expressions for this projection are given as

$$T(Na^+) = \frac{x_{Na^+}}{x_{Na^+} + x_{Cl^-} + x_{Lys13-}} \tag{5.4}$$

$$T(Cl^-) = \frac{x_{Cl^-}}{x_{Na^+} + x_{Cl^-} + x_{Lys13-}} \tag{5.5}$$

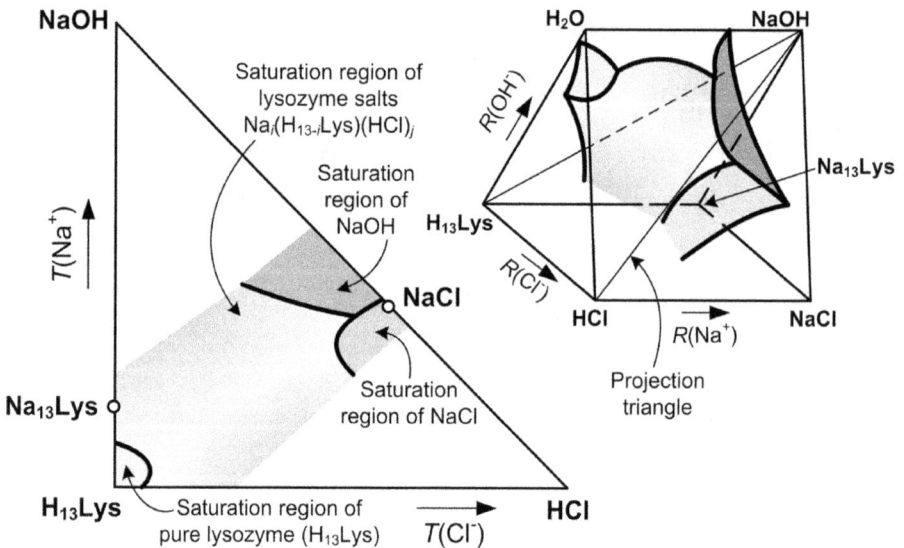

Figure 5.29: Conceptual phase diagram of lysozyme/NaOH/HCl system [150].

Essentially, all saturation surfaces in the 3D diagram are projected to the triangle $H_{13}Lys$-NaOH-HCl. The projection features saturation regions of pure lysozyme ($H_{13}Lys$), NaOH, NaCl, and lysozyme salts. The region of lysozyme salts is actually an aggregate of many regions of individual salts with different contents of counterions Cl^-, OH^-, H^+, and Na^+. Only the areas near the center of the triangle, where most of the process points are expected to be, are shown in the projection. The blank areas close to the NaOH and HCl vertices are not of interest, because extreme pH conditions would destroy the protein.

Figure 5.30 shows an enlarged section of the projection near the NaCl vertex, along with the process path. The contour lines represent different values of $(H^+ + OH^-)$ mole fractions in the solution, and the gradient shading represents the solid composition in equilibrium with the solution, all of which are plotted based on experimental data. A darker shade corresponds to a higher content of Na^+ and Cl^- in the solids.

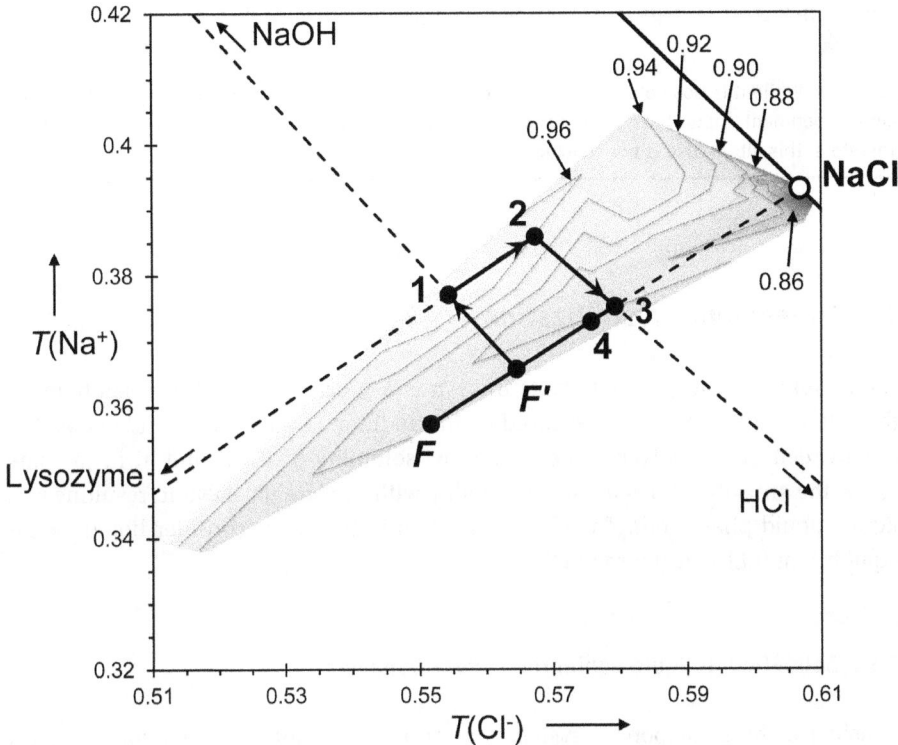

Figure 5.30: Process path for separating lysozyme from chicken egg white [150].

The process flowsheet is depicted in Figure 5.31. Crude lysozyme solution, which is a dilute lysozyme solution with sodium chloride (point F), is mixed with a recycle stream (point 4) to give point F'. By adding NaOH to reach point 1, supersaturation is generated because of decreasing solubility in this direction. As a result, lysozyme with low NaCl content is crystallized out, and a liquid phase with a composition represented at point 2 is obtained. HCl is then added to give point 3. As the saturation region of NaCl is very small, removal of NaCl by crystallization is impossible, and ultrafiltration is used to remove a concentrated NaCl solution (point 5) from the system. The retentate (point 4) is recycled back to the feed stream. A portion of stream 2 is purged to prevent accumulation of impurities.

The first step for recovering lysozyme in which NaOH is added to generate supersaturation is consistent with a patented process for recovering lysozyme [152]. To increase the yield of lysozyme, the patented process involves lowering the temperature after adding NaOH, to further decrease the

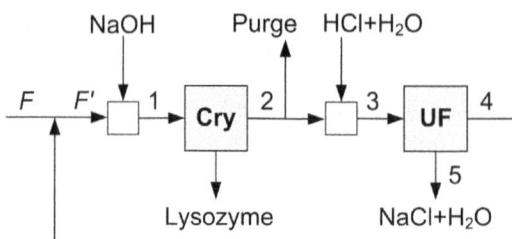

Figure 5.31: Process for separating lysozyme from chicken egg white.

solubility. While it is possible to generate such a process alternative involving temperature swing, more experimental data have to be used to generate the phase diagram at a lower temperature. Therefore, this alternative is not considered here.

5.5 Antisolvent Crystallization

Antisolvent or drowning-out crystallization is a process in which the solubility of the solute in the solvent is decreased by the addition of an MSA. The added MSA is often called an antisolvent, due to the low solubility of the solute in it. In some cases, the antisolvent has limited miscibility with the original solvent, resulting in a liquid–liquid phase split. Therefore, it is often necessary to consider liquid–liquid equilibrium (LLE) together with SLE.

5.5.1 Solid–liquid–liquid equilibrium phase behavior

Consider a three-component system consisting of a solute and two solvents with simple eutectic behavior, as depicted in Figure 5.32. At the temperature of interest, A has a much higher solubility in S than in D, so that S and D are the solvent and antisolvent, respectively. In this simplest scenario, there is a gradual decrease in solubility of A in a mixture of S and D, and there is no liquid–liquid phase split at any composition.

However, since the solvent and antisolvent often have very different physical properties, partial immiscibility in the liquid phase is not uncommon. Figure 5.33 illustrates a situation in which S and D are only partially miscible at some temperatures, and the liquid–liquid immiscibility region intersects the solubility surfaces. An isothermal cut at the temperature of interest features two separate regions where pure solid A coexists with a single liquid phase ($A(s)$+Liq), with a three-phase region ($A(s)$+L_1+L_2) between them. Points L_1 and L_2 are invariant points at the

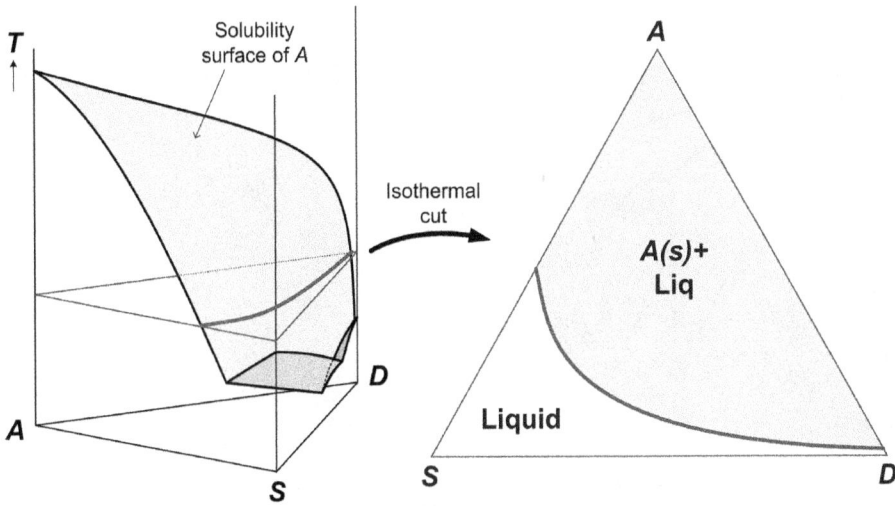

Figure 5.32: SLE phase behavior of a ternary system involving a solute, a solvent, and an antisolvent.

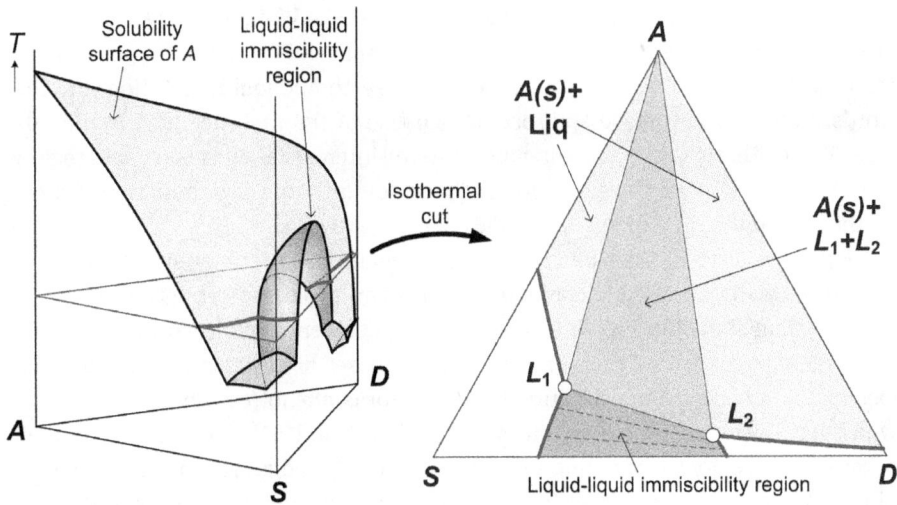

Figure 5.33: SLE phase behavior of a ternary system involving two solvents with limited miscibility.

given temperature, as any mixture within the three-phase region phase splits into pure solid A and two liquid phases, with compositions corresponding to these two points. From the 3D diagram, it can be deduced that the location of these points changes with temperature. If the temperature is increased above a certain limit, the three-phase region would disappear altogether, and the isothermal cut would simply look like the one in Figure 5.32.

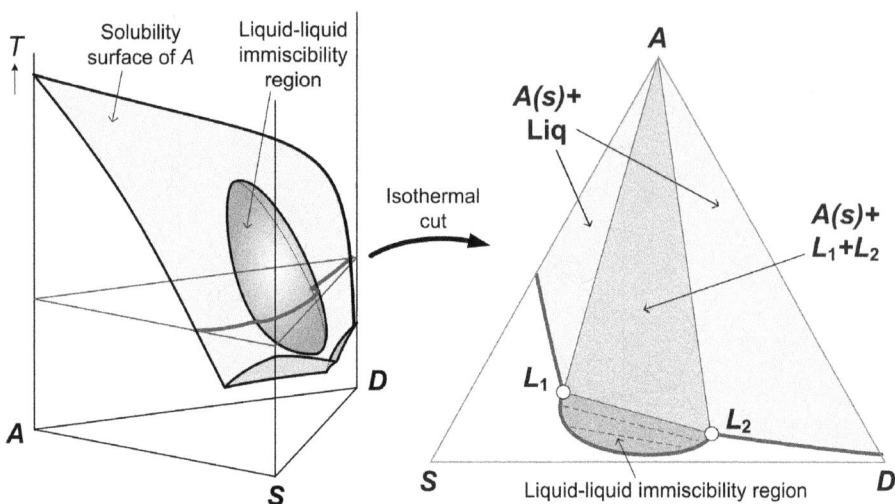

Figure 5.34: SLE phase behavior of a ternary system involving liquid–liquid immiscibility region resulting from three-component interactions.

Another possible scenario is illustrated in Figure 5.34. While S and D are completely miscible at all temperatures, a liquid–liquid immiscibility region appears in the interior of the phase diagram as a result of three-component interaction. In other words, liquid–liquid phase split occurs because of the presence of A in the mixture. The isothermal cut at the temperature of interest features the same regions as in Figure 5.33, with the exception that the liquid–liquid region does not extend all the way to the S-D edge.

Figure 5.35 depicts yet another scenario, featuring a ternary system with two pairs of partially immiscible components. This can be considered as a special case of the system shown in Figure 5.33, with the antisolvent D replaced by component B, whose mixture with A can form two liquid phases at elevated temperatures. An example of a real system exhibiting this behavior is phenol/catechol/water system, which has been experimentally measured by Lai et al. [153]. The two liquid–liquid immiscibility regions emanating from the A-B and S-B edges sit on top of the solubility surfaces and intersect at the interior of the phase diagram. Cuts at different fixed solvent compositions (S/B ratio), one of which is shown in Figure 5.35b, are easier to be measured. By combining the information from the various cuts, it is possible to deduce the overall phase behavior, as well as the isothermal cut at a temperature of interest, shown in Figure 5.35c.

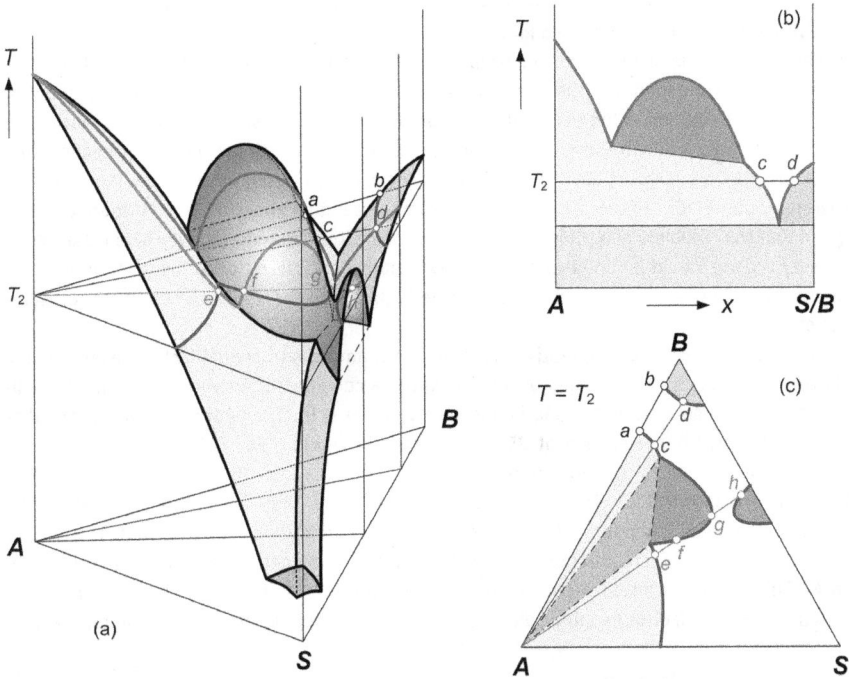

Figure 5.35: SLE phase behavior of a ternary system with two liquid–liquid immiscibility regions [153]: (a) T–x diagram, (b) cut at constant S/B ratio, and (c) isothermal cut at T_2.

5.5.2 Process synthesis

Antisolvent crystallization process can be properly synthesized with the help of the phase diagram. A suitable MSA needs to be selected, considering the overall SLE phase behavior, such as how much decrease in solubility can be effected by antisolvent addition. The possibility of hydrate and solvate formation also needs to be considered [154]. It is also important to consider the presence of liquid–liquid immiscibility region, as crystallization in the presence of a second liquid phase may be undesirable. Formation of liquid bubbles may interfere with crystal growth, giving oily fine crystals or even "dumplings" due to inclusion of the oil phase in the crystals. Highly inhomogeneous crystals containing hard agglomerates and partially amorphous materials have also been reported [155]. However, if crystallization can be performed with no problem, operation in the solid–liquid–liquid region may be advantageous as the mother liquor can be readily separated using a decanter, and the two liquid phases can be sent to different destinations.

Example 5.9 – Production of salt from a mine

Sodium chloride, NaCl, is mainly produced from either seawater or deposits on land [156]. In the latter method, water is pumped into the mine to dissolve the solid salt as well as a small amount of impurities, and the solution is then crystallized to produce salt crystals. The phase behavior of the NaCl-H_2O system features the presence of a dihydrate (NaCl.$2H_2O$), which has an incongruent melting point at 0.15 °C [157], which implies that dihydrate crystallization would occur below this temperature. Therefore, pure NaCl can only be precipitated from the solution above 0.15 °C. Furthermore, it is well known that the solubility of NaCl in water is rather insensitive to temperature, which means crystallization by cooling would be infeasible. Hence, evaporative crystallization is the most straightforward alternative, although the operating cost can be high due to the energy requirement for water evaporation.

As a process alternative to evaporative crystallization, antisolvent crystallization using organic solvents that can reduce the solubility of NaCl in water is considered. Among the potential candidates is ethanol, which does not cause liquid–liquid phase split. The isothermal SLE phase diagram of NaCl/water/ethanol system at 25 °C is shown in Figure 5.36. Solubility data from the literature [158] are shown as triangles on the diagram. A recycle stream (stream 5′) is injected into the mine (M) to dissolve NaCl. The NaCl solution (stream 1) is first mixed with an ethanol-rich stream 4′ to give stream 2 inside the two-phase region, causing NaCl to crystallize out. The mother liquor (stream 3) is sent to a distillation column to recover ethanol-water azeotrope as the distillate (stream 4). This stream is mixed with make-up ethanol to produce stream 4′. The water-rich bottoms (stream 5) is recycled to the mine, with a partial purge to avoid accumulation of impurities.

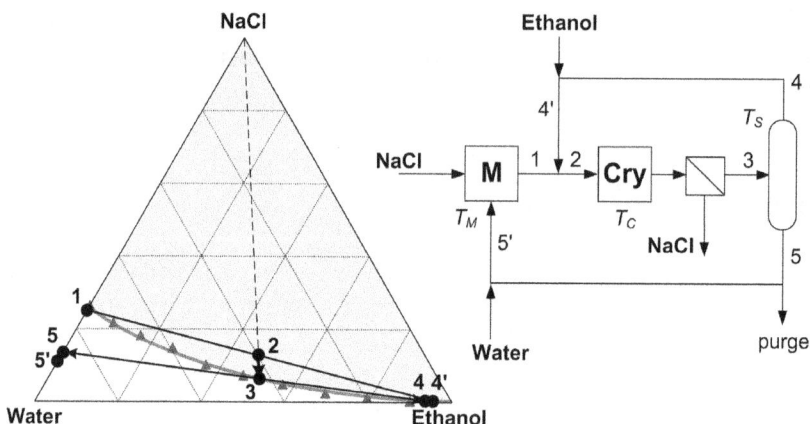

Figure 5.36: Process alternative for salt crystallization using ethanol as antisolvent.

Another alternative is to use an antisolvent that creates liquid–liquid immiscibility. One such solvent is 2-propanol, which leads to the phase behavior shown in Figure 5.37. The solubility data (triangles) as well as liquid–liquid tie-line data (circles) at 25 °C are taken from the literature [159]. Instead of a distillation column, a decanter can be used to separate the mother liquor (stream 3) into a 2-propanol-rich mixture (stream 4) and a water-rich mixture (stream 5), which is partially purged to avoid accumulation of impurities. To balance the loss to the purge stream, make-up 2-propanol and water are added to streams 4 and 5, respectively, resulting in streams 4′ and 5′, which are recycled. The use of decanter instead of distillation column means less energy is required for separation. Note that in the process illustrated in Figure 5.37, the crystallizer is operated at a lower temperature than the decanter temperature (T_D=25 °C), so that point 3 is located underneath the LLE region at 25 °C. When this stream is heated up to 25 °C, the liquid compositions (streams 4 and 5) follow one of the tie-lines in the LLE region. The composition of stream 1 is determined by the solubility of NaCl in the mixed solvent at T_M.

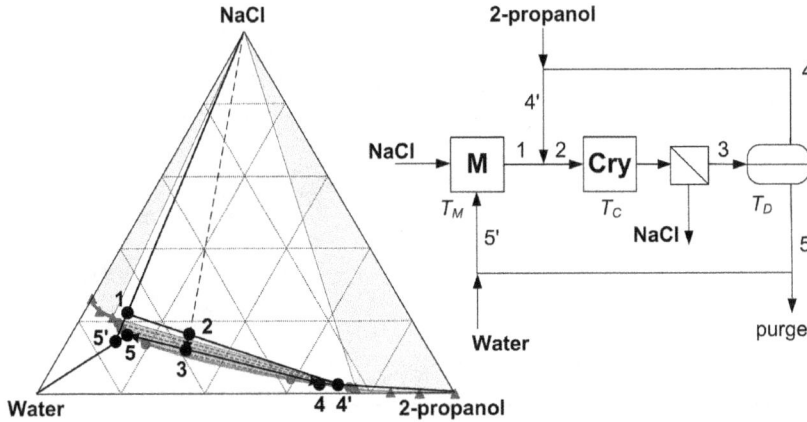

Figure 5.37: Process alternative for salt crystallization using 2-propanol as antisolvent.

Another possible solvent is diisopropylamine (DiPA). The phase behavior of NaCl/water/DiPA at 25 °C is also reported in the literature [160] and features limited miscibility between water and DiPA (similar to Figure 5.33). The process flowsheet is exactly the same as Figure 5.37, but the mass balance and process economics would be different due to different locations of points 4 and 5 [100].

Example 5.10 – Preventing oiling out during antisolvent crystallization

Kim et al. [155] reported some of the challenges during the development of the final crystallization step for Bristol-Myers Squibb drug candidate A, which involves a drowning-out crystallization process. Experimental observations show that crystallization of compound A in solvent S at 50 °C by adding antisolvent D leads to oiling out; however, when the drowning-out crystallization process takes place at 35 °C, no oiling out is observed. Based on this information, it can be deduced that there exists a liquid–liquid immiscibility region that interferes with the solid-liquid region at 50 °C, but disappears at 35 °C. This is consistent with the phase behavior shown in Figure 5.34.

Process paths for two alternatives are depicted on isothermal cuts at 50 °C and 35 °C in Figure 5.38. Both alternatives start with the same feed (point 1), which is a nearly saturated solution of A in S. In the first alternative (Figure 5.38a), antisolvent D is added to the feed at 50 °C, resulting in a mixture inside the three-phase region (point 2). Therefore, crystallization of A gives mother liquor (point 3) that splits into two liquid phases (oiling out). The second alternative (Figure 5.38b) starts with cooling to 35 °C, causing some A to crystallize. Addition of D to the mother liquor (point 2), at this temperature, produces a composition given by point 3, leading to additional crystallization of A. Since there is no liquid–liquid immiscibility region at 35 °C, no oiling out occurs. Basically, the second process alternative circumvents the liquid–liquid immiscibility region in order to avoid the problems caused by oiling out.

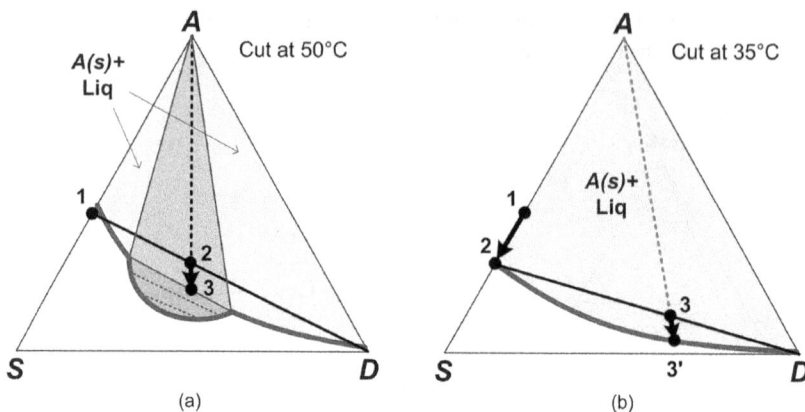

Figure 5.38: Avoiding oiling out in the crystallization of a drug substance: (a) antisolvent crystallization at 50 °C and (b) cooling followed by antisolvent crystallization at 35 °C.

5.6 Supercritical Fluid Crystallization

Crystallization using supercritical fluid has increasingly become a suitable alternative to conventional processes such as evaporative crystallization and antisolvent crystallization. Carbon dioxide (CO_2), which is inexpensive, generally recognized as safe, and has a relatively low critical temperature, is particularly attractive for food and pharmaceutical applications [161, 162]. Supercritical crystallization is capable of producing micro- and nanoparticles, usually with a relatively narrow size distribution and a different morphology. Furthermore, separation of the crystals from the supercritical fluid is relatively easy and the products can be free from organic solvent residues.

5.6.1 Solid–fluid equilibrium phase behavior

In contrast to solid–liquid systems, pressure change often plays a significant role in systems involving supercritical fluid. Therefore, it is often necessary to consider a polybaric phase diagram for process representation alongside the isobaric, polythermal phase diagrams that are widely used for SLE representation. Figure 5.39a shows the polythermal, polybaric solid-fluid equilibrium phase diagrams for a binary system of a typical component, A, and CO_2, for which a more detailed explanation can be found in Harjo et al. [163]. Note that different possible behaviors of solid-supercritical fluid systems have been classified by several authors based on various characteristics

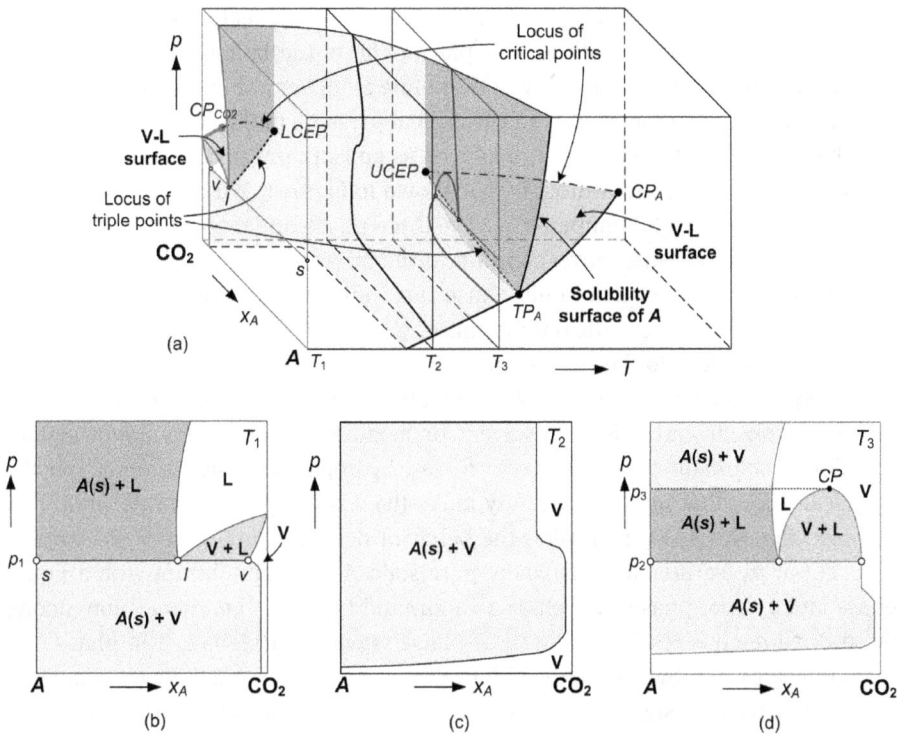

Figure 5.39: Solid-fluid equilibrium phase diagram of binary system A-CO_2 [163]: (a) polybaric polythermal diagram, (b) isothermal cut at T_1, (c) isothermal cut at T_2, and (d) isothermal cut at T_3.

[164, 165]. The discussion here focuses on a particular type characterized by a discontinuous critical mixture curve (the locus of mixture critical points running between the pure component critical points), which is typical of systems consisting of components that show considerably different molecular size, shape, and polarity [165]. One branch of the curve runs from the critical point of CO_2 (CP_{CO2}, on the back face) and intersects the locus of triple points at the lower critical end point (LCEP). Another branch runs from the critical point of A (CP_A, on the front face) and intersects the locus of triple points at the upper critical end point (UCEP).

Figure 5.39b shows an isothermal cut at T_1, which is below the critical temperature of pure CO_2. This cut features regions where solid A is in equilibrium with a liquid phase (at high pressure) and with a vapor phase (at low pressure), as well as a vapor–liquid region. At pressure p_1, solid A is in equilibrium with both liquid and vapor phases. This implies that any mixture of A and CO_2 at T_1 and p_1 would split into three phases: pure solid A (point s), a liquid phase with a composition given by point l, and a vapor phase with a composition given by point v.

When the temperature is increased to T_2, which is between the LCEP and UCEP temperatures, the boundary between vapor and liquid phases disappears.

The isothermal cut at this temperature (Figure 5.39c) only features a region where solid A is in equilibrium with a fluid phase. The phase boundary illustrates the typical behavior that the solubility of the solute is higher at high pressures, with a significant change in the vicinity of the critical pressure of CO_2. It is this portion of the phase diagram that is usually displayed as an isotherm curve in solubility diagrams reported in the literature. The cut shown in Figure 5.39d is taken at temperature T_3 above the UCEP temperature. The three-phase line reappears at p_2. Above p_3, which is the pressure corresponding to the mixture critical point at this temperature (CP), the vapor and liquid phases are indistinguishable, and solid A is in equilibrium with a supercritical fluid phase.

Figure 5.40 depicts the isothermal solid–fluid equilibrium phase diagram of a ternary system involving a solute A, an organic solvent C, and CO_2. Here, CO_2 acts as an antisolvent and C is a co-solvent or modifier, which brings about certain chemical functionalities such as polar forces, hydrogen bonding, or other specific chemical forces that may significantly affect the solubility of the solute [166]. It is assumed that the A-CO_2 pair takes the behavior described in Figure 5.39c. An isobaric cut at p_4 features regions where pure solid A is in equilibrium with a liquid phase and a vapor phase, as well as a vapor-liquid region. The intersection among these three regions results in the three-phase region of A(s)+V+L. The phase diagram indicates that the solubility of A in CO_2 is, in general, much lower compared to that in solvent C, signifying the role of CO_2 as an antisolvent. Furthermore, the solubility of A in CO_2 increases with the addition of co-solvent C. When the pressure

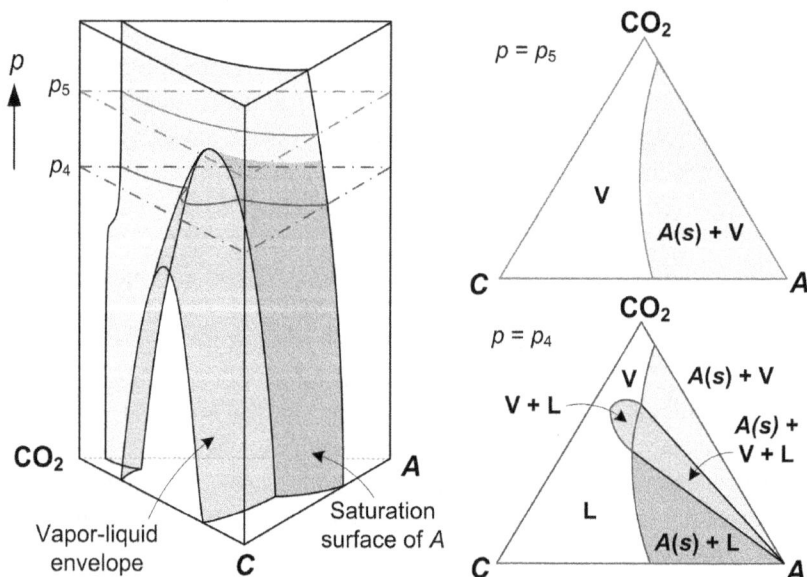

Figure 5.40: Solid–fluid phase diagram of ternary system involving solute A, co-solvent C, and CO_2.

is increased to p_5, all the liquid regions disappear and solid A is in equilibrium with a supercritical fluid phase.

Another ternary system of interest is a system involving two solutes and CO_2, which is relevant to the design of a depressurization crystallization process. One of the possible phase behaviors for such a system is shown in Figure 5.41. To illustrate the presence of different phase behaviors, it is assumed that at this temperature (above the critical temperature of pure CO_2), the A-CO_2 pair exhibits the behavior described in Figure 5.39c, while the B-CO_2 pair exhibits the behavior described in Figure 5.39d. Such phase behavior may be possible when the UCEP of A-CO_2 is significantly different from that of B-CO_2. An isobaric cut at pressure p_6 below the critical point of the mixture features a solid–vapor region extending from the A-CO_2 edge and both vapor–liquid and solid–liquid regions from the B-CO_2 edge. The solid A–vapor region intersects with the vapor–liquid region inside the triangle, resulting in the three-phase region of $A(s)$+V+L. Next to this region is the solid A–liquid region, which intersects the solid B–liquid region to form the three-phase region of $A(s)$+$B(s)$+L. Increasing the pressure to p_7 leads to the disappearance of the liquid region, as the fluid phase is now supercritical in all proportions. The three saturation regions corresponding to solid A, solid B, and A+B mixture are similar to those observed in an isothermal isobaric cut of an SLE phase diagram of a simple eutectic system. Also, notice that a cut at a very low pressure would also look similar, but the fluid phase would be subcritical vapor and the solubility of both A and B in the fluid would be much lower than at p_7.

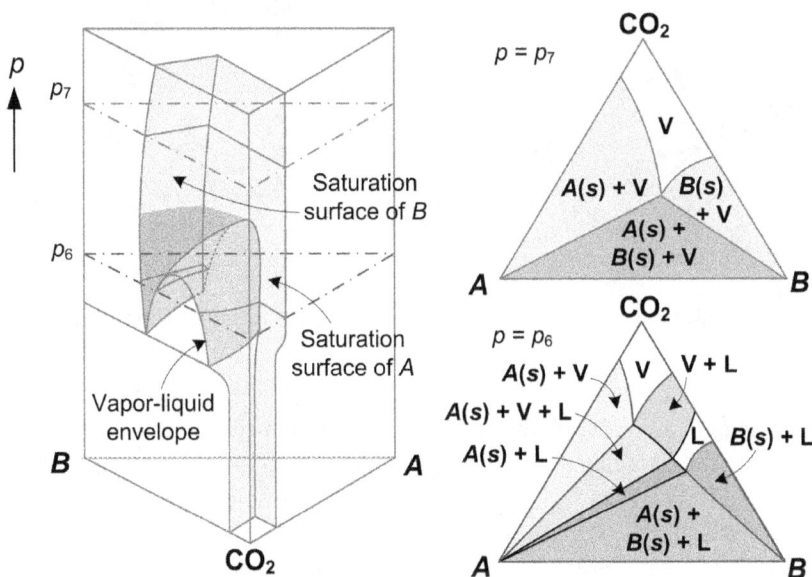

Figure 5.41: Solid-fluid phase diagram of ternary system involving solute A, solute B, and CO_2.

5.6.2 Process synthesis

The phase diagram can be used to guide process synthesis and determine suitable operating conditions of a supercritical crystallization process. One common process is *precipitation with compressed antisolvent* (PCA), in which the solute is first dissolved in an organic solvent, then, a supercritical solvent is added to induce crystallization [167]. As indicated in the phase diagram in Figure 5.42a, the solution of solute A in solvent C (point F_1) is mixed with supercritical CO_2 with a high CO_2-to-feed ratio, so that the operating point r falls in the solid–vapor region. Operating in this region is desirable because the particles can be readily separated from the supercritical fluid, without the need of a drying step. One way to achieve this is to spray the feed solution into a crystallizer where CO_2 passes through at constant pressure, temperature, and flow rate, as illustrated in Figure 5.42b, so that the liquid droplets formed inside the crystallizer are in contact with a large excess of CO_2. The organic solvent and CO_2 in the saturated vapor, which would have a composition along the saturation curve such as point 1, can be separated in a low-pressure tank and recycled. This process is done in a semibatch fashion; the particles accumulate in the crystallizer and are collected at the end of the batch.

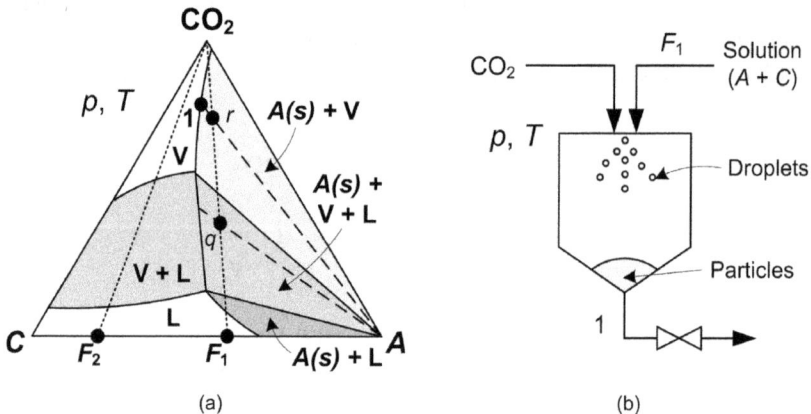

Figure 5.42: Precipitation with compressed antisolvent process: (a) process paths (b) process schematics.

With the phase diagram in hand, the suitable feed solution concentration and CO_2-to-feed ratio can be determined. For example, a lower CO_2-to-feed ratio may result in point q in the three-phase region, and a liquid phase would form in the crystallizer. If the feed solution is too dilute (point F_2), there would be a single liquid phase, vapor–liquid equilibrium, or a single vapor phase, depending on the CO_2 loading. No solid can be collected, regardless of the CO_2-to-feed ratio for this feed.

Another common process is depressurization crystallization, in which solutes are dissolved at high pressure and then crystallized out by depressurization. One way to dissolve the solute is to make use of solid–vapor equilibrium at high pressures, such as at p_2 and T_2, as illustrated in Figure 5.43a. Note that the solubility curves are greatly exaggerated for clarity; in reality, they are usually located very close to the CO_2 vertex due to the relatively low solubility of the solutes in CO_2. The feed (point F), a mixture of solid A and solid B, is mixed with CO_2 to give a composition along the dotted line. If crystallization of pure A is desired, the CO_2-to-feed ratio should be chosen such that the overall mixture during the dissolution process at p_2 and T_2 (point 1) is located inside the A saturation region under the crystallization condition at p_1 and T_1. Therefore, pure A crystallizes out in the crystallizer, leaving behind a saturated vapor with a composition given by point 2, as schematically shown in Figure 5.43b. Such a process is generally known as the *rapid expansion of supercritical solution* (RESS) [168].

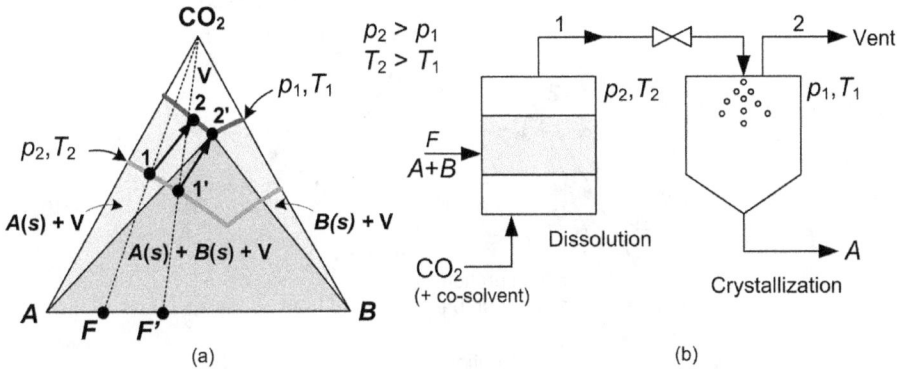

Figure 5.43: Depressurization crystallization process: (a) process paths on phase diagram (b) process schematic.

Sometimes co-crystallization of both solids is desired, such as in the production of microcapsules for controlled-release dosage forms [169]. By starting from a different feed composition (point F' in Figure 5.43a) and choosing a CO_2-to-feed ratio in the dissolution process at p_2 and T_2, to produce point $1'$ inside the double saturation region under the crystallization condition at p_1 and T_1, A and B can be crystallized out simultaneously in the crystallizer. The saturated vapor in the crystallizer would have the composition of the double saturation point at p_1 and T_1 (point $2'$). Clearly, depending on the process objective, a different region on the phase diagram is used in synthesizing the process.

Example 5.11 – Salicylic acid production using PCA process

Precipitation with compressed antisolvent (PCA) using supercritical CO_2 and ethanol as co-solvent was used in producing fine particles of salicylic acid (SA) [170, 171]. To identify feasible operating conditions, particularly the CO_2-to-feed ratio and feed concentration, the relevant phase diagram was constructed using solubility data from the literature [172] to guide the effort, rather than performing a series of bench-scale experiments in a trial-and-error manner.

Figure 5.44 shows the isothermal phase diagram of the SA-ethanol-CO_2 at 45 °C and two different pressures (96 and 157 bar). As the literature data (shown as triangles and squares) concentrate near the CO_2 vertex due to the very low solubility and high CO_2-to-feed ratio, the top portion of the triangle is zoomed for clarity. The phase behavior at this temperature and pressure only features two regions (vapor and solid SA+vapor), as the pressure is above the critical pressure of CO_2. It can be seen that the solubility of SA is higher at the higher pressure (157 bar) compared to that at the lower pressure (96 bar).

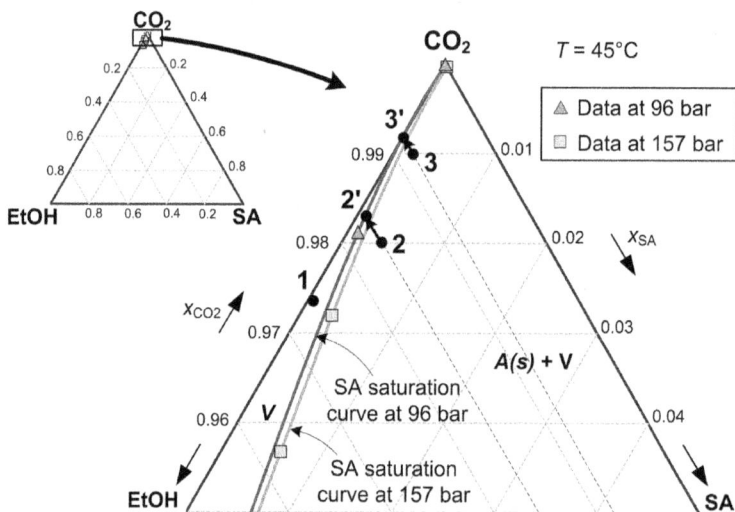

Figure 5.44: Ternary phase diagram of salicylic acid-ethanol-CO_2 at 45 °C, showing various operating points.

Three conditions were examined and tested experimentally. The first run used a feed concentration (SA in ethanol) of 0.05 g/mL and a CO_2 (g/min)-to-feed (mL/min) ratio of 28.3, giving an overall composition at point 1. As expected, no solid particles were collected, as point 1 is located in the vapor region. In order to obtain solid particles, point 1 has to move to the right to get inside the *A (s)+V* region. This can be achieved by using a more concentrated feed solution. Therefore, a feed concentration of 0.35 g/mL was used in the next run, with the same ratio of CO_2 mass flow to feed volumetric flow rate. The overall composition is at point 2 inside the solid–vapor region at 96 bar. A third run with a feed concentration of 0.35 g/mL and CO_2 (g/min)-to-feed (mL/min) ratio of 56.6 was also performed, giving an overall composition indicated by point 3. From both the second and third runs, solid particles could be successfully recovered [170].

The crystallization process in the last two runs can be represented on the phase diagram. Assuming equilibrium is achieved, points 2′ and 3′ (on the saturation curve of SA at 96 bar) mark the saturated vapor composition obtained from the second run and third run, respectively. By mass

balance (assuming no loss), the solid recovery is almost 100% for the third run because the saturated vapor (point 3′) has a very low SA concentration. On the other hand, the solid recovery for
the second run is only about 77%, but the amount of solid produced per g of CO_2 is higher than in
the third run, because the CO_2-to-feed ratio is lower.

Example 5.12 – Separation of palmitic acid and tripalmitin
Palmitic acid (PA) and its triglyceride, triplamitin (TP), are main components in palm oil and palm
kernel oil, both of which are edible vegetable oils. The use of supercritical CO_2 for extraction of
these oils from the fruit and kernel of palm tree has been investigated [173]. Since the fractions
obtained from such an extraction process are usually not pure, it is interesting to explore the possible use of supercritical crystallization process, particularly depressurization crystallization using
RESS, for separation of individual components.

Figure 5.45 shows the phase diagram of the ternary system PA-TP-CO_2. The four saturation
curves at two temperatures (35 and 50 °C) and two pressures (110 and 130 bar), determined based
on the solubility data calculated by Mukhopadhyay and Rao [174], are plotted on the phase diagram. Because of the very low solubility of both PA and TP at these conditions, only a zoomed section near the CO_2 vertex is shown. Furthermore, as the solubility of TP is more than an order of
magnitude lower compared to that of PA, the scale on the x-axis is much smaller than on the y-axis.
The lightly shaded region is the TP saturation region at 50 °C and 130 bar, while the darker region
is the PA saturation region at 35 °C and 110 bar. These two regions overlap because the two sets of
saturation curves at 50 °C, 130 bar and 35 °C, 110 bar have a crossover behavior; that is, if the
temperature and pressure are simultaneously decreased from 50 to 35 °C and 130 to 110 bar, respectively, the solubility of PA in CO_2 decreases, while that of TP in CO_2 increases.

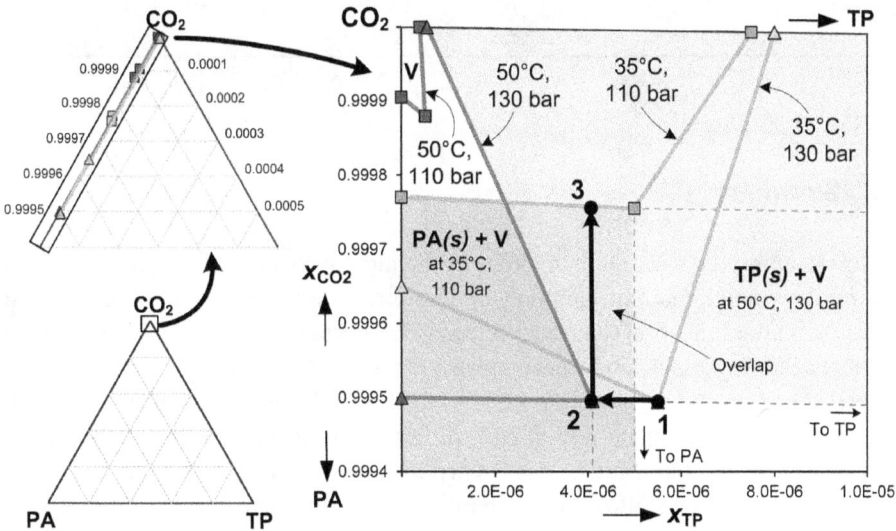

Figure 5.45: Ternary phase diagram of palmitic acid-triplamitin-CO_2 system.

Harjo et al. [163] proposed a process for obtaining both PA and TP in substantially pure form, starting with a feed mixture containing equal amount of PA and TP, as shown in Figure 5.46. The corresponding process paths are depicted on the phase diagram in Figure 5.45. The feed (not shown) is

mixed with CO_2 in the dissolution vessel at 35 °C and 130 bar, at which both solutes have relatively high solubility. The saturated vapor composition (point 1) is at the double saturation point because both PA and TP solids are in excess in this vessel. As point 1 is located in the TP(s)+V region at 50 °C and 130 bar, TP is first crystallized out by isobarically increasing the temperature to 50 °C in the first crystallizer, giving a saturated vapor with a composition given by point 2, which is close to the double saturation point at this condition. Next, to crystallize PA, both the temperature and pressure are reduced to 35 °C and 110 bar, respectively, in the second crystallizer, because point 2 is inside the PA(s)+V region at this condition. The saturated vapor (point 3) is recycled (with a partial purge if required to avoid accumulation of impurities) by recompression to 130 bar. Fresh CO_2 can be added as make-up.

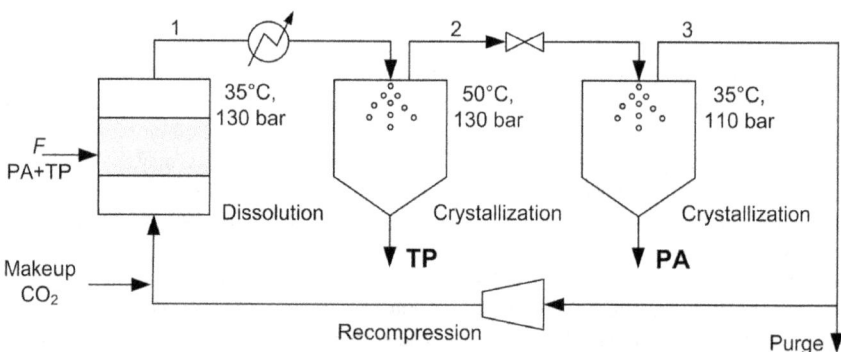

Figure 5.46: Process for separating palmitic acid and tripalmitin using depressurization crystallization.

5.7 Summary

In this chapter, the separation of different chemical systems, which include adduct, enantiomer, solid solution, ampholyte, amino acid, antisolvent, and supercritical fluid, is studied. Each system has its own idiosyncrasies. An adduct such as the phenol-acetone adduct can be decomposed to recover the desirable compound. The separation of enantiomers requires the conversion of the enantiomers with a resolving agent to diastereomers, which have different physical properties. A solid solution has to be purified using a series of crystallization stages. Both acids and bases are needed in the crystallization of amino acids and proteins because of their amphoteric nature. Supercritical crystallization has limited productivity because the solubility in supercritical fluid is very low for most solutes. Despite their significantly different solid-liquid phase behavior, the same techniques in the form of movements in SLE phase diagrams – heating/cooling, solvent addition/removal, and stream combination/splitting are used to effect separation by crystallization.

Exercises

5.1. A solid product C is an adduct that can be formed by mixing A and B in solvent D. In a laboratory test for developing the process, 2 mol of A is dissolved in 8 mol of D at 80 °C, followed by mixing with 2 mol of B. The mixture is then cooled to 40 °C to obtain a solid product. The SLE phase diagram of the system (in mole fraction), which features incongruent melting of C, is shown in Figure 5.47. An isothermal cut at 40 °C is indicated in the figure.
a. Locate the composition of the dissolver outlet and crystallizer feed on the diagram.
b. What is the composition of the solid product?
c. Explain why the solid product is not pure C, and suggest a way to obtain pure C by crystallization at 40 °C based on the SLE phase diagram.

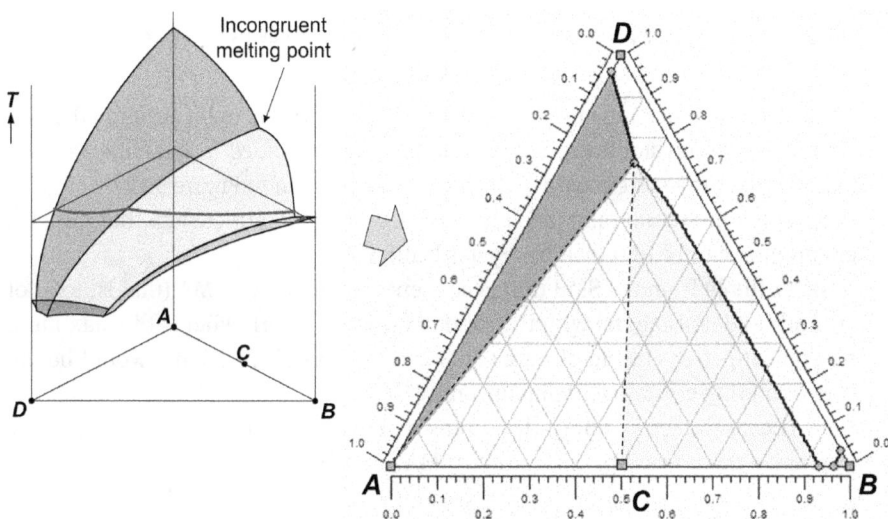

Figure 5.47: SLE phase diagram of a ternary system involving an adduct (problem 5.1).

5.2. Consider the process for separating cresol isomers using t-butyl alcohol (TBA) as an MSA, which is discussed in Example 5.2. With the aid of the phase diagram in Figure 5.48, synthesize a process alternative that features crystallization of m-cresol (B) before crystallization of the adduct (AS) between p-cresol and TBA, starting from a feed containing an equal amount of A and B, as indicated by point F on the phase diagram. Show the process path on the phase diagram and draw the corresponding flowsheet.

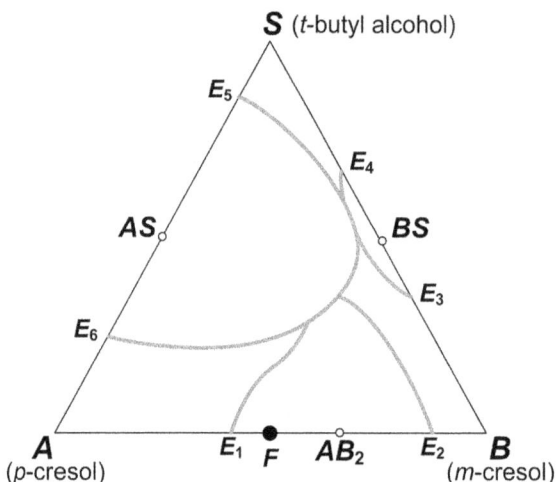

Figure 5.48: Polythermal SLE phase diagram of p-cresol/m-cresol/t-buytl alcohol (problem 5.2).

5.3. Resolution of a racemic mixture of mandelic acid (MA) is to be performed using a chromatography/crystallization hybrid process, as discussed in Example 5.4. The phase diagram of the system with water as solvent is given in Figure 5.15.

a. What is the minimum ratio of (+) to (–)-MA in the SMB outlet such that pure (+) enantiomer can be obtained by crystallization?

b. If the raffinate from the SMB unit is concentrated to 50 wt% MA (that is, total of (+) and (–) enantiomers) with the ratio of (+) to (–) of 9:1, what is the maximum theoretical per-pass yield of pure (+)-MA in the crystallizer? What would be the corresponding crystallizer temperature?

   Hint: Locate the feed composition to the crystallizer on the phase diagram, and then trace the process path as pure (+)-MA crystallizes.

5.4. The curvature of the eutectic trough in some systems involving enantiomers that form racemic compound can be exploited to obtain pure enantiomers via a two-step crystallization process, as proposed by Lorenz and coworkers [128, 129, 130]. An example is DL-methionine/water system. A sketch of the ternary phase diagram, which is roughly based on the data of Polenske and Lorenz [175] is shown in Figure 5.49. The first step of the process involves crystallization of the racemic compound at a low temperature from a solution that is enriched in the enantiomer to be purified. The mother liquor from the first step is processed in the second step, which involves solvent evaporation at a high temperature to selectively crystallize the desired enantiomer. Using the phase diagram, explain how the purification process works. The starting material is given as point 1 on the diagram, and the final product is pure L-Methionine.

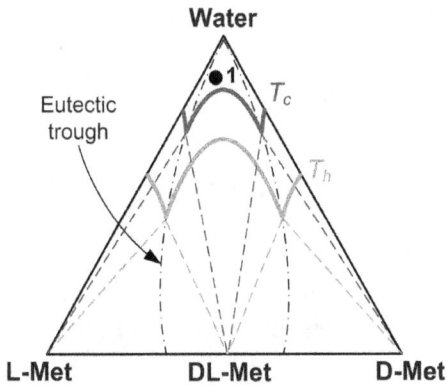

Figure 5.49: SLE phase diagram of the ternary system L-Met/D-Met/Water (problem 5.4).

5.5. In the isolation and purification of Schisandrin B discussed in Example 5.5, separation is achieved by first crystallizing a (−)Sch B-enriched solid in the first crystallizer (see Figure 5.22). According to Table 5.1, the (−)Sch B content in the feed, liquid, and solid (on a solvent-free basis) is 83.3 wt%, 76.4 wt%, and 91.6 wt%, respectively. Based on this information, calculate the crystallization yield of (−)Sch B, defined as the fraction of (−)Sch B in the feed that ends up in the solid.

5.6. The purification process of C_{60} fullerene using o-xylene discussed in Example 5.6 can alternatively be done via a continuous countercurrent four-stage crystallization

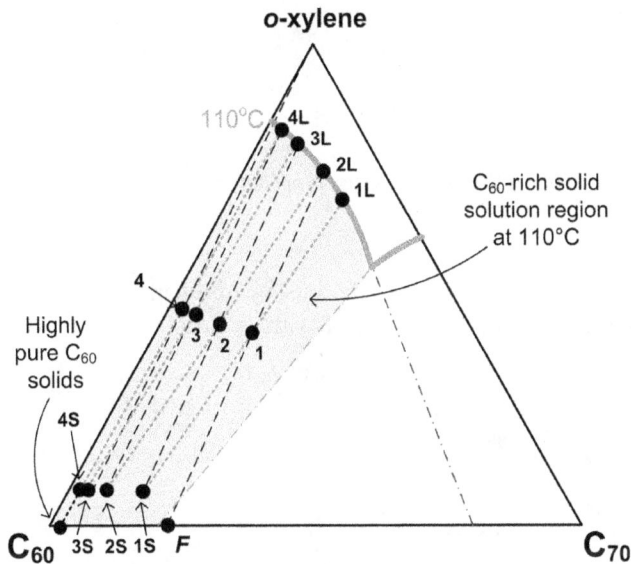

Figure 5.50: Continuous C_{60} purification process by multistage crystallization (problem 5.6).

process, with the fresh feed entering the first stage and pure solvent (*o*-xylene) being introduced to the last stage. Based on the plot of the process points on the phase diagram as shown in Figure 5.50, draw the flowsheet of this continuous purification process.

5.7. A sketch of the solid–liquid–liquid phase diagram of the ternary system LiCl/water/1-propanol at 25 °C, based on data reported by Gomis et al. [176] is shown in Figure 5.51. Using this phase diagram, synthesize a process for recovering LiCl from a land deposit, which involves dissolution of LiCl in a mine, crystallization of LiCl·H_2O using 1-propanol as an antisolvent, and separation of organic and aqueous phases in a decanter. Both the crystallizer and the decanter are operated at 25 °C.

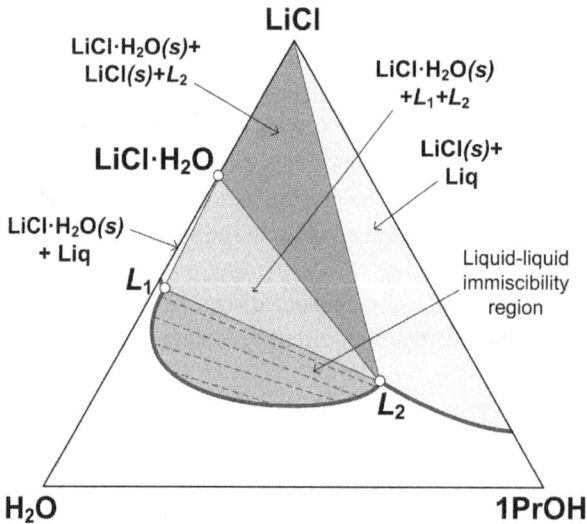

Figure 5.51: Solid–liquid–liquid equilibrium (SLLE) phase diagram of LiCl/water/1-propanol system at 25 °C (problem 5.7).

5.8. Consider the drowning-out crystallization of NaCl using 2-propanol as an antisolvent, which is discussed in Example 5.9. The compositions of points 4 and 5 are given as follows:

Point	NaCl (wt%)	Water (%)	2-propanol (wt%)
4	2.9	30.6	66.5
5	16.4	70.0	13.6

Assume that stream 1 is saturated with NaCl, and its composition (determined by the solubility of NaCl in a mixture of water and 2-propanol at T_M) can be described by $x_{NaCl,1}=0.27-0.5\ x_{D,1}$, where x is mass fraction and D represents 2-propanol. The product stream (NaCl from the crystallizer) is 100 kg/h and contains no water or 2-propanol.

a. From mass balances around the mine (block M) as well as for the overall process, determine the product and aqueous recycle stream flow rates as functions of purge flow rate.

b. If the purge flow rate=20 kg/h, what is the aqueous recycle flow rate and the composition of stream 1?

c. How would the aqueous recycle stream, product, and make-up solvent flow rates change if the purge flow rate is increased? Describe the potential economic tradeoff.

6 Impact of Kinetics and Mass Transfer on Crystallization

6.1 Kinetic Effects in Crystallization Process

Thermodynamics (SLE) alone does not provide a complete picture in designing a crystallization process. Kinetic and mass transfer issues can also have a significant effect on the outcome of the crystallization process. This chapter focuses on the opportunities provided by operation under nonequilibrium conditions for selectively obtaining the desired product.

6.1.1 Supersaturation

During crystallization, material from the liquid phase turns into solid phase via two different mechanisms: *nucleation*, the formation of solid particles from liquid solution, and *growth*, the deposition of additional substance on existing particles, as illustrated in Figure 6.1. Nucleation can be of two types: primary and secondary. Primary nucleation is the initial formation of a crystal where there are no other crystals present. It can be homogeneous, if it is not influenced in any way by solids; or heterogeneous, if it is affected by other solids such as crystallizer walls or dust. Secondary nucleation is induced by the presence of crystals that are already present in the mixture.

Both nucleation and growth rates depend on *supersaturation*, which is defined as the difference between the actual concentration of the crystallizing substance in the solution and its solubility. The idea of supersaturation is usually explained using a conventional solubility diagram, similar to the one shown in Figure 6.2a. The solubility as a function of temperature is represented by the saturation curve. Any point above this curve, such as point P, is said to be *supersaturated*. The supersaturation (Δc) is defined as $c - c^*$, where c^* is the solubility at temperature T. Also, *supercooling* ($\Delta T = T^* - T$, where T^* is the temperature at which a solution with concentration c becomes saturated) is commonly used as an alternative to represent the extent of supersaturation. The two quantities are related through the local slope of the solubility curve (dc^*/dT):

$$\Delta c = \left(\frac{dc^*}{dT}\right)\Delta T \tag{6.1}$$

The dashed curve in Figure 6.2a is the *metastable limit*, or sometimes called the supersolubility curve [157]. Together with the saturation curve, it bounds the *metastable zone*, in which spontaneous nucleation is unlikely, but growth can still occur

https://doi.org/10.1515/9781501519901-006

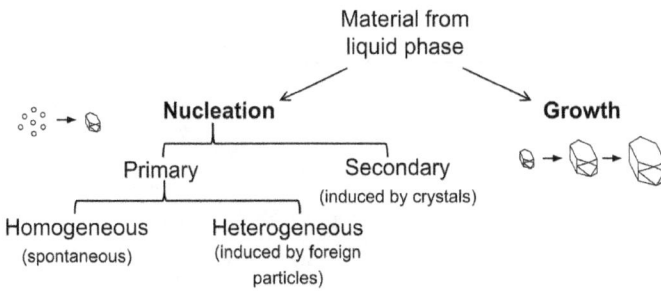

Figure 6.1: Nucleation and growth.

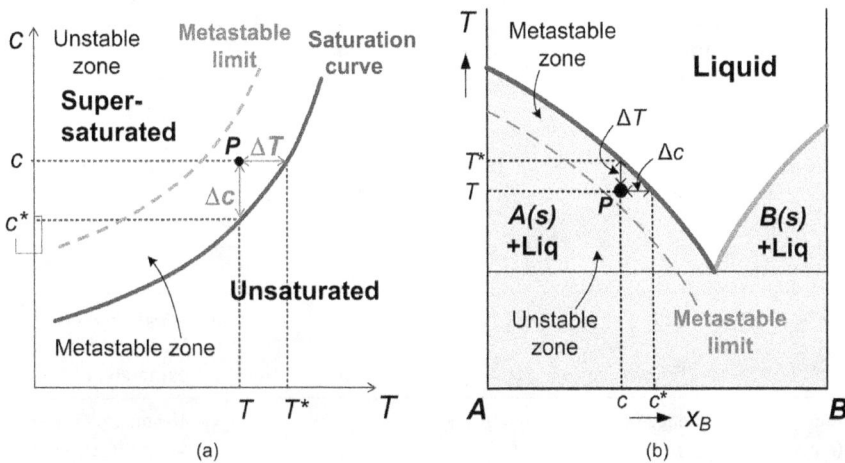

(a)

(b)

Figure 6.2: Supersaturation and metastable limit: (a) conventional representation and (b) phase diagram representation.

if crystals are already present. On the other side of the metastable limit lies the *unstable zone*, where nucleation occurs spontaneously. In light of the previous discussion on the use of SLE phase diagrams, Figure 6.2b depicts the representation of the same idea on the phase diagram. Here, concentrations are expressed in terms of mass or mole fraction.

6.1.2 Nucleation and growth models

Various models have been proposed to describe the dependence of nucleation and growth rates on supersaturation, as summarized in Table 6.1. Some models are based on classical theories, which account for individual mechanisms of nuclei formation and crystal growth. For example, the classical nucleation theory considers the chemical potential difference between a molecule in solution and in the bulk of

Table 6.1: Summary of nucleation and growth models.

Theoretical models		
Nucleation rate from homogeneous supersaturated solution [179]	$$B = A \exp\left[-\frac{16\pi\sigma^3 v^2}{3k^3 T^3 (\ln S)^2}\right] \quad (6.2)$$	A=pre-exponential factor σ=surface energy per unit area v=molar volume k=Boltzmann constant T=temperature S=supersaturation ratio $(= c/c^*)$
Growth rate based on screw dislocation mechanism for crystal growth [180]	$$G = C\left(\frac{\varepsilon \Delta c^2}{\sigma_1'}\right)\tanh\left(\frac{\sigma_1'}{\varepsilon \Delta c}\right) \quad (6.3)$$	C=constant ε=screw dislocation activity σ_1'=a function of temperature Δc=supersaturation $(c - c^*)$
Empirical models		
Nucleation rate (power law)	$$B = k_b \Delta c^b \quad (6.4)$$	Δc=supersaturation k_b, b=empirical parameters
Nucleation rate dependence on slurry (magma) density and agitation rate [181]	$$k_b = k_b' M_T^i N^j \quad (6.5)$$	M_T=magma density N=agitator speed k_b', i, j=empirical parameters
Growth rate (power law)	$$G = k_g (\Delta c)^g \quad (6.6)$$	Δc=supersaturation k_g, g=empirical parameters
Growth rate dependence on particle size [182]	$$k_g = k_{g,0}(1 + k_l L)^l \quad (6.7)$$	L=crystal size $k_{g,0}$, k_l, l=empirical parameters
Growth rate in the presence of impurities [183]	$$G = G_0\left(1 - \frac{\alpha K_{imp} c_{imp,L}}{1 + K_{imp} c_{imp,L}}\right) \quad (6.8)$$	G_0=growth rate in clean solution α=effectiveness factor K_{imp}=constant in Langmuir adsorption isotherm $c_{imp,L}$=impurity concentration in solution

the crystal phase as the driving force for nucleation [177]. While the theoretical models are supposed to be predictive, some of the parameters are often difficult to obtain, rendering them impractical for industrial crystallization. Therefore, empirical models that capture the overall effect of various mechanisms are widely used in practice. These empirical models can sometimes be obtained from simplification of the fundamental models, but the parameters must be determined experimentally. Parameter values for some common systems have been reported in the literature [178], although the results obtained by different researchers are generally not very consistent.

6.2 Actual Process Path

In reality, crystallization cannot start at equilibrium condition because supersaturation is needed as a driving force. Supersaturation has to be first generated by cooling, solvent removal, addition of MSA, or reaction. Nucleation and growth will then occur, and the solution composition approaches an equilibrium condition as the supersaturation diminishes. Therefore, the actual process path or locus of liquid compositions does not actually coincide with the saturation curve as has been assumed in previous chapters.

The actual process path for a case of batch cooling crystallization in a binary mixture of A and B is illustrated in Figure 6.3a. Starting with an unsaturated liquid (point 1), the solution becomes saturated when the temperature reaches T^* at point 2. Upon further cooling, the mixture enters the metastable zone, in which the concentration of

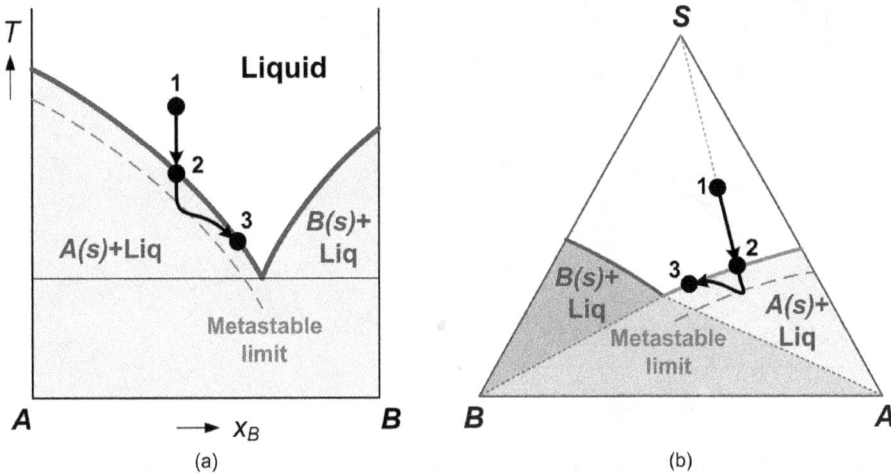

Figure 6.3: Actual process path during crystallization: (a) cooling crystallization in a binary mixture, (b) evaporative crystallization in a ternary mixture.

the solution stays constant in the absence of nucleation and growth, until the composition reaches the metastable limit. As further cooling brings the mixture into the unstable zone, crystallization starts, and the process path turns away from *A* and approaches the saturation curve, eventually reaching equilibrium at point 3. The exact path depends on the balance between cooling rate and nucleation and growth rates. Similarly, Figure 6.3b illustrates the process path for a batch evaporative crystallization in a ternary mixture. As more solvent is removed beyond point 2, the process path continues to venture inside the saturation region of *A* without any crystallization, until it reaches the metastable limit. The path takes a turn once *A* starts to crystallize out, and approaches point 3 on the saturation curve. In reality, it would take infinite time to achieve true equilibrium condition, as the supersaturation would become smaller and smaller as the composition gets closer to the equilibrium composition. Therefore, the final composition is not exactly located on the saturation curve, since the system can only be infinitesimally close to equilibrium at most.

Example 6.1 – Crystallization of nitroguanidine

Nitroguanidine (NQ) is a low-sensitivity explosive with high detonation velocity, widely used as a propellant. However, only high bulk density NQ is useful for this application [184]. In an effort to obtain high bulk density NQ crystals, crystallization of NQ from its solution in water has been studied. The solubility curve of NQ, established from experimental solubility data at different temperatures, is shown in Figure 6.4. Also shown on the same diagram is the trajectory of NQ concentration in the solution during cooling crystallization, with a cooling rate of 2.5 °C/min. Starting with a solution containing 3.8 g NQ/100 mL water at 90 °C (point 1), the concentration of NQ in the solution remains unchanged even after cooling below the dissolution temperature, indicating no crystallization. It only starts to decrease sharply once the temperature reaches about

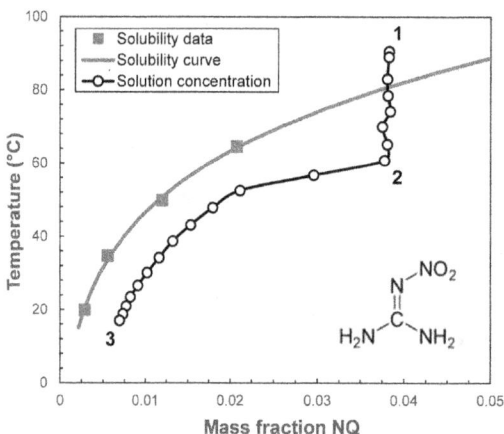

Figure 6.4: Solution concentration profile during cooling crystallization of nitroguanidine from water solution with a cooling rate of 2.5 °C/min (data from [184]).

60 °C (point 2), corresponding to a substantial degree of supersaturation before nucleation starts to occur. Although the supersaturation decreases as crystals are formed upon further cooling, the solution at the end of cooling (point 3) appears to be still supersaturated. Crystals with different attributes can be obtained by controlling the supersaturation. The degree of supersaturation can be controlled by various means, including changing the cooling rate, adding seeds, or using additives that restrict crystal growth. The study found that adding methylcellulose and polyvinyl alcohol as growth inhibitors results in a different crystal shape and higher bulk density.

6.3 Preferential Crystallization

Kinetically controlled crystallization can also be an attractive alternative for separating chiral enantiomers. In particular, preferential crystallization has been commonly employed in industrial production of antibiotics chloramphenicol and thiamphenicol, among others [185]. This technique relies on crystallization of pure enantiomer inside the metastable region by introducing the appropriate seeds. To understand how the process works, consider an isothermal cut of the SLE phase diagram of a ternary system involving two enantiomers and a solvent, as shown in Figure 6.5. The metastable zone of R and S can be obtained by extending the solubility curves and identifying the metastable limits, as indicated by the dashed curves.

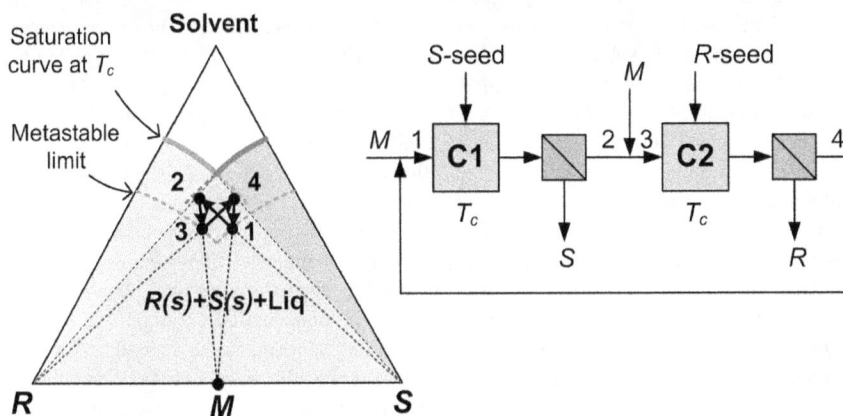

Figure 6.5: Preferential crystallization to separate R and S isomers.

Beginning with a solution that lies within the metastable zone of both enantiomers (indicated by point 1), seeds of S are introduced to induce crystallization of S, so that the solution composition moves to point 2 on the extension of the S solubility curve. Fresh feed of racemate mixture (point M) is then added to stream 2 at a higher temperature to completely dissolve the solids. The resulting solution is given by point 3, which is located above the solubility curves (not shown) of the higher temperature. Then, stream 3 is introduced to C2 at T_c and seeds of R are introduced, so that R crystallizes out and the composition moves to point 4. Again, fresh feed is introduced at the higher temperature to obtain a solution with composition given by point 1, and the process repeats itself. Note that the entire process path is located inside the two solids region where thermodynamically both R and S would crystallize, but appropriate seeding induces the crystallization of just one form. Therefore, this technique is only suitable when there is an overlap between the R and S metastable zones, which is observed only when the system exhibits a conglomerate or simple eutectic phase behavior.

When preferential crystallization is conducted as a continuous process in an industrial setting, adding seeds can be problematic. Therefore, Vetter et al. [186] proposed a process alternative in which a portion of the product crystals is continuously withdrawn, milled in a suspension mill, and returned to the crystallizer to provide sufficient surface area for crystal growth. Different configurations of preferential crystallization process for enantiomer separation have been reviewed by Rougeot and Hein [187], who classified the configurations based on batch or continuous process and the use of multiple stirred tanks or fluidized beds as crystallizers.

Example 6.2 – Preferential crystallization of L-threonine
Alvarez Rodrigo et al. [188] experimentally demonstrated the feasibility of preferential crystallization in the resolution of threonine enantiomers (D-threonine and L-threonine) following the process shown in Figure 6.5. The change in solution concentration during crystallization was monitored online by polarimetric measurement, which was combined with refractometry to determine the liquid phase composition at any time. Figure 6.6a shows the SLE phase diagram of D-threonine/L-threonine/water ternary system, with isothermal cuts at different temperatures obtained from experimental data.

The process path during a preferential crystallization experiment is shown on the enlarged section of the phase diagram in Figure 6.6b. The experiment starts with a saturated solution at 40 °C, containing a slight excess of L-threonine (L-Thr), as indicated by point 1. The solution is cooled down to 33 °C and seeds of L-Thr are added. The L-Thr content in the solution decreases as a result of preferential crystallization of L-Thr. The process path practically follows a straight line with constant D-Thr/H₂O ratio until the composition reaches point 2, indicating that only L-Thr is crystallized. The composition takes a turn beyond point 2, indicating spontaneous co-crystallization of D-Thr. When the mixture is left to equilibrate, the solution composition eventually reaches point 3, which is the double saturation point at 33 °C. To ensure that the solid product contains only L-Thr, crystallization needs to be stopped at point 2 by isolating the crystals and sending the mother liquor to the next step of the process, as depicted in Figure 6.5.

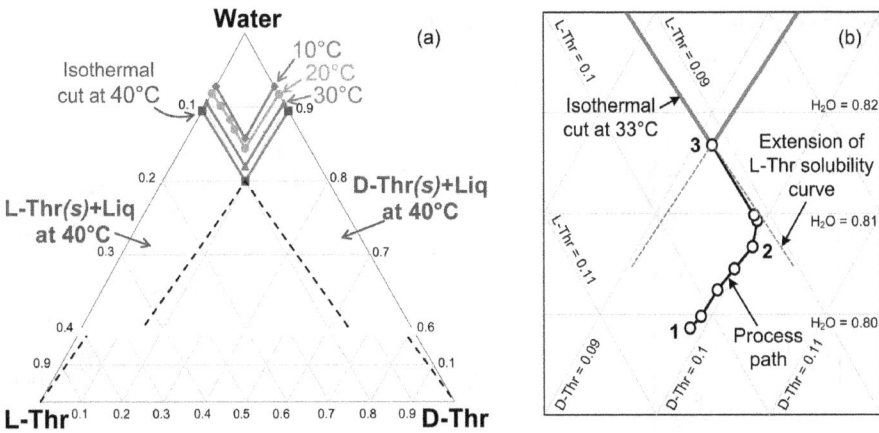

Figure 6.6: D-Threonine/L-threonine/water system (data from [188]): (a) SLE phase diagram and (b) process path for preferential crystallization of L-threonine.

Example 6.3 – Kinetic-based alternative for ibuprofen resolution

As an alternative to the thermodynamic-based process discussed in Example 5.3, NMDG-(S)-Ibuprofen can also be recovered under kinetically controlled operation by seeding [134]. The process flowsheet along with the process paths represented on a Jänecke projection is shown in Figure 6.7. Racemic ibuprofen (stream F) is mixed with a recycle stream to form stream 1. It reacts with NMDG to form diastereomeric salts in the presence of solvent. The resulting solution (stream 2) is mixed with the recycle stream (stream 5) to give stream 3. As in the equilibrium-based process, NMDG-(R)-Ibuprofen is crystallized out from C1 under equilibrium condition, and evaporation is stopped when the mother liquor is about to reach the double saturation composition (stream 4).

In the second crystallizer (C2), the solution is subcooled to 5 °C and a small amount of NMDG-(S)-Ibuprofen crystals is introduced to the crystallizer for seeding. Note that the location of stream 4 can be controlled by adjusting the amount of NMDG fed into R1. Results from the seeding experiments indicate that it is desirable to carry out seeding at the double saturation composition with R(NMDG-H^+) = 0.6. By allowing crystallization to proceed for 30 minutes, after which the solids and mother liquor are promptly separated, and pure NMDG-(S)-Ibuprofen can be recovered with about 20% yield. Although the composition of the mother liquor (stream 5) appears to be well inside the saturation region of the (R) isomer, it is actually still located within the metastable limit of both (R) and (S) isomers. This process alternative is simpler compared to the equilibrium-based alternative because there is no need to release free ibuprofen in between the two crystallizers, but a more elaborate control is required for such a kinetically controlled process.

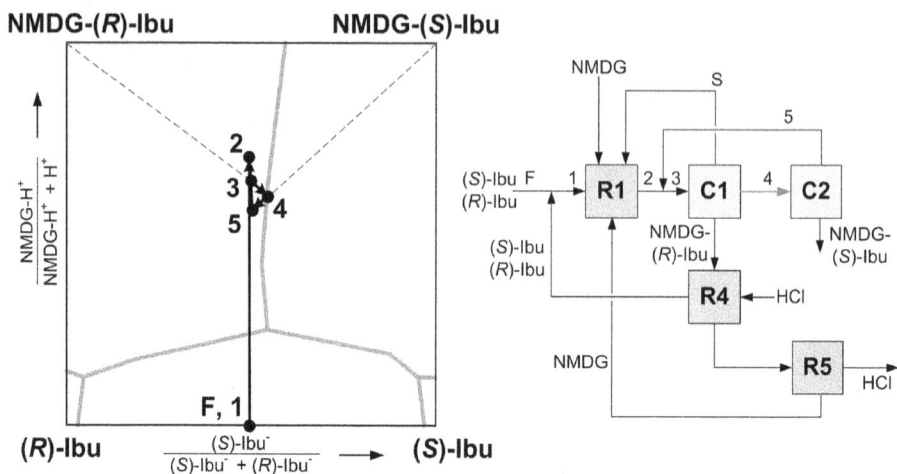

Figure 6.7: Kinetic-based process alternative for resolution of ibuprofen using NMDG as resolving agent [134].

6.4 Kinetically Controlled Reactive Crystallization

Chemical reactions in the liquid and/or solid phase can occur simultaneously with the crystallization process, resulting in a final composition that is not necessarily in phase equilibrium or reaction equilibrium. The outcome of the process depends on the interplay between reaction and crystallization kinetics, particularly the relative rates of reaction and crystallization. As an illustration, consider an *asymmetric transformation* process, where racemization reaction proceeds simultaneously with crystallization to improve the yield and purity of the desired enantiomer. Asymmetric transformation has been used to obtain a pure enantiomer from a racemic mixture, as well as to convert one enantiomer to the other. An industrial example is the resolution of (−)-narwedine in the production of the anti-Alzheimer drug (−)-galanthamine [189, 190]. Note that asymmetric transformation is different from an asymmetric synthesis process, which produces a single enantiomer from some other intermediates by chemical or biological reaction and does not necessarily involve a racemization reaction or crystallization.

Figure 6.8 shows four possible scenarios for an asymmetric transformation process for resolution of a racemic mixture of R and S enantiomers by batch crystallization. The process path as well as the amounts of R and S crystals as a function of time during the course of the process, obtained from simulation using a model that accounts for

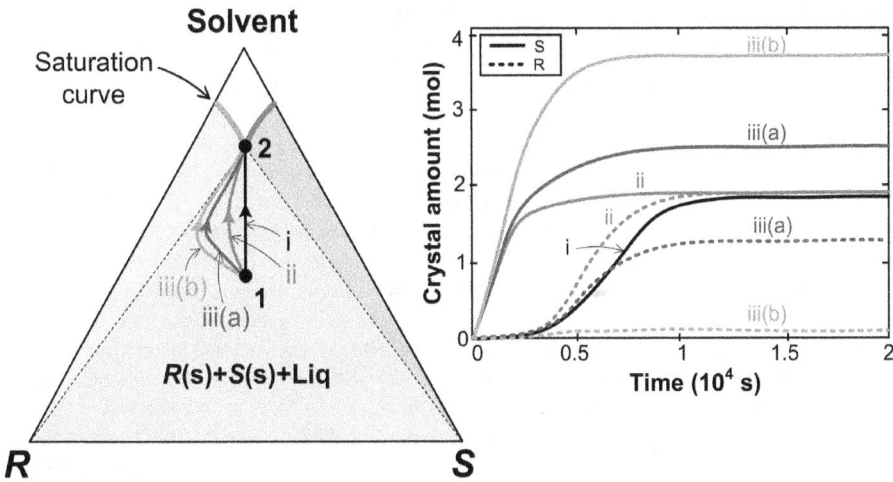

Figure 6.8: Process paths and crystal amounts for various operating policies.

racemization reaction, crystal growth, nucleation, and dissolution [191] is plotted for each scenario. Different seeding policies and relative rates between crystal growth and racemization are simulated by changing the values of the model parameters.

The starting point for all scenarios is an initial solute composition of a 50:50 mixture of the R and S enantiomers (point 1). An unseeded crystallization without racemization reaction is considered as the base-case scenario. The process path (i) moves directly towards the double saturation point (point 2), since both enantiomers nucleate and crystallize at the same time. The same amount of R and S is present in the crystal product at all times. In the second scenario, seeds of S are added at the start of the batch, but no reaction occurs. As seeding enhances the crystallization of S, the process path (ii) initially moves away from the S vertex. However, as R nucleates and begins to crystallize, the path turns towards the double saturation point and the ratio of R to S crystals increases and approaches unity after about 10^4 s. Therefore, some enrichment in S is achieved if the crystallization is stopped before this point. The product purity can be controlled by adjusting the time at which the crystallization is stopped.

Asymmetric transformation process, with racemization reaction occurring in conjunction with seeding of the S enantiomer, is considered in the last two scenarios. The difference between them is the racemization reaction rates, the reaction rate in iii(b) being 10 times faster than in iii(a). As the racemization reaction consumes the excess R in the mother liquor as S is being crystallized, the effect is reducing the supersaturation of R, and ultimately slowing down the crystallization of R. Therefore, the process path ventures further away from the S side, although eventually R still crystallizes and drives the process path towards point 2. Despite the fact that the final liquid composition is still the double saturation point, the final crystal product is enriched in S, because the reaction has converted some of the R in the feed into S. The faster the racemization reaction, the higher the S content that can be obtained in the final product.

Example 6.4(I) – Reactive crystallization of cycloaliphatic diacid
A cycloaliphatic diacid is typically produced as a mixture of two isomers (*cis* and *trans*) with a *cis*/*trans* ratio of approximately 70/30. Since high *trans* content is required to achieve many of the desirable end-use properties, a process to produce high-purity *trans* isomer is to be developed. The melting point of the two isomers differs significantly (*trans*: 312 °C, *cis*: 172 °C), and isomerization of *cis* to *trans* can occur in the melt phase in the presence of a catalyst. Due to these facts, a reactive crystallization process can be performed in a kiln, where the feed is heated until it partially melts and undergoes isomerization reaction to produce *trans* isomer, which crystallizes out as the mixture cools down towards the end of the kiln. Since heating is provided by electrical heaters, the temperature profile inside the kiln can be controlled by adjusting the power to the heaters. To understand the effect of the temperature profile inside the kiln on the product composition, a model accounting for crystallization and reaction kinetics as well as heat transfer inside the kiln was developed. While the model is not presented here due to its complexity, a set of results that is consistent with available data (temperature at several locations inside the kiln and the final product composition) is shown in Figure 6.9. The temperature and composition profiles in the kiln are shown on the left-hand side, while an image of the process path on the SLE phase diagram is shown on the right-hand side to provide insights to the various mechanisms occurring inside the kiln. Note that although it was difficult to accurately determine the solubility curve due to the presence of reaction, the simple eutectic behavior of the *cis*/*trans* system was confirmed by the results of experimental tests.

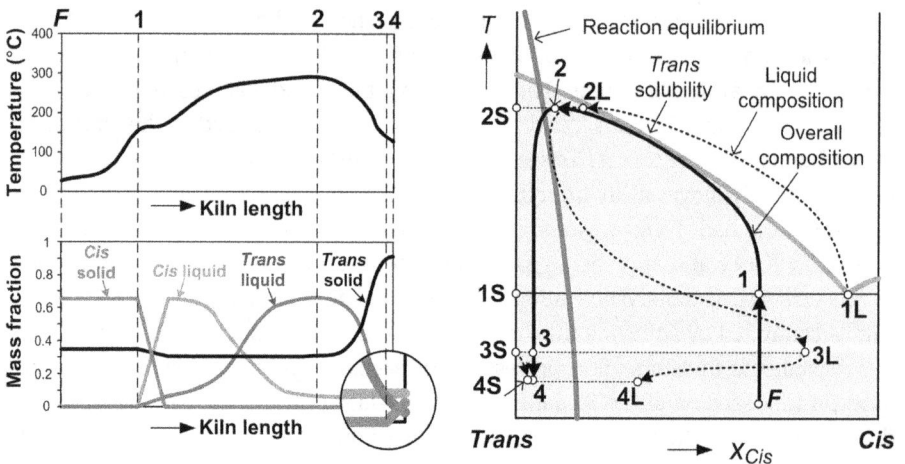

Figure 6.9: Temperature and composition profiles in the kiln and the image of process path.

The feed enters the kiln as a solid mixture (point *F*). Partial melting starts to occur, once the mixture is heated past the eutectic temperature (point 1), until all *cis* in the original mixture melts. The remaining solid (point 1S) contains 100% *trans*, which is the *trans* in the original feed mixture that has not melted. As heating continues, isomerization reaction occurs in the liquid phase to convert *cis* to *trans*, as illustrated by the liquid composition trajectory turning toward *trans* from point 1L to point 2L. Since the reaction rate is relatively slow, the liquid composition stays outside of the *trans* saturation region, so that no crystallization of *trans* occurs during heating. After passing the maximum temperature (point 2), the mixture cools down while reaction continues in the liquid phase to approach equilibrium. As cooling continues, *trans* eventually crystallizes out from the liquid, causing the liquid composition to turn away from the reaction equilibrium curve. Due to slow crystallization rate, the trajectory of the liquid composition stays well inside the *trans* saturation region, as opposed to closely following the *trans* solubility curve. Eventually, the temperature is sufficiently low that *cis* also becomes saturated and starts to crystallize out (point 3). Therefore, the final solid exiting the kiln (point 4S) contains some *cis*, as can be seen in the magnified portion of the composition profile diagram in Figure 6.9. As *cis* crystallizes faster than *trans* during the final stretch of cooling, the liquid composition trajectory takes a turn toward *trans*, resulting in a final liquid composition (point 4L) that is rich in *trans*.

The exact process path is determined by the balance between reaction rate and heating/cooling rate. The final product composition highly depends on the liquid composition when the mixture begins to cool down, as the overall composition does not change much during the later part of cooling. Higher kiln temperature in the middle section (between point 1 and point 2) would result in faster heating and faster reaction, so that the liquid would be richer in *trans* and the final solid product would also have a higher *trans* content. On the contrary, not enough heating in either the beginning or middle section of the kiln would result in lower temperature and slower reaction, and the final product would have a lower *trans* content.

In reactive crystallization or precipitation, there is less control on degree of supersaturation, and small crystals are usually produced. Many nonequilibrium issues such as reaction kinetics, mixing and mass transfer, rapid nucleation, and crystal growth, as well as possible secondary processes such as aging, ripening, breakage, and agglomeration determine the actual process paths as crystallization occurs, as well as the final particle size distribution (PSD). The relative importance of individual steps can be characterized by various dimensionless numbers, as summarized in Table 6.2 [192]. The mathematical expression for these dimensionless numbers depends on the relevant equations of reaction kinetics, nucleation, and growth rate.

The relative values of these dimensionless numbers for a particular system can indicate the regime of operation, characterized by the dominating step in the reactive crystallization process. A good balance among the various steps leads to the desirable outcome, which generally features large median crystal size and high crystal yield. This is characterized by high values of $N_{Da,R}$ and N_{Gr}, as well as low

Table 6.2: Dimensionless numbers characterizing individual steps in a reactive crystallization process.

Step	Dimensionless number	Expression
Reaction	Damköhler number	$N_{Da,R} = \dfrac{\text{Reaction rate}}{\text{Reactor throughput}}$
Dissolution	Damköhler number	$N_{Da,D} = \dfrac{\text{Reaction rate}}{\text{Dissolution rate}}$
Mass transfer	Damköhler number	$N_{Da,M} = \dfrac{\text{Reaction rate}}{\text{Mass transfer rate from bulk to crystal face}}$
Nucleation	Nucleation number	$N_{Nu} = \dfrac{\text{Nuclei generation rate}}{\text{Reaction rate}}$
Growth	Growth number	$N_{Gr} = \dfrac{\text{Surface integration rate}}{\text{Mass transfer rate from bulk to crystal face}}$

values of $N_{Da,D}$, $N_{Da,M}$, and N_{Nu}. Table 6.3 summarizes different situations where a certain step, as captured by one of the dimensionless numbers, becomes too dominant (italicized), leading to deviations from the desirable performance [193]. By understanding the physical meaning of each dimensionless number, it is also possible to figure out the appropriate operational changes to improve the performance. For example, a low value of $N_{Da,R}$ (second column) indicates that the reaction is the dominant step characterized by a slow reaction rate. This results in low crystal yield as there is not enough product formed with the given reactor throughput. Actions such as increasing temperature, batch time, and/or reactant concentration, which increase the value of $N_{Da,R}$, should be considered to improve the performance. On the other hand, when the mass transfer of the solute from the bulk to the liquid–crystal interface and vice versa is much slower than the reaction (high $N_{Da,D}$ and $N_{Da,M}$) (third and fourth columns), it results in large supersaturation and excessive nucleation, eventually leading to more fines. This can be resolved by increasing mixing to speed up mass transfer, or decreasing the reactant concentration to slow down the reaction. Operation at high nucleation rate (high N_{Nu}) (fifth column) means that whatever supersaturation is generated by the reaction is quickly relieved by the formation of new nuclei, leading to low supersaturation and slow growth rate. This effect can be dampened by reducing the supersaturation generation rate, by slowing down the reaction or improving mixing. Finally, very slow crystal growth (low N_{Gr}) (sixth column) leads to small crystal size and low crystal yield. This can be improved by increasing the batch time to allow more time for crystal growth.

The impact of the dimensionless numbers on the process path is illustrated in Figure 6.10a, taking the formation of a solid product C from liquid reactants A

Table 6.3: Regimes of operation and suggested changes for improving process performance.

Dominating step	$N_{Da,R}$	$N_{Da,D}$	$N_{Da,M}$	N_{Nu}	N_{Gr}	Deviation from desirable crystallizer performance	Operational changes to improve performance
Reaction kinetics	Low	Low	Low	Low	High	Low crystal yield	Increase temperature, batch time, and/or reactant concentration
Solid-to-liquid or liquid-to-solid mass transfer	High	High	High	Low	High	Low crystal yield for solid reactant systems Large supersaturation, leading to more fines	Increase mixing and/or batch time Decrease reactant concentration
Nucleation	High	Low	Low	High	High	Nucleation rate is faster than growth rate, leading to a lot of fines	Reduce reactant concentration and/or temperature Reduce supersaturation by improving mixing
Crystal growth	High	Low	Low	Low	Low	Low growth rate leading to small crystal size and low crystal yield	Increase batch time

and B as an example. The effect on PSD is shown in Figure 6.10b. N_{Gr} characterizes how fast the supersaturation generated by reaction is consumed by growth,
thereby significantly affecting the supersaturation level. Starting from a mixture
of A and B that contains slightly more A than B (point 1), reaction brings the
composition into the saturation region of C, creating a supersaturation. As the
reaction approaches equilibrium and C starts to precipitate, the process path
turns toward point 2, where both reaction and solid–liquid equilibria are simultaneously satisfied. If the growth rate is slow relative to the mass transfer rate
(small N_{Gr}), supersaturation is consumed more slowly and the process path ventures deeper into the saturation region of C. This leads a higher supersaturation
level during crystallization, eventually resulting in smaller crystals. Faster
growth rate (increasing N_{Gr}) causes the process path to turn quicker, leading to

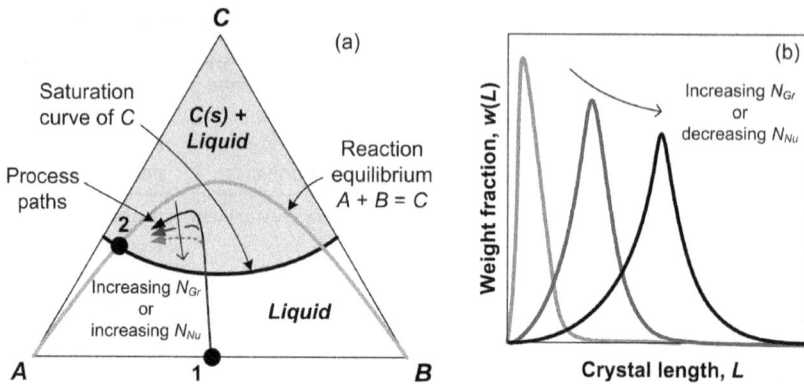

Figure 6.10: Kinetic effects in reactive crystallization: (a) process paths, (b) particle size distribution of the crystals.

lower supersaturation level and bigger crystals. In contrast, N_{Nu} characterizes how the supersaturation generated by reaction translates to new nuclei. Faster nucleation (increasing N_{Nu}) means generation of more nuclei, allowing supersaturation to be consumed more quickly, so that the process path turns quicker and the supersaturation is maintained at a lower level. However, the presence of more nuclei leads to smaller crystals as the crystallized mass has to be distributed over a higher number of particles. Therefore, the effect of N_{Nu} on the PSD is opposite that of N_{Gr}. Note that although these effects are illustrated on the same plots in Figure 6.10, in general, the impact of N_{Nu} on supersaturation and PSD is not of the same magnitude as that of N_{Gr}. Furthermore, the actual shape of the PSD, which depends on operating conditions such as cooling profile, reactant addition rate, and so on may not be the same as those in the figure.

6.5 Polymorphic Crystallization

One of the most important issues in the development of crystallization processes for active pharmaceutical ingredients (APIs) is to obtain the desired polymorph of the product. Polymorphs are different crystalline structures of the same compound; they are identical in chemical composition but have different molecular arrangements in the crystal lattice. Polymorphs often exhibit markedly different physical and chemical properties. For example, their solubility can be very different, affecting the bioavailability and efficacy of a drug [194]. Polymorphs can have different shapes ranging from spheres to needles, which, in turn, exert considerable influence on downstream processing steps such as filtration, washing, bulk solids handling, and product formulation. Furthermore, finding all possible polymorphic forms of a compound is a difficult task, and interconversion among polymorphs may also occur during and after the manufacturing process. An example is Ritonavir, an HIV drug discovered by Abbott Laboratories [195]. When it was introduced in the market in 1998, many batches of its semisolid capsule formulation failed the dissolution test. An investigation revealed a previously unknown, thermodynamically more stable and less soluble crystalline form. Indeed, polymorphs are patentable as they are considered as distinct products. Therefore, selecting the right polymorph for formulation and being able to produce it in a consistent manner are exceedingly important in pharmaceutical processing.

6.5.1 SLE phase behavior of polymorphic systems

Polymorphic behavior also adds complexity to the phase diagram. Basically, each polymorphic form has its own solubility curve (or surface). At any temperature, the form with the lowest solubility is the most stable form. One possible relationship between polymorphic forms is *monotropic* behavior, where one form is always more stable than the other, as shown in Figure 6.11a. In this case, the entire solubility curve of the stable form β is above that of the unstable form α on the phase diagram. The other possibility is *enantiotropic* behavior, in which the solubility curves cross at a transition temperature, as illustrated in Figure 6.11b. Above the transition or crossover temperature, form α has lower solubility than form β, making it the more stable form. Below the crossover temperature, form β is more stable. This leads to the existence of two SLE regions for component A: that of form α above the crossover temperature, and another of form β below the crossover temperature.

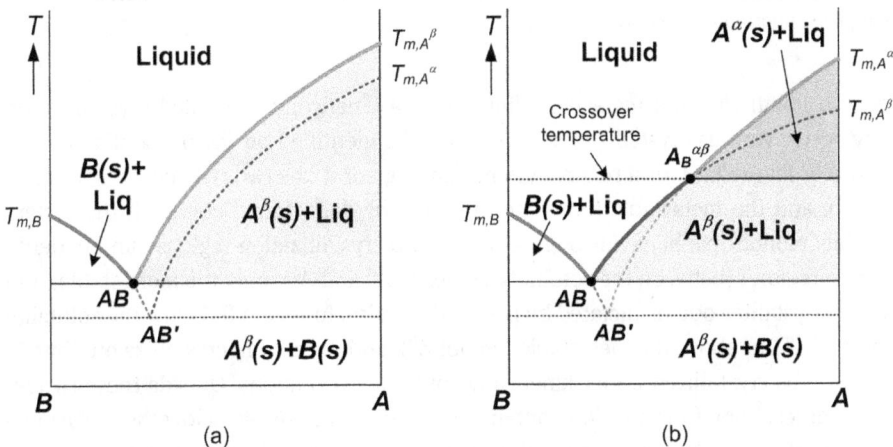

Figure 6.11: SLE phase diagram of a binary polymorphic system: (a) monotropic behavior and (b) enantiotropic behavior.

Similarly, the SLE phase diagram of a ternary system involving one component, A, having two polymorphs, and two solvents, D and S, are shown in Figure 6.12. If the two forms exhibit monotropic behavior, the solubility surface of one form lies entirely below that of the other form. With enantiotropic behavior, the solubility surfaces intersect at the crossover temperature. Note that the crossover temperature is a property of the polymorphic system and does not depend on the solvent.

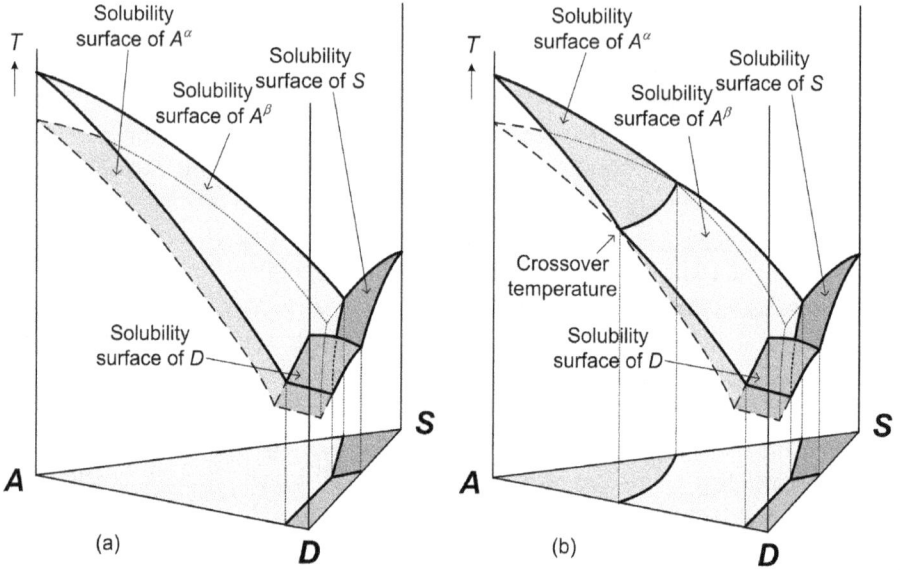

Figure 6.12: SLE phase diagram of a ternary polymorphic system: (a) monotropic behavior and (b) enantiotropic behavior.

By also identifying the metastable limits where kinetically controlled crystallization may occur, various crystallization regions can be identified on the phase diagram, as shown in Figure 6.13 for a binary system. Component A has two polymorphic forms (α and β), and the metastable limits are indicated by dash-dotted curves. If the system exhibits monotropic behavior (Figure 6.13a), two crystallization regions can be identified. In region I (between the solubility curves of A^α and A^β), only the more stable form A^β can crystallize out, as dictated by thermodynamics. In region II (below the solubility curve of A^α but above the metastable limit of A^β), both forms can crystallize out, but A^α can only be crystallized under kinetically controlled conditions. Outside these two regions, at least one form would spontaneously crystallize out (A^β below the metastable limit of A^β, and both forms below the metastable limit of A^α), so it is not possible to control the product form.

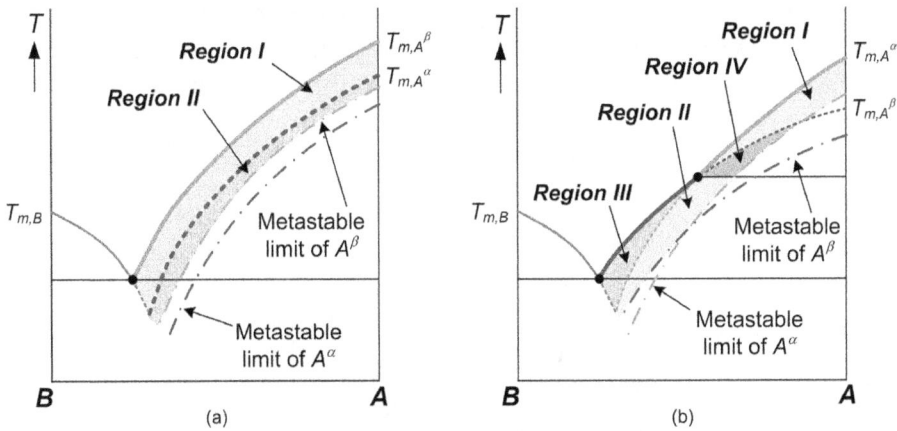

Figure 6.13: Various crystallization regions on the SLE phase diagram of a binary polymorphic system: (a) monotropic behavior and (b) enantiotropic behavior.

More regions can be identified if the system exhibits enantiotropic behavior (Figure 6.13b) instead. In region I, which is located below the solubility curve of A^α but entirely above the unstable branch of the solubility curve of A^β, thermodynamics dictates that only A^α can crystallize out. Therefore, it is desirable to operate within this region if A^α is the target product. On the other hand, only A^β can crystallize out in region III (below the solubility curve of A^β but above the solubility curve of A^α), making this the ideal operating region if A^β is the desired product. Regions II and IV are located within the metastable zone of both forms, so thermodynamically, crystals of both forms can be obtained. With proper seeding, the desired form can be preferentially crystallized out. For example, region II would be a good choice for obtaining A^α if the temperature for operation in region I is considered too high. Seeds of A^α can be introduced to induce the initial crystallization, or if A^α nucleates faster than A^β, an unseeded operation may also be suitable. Operation outside these four regions, which is beyond the metastable limit of any form, is generally undesirable due to spontaneous nucleation.

6.5.2 Process synthesis

In synthesizing a polymorphic crystallization process, it is important to understand how the process path can affect the outcome of the process. Figure 6.14 shows the SLE phase diagram of a binary system involving a compound A, which has two polymorphic forms with enantiotropic behavior, and a solvent B. The solubility curves are shown as thick curves, with dotted portions indicating the unstable branches. The metastable limits are indicated by the dashed curve. Two process paths with the same starting and ending points are considered.

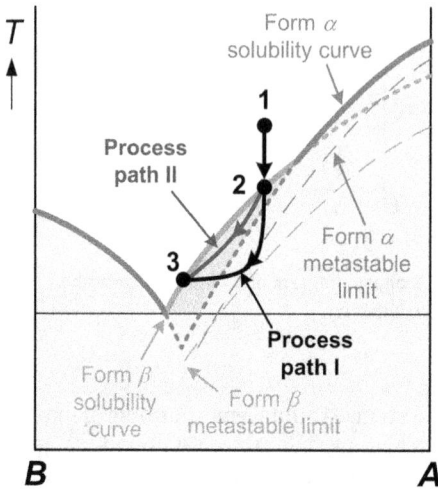

Figure 6.14: Process paths in polymorphic crystallization.

Process path I begins with cooling of an unsaturated solution (point 1) and cools down to point 2, at which it becomes saturated with form β. Further cooling would bring the composition into the metastable region, where form β may not crystallize out yet. If nucleation does not start until the composition goes below the metastable limit of form α (thus entering the unstable region of this form), spontaneous crystallization of form α would occur. As a result, the path would turn away from A as crystallization proceeds. If the formed solids are quickly isolated before any transformation to the more stable form β occurs, it is possible to obtain form α as the product. Otherwise, if the mixture is left indefinitely, form β eventually crystallizes out and the final solution composition would approach point 3 on the solubility curve of form β.

A different scenario occurs when the process follows path II. If crystallization of form β is initiated soon after the composition enters the metastable zone (e.g., by introducing seeds of form β), the process path would immediately turn away from A, closely following the solubility curve of form β. Since it never ventures below the solubility curve of form α, thermodynamically, only form β can crystallize out of the solution, and no form α would be present in the solids. The final solution composition would also be at point 3, but the obtained product is totally different compared to process path I.

Clearly, both thermodynamics and kinetics play important roles in designing a crystallization process for obtaining the desired polymorphic form. By ensuring that the process path stays within certain regions of the phase diagram, which are bounded by saturation curves and metastable limits, the outcome of the crystallization process can be controlled. To illustrate the application of this concept, a few examples are presented.

Example 6.5 – Crystallization of p-aminobenzoic acid

Lin et al. [196] discussed the crystallization process of p-aminobenzoic acid (PABA), which was widely used in sunscreens as UV filter before it was found to be responsible for some allergic reactions. PABA has two polymorphic forms: the hexagonal form α and the needle-shaped form β. They exhibit enantiotropic behavior with a crossover temperature of 25 °C. Below this temperature, form β is the stable polymorph. No solvate is formed when water is used as a solvent for crystallization. Figure 6.15 shows a section of the phase diagram of PABA/water system, with temperature ranging from 0 to 50 °C. The open circles and triangles represent the solubility data reported by Gracin and Rasmusson [197].

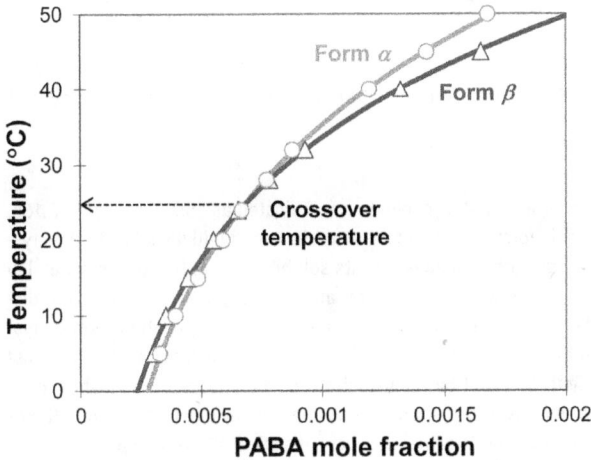

Figure 6.15: Solubility of PABA polymorphic forms.

With its focus on operation within region II (in Figure 6.13b), Figure 6.16 shows the effect of different cooling rates on the outcome of the crystallization process. The cooling starts at 20 °C and terminates at 5 °C, and no seeding is applied. The process paths are shown on the phase diagram (Figure 6.16a), and the relative amount of forms α and β produced in each case are shown as functions of time (Figure 6.16b).

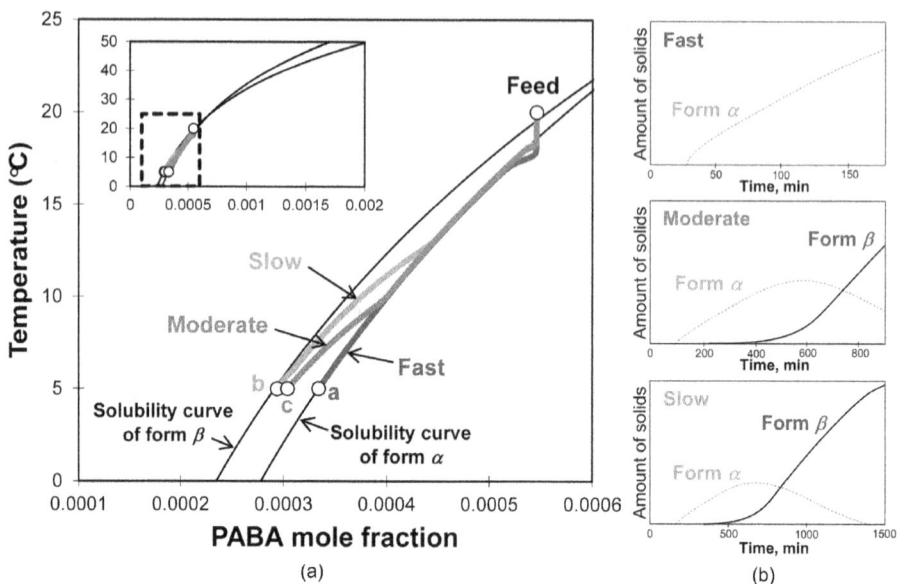

Figure 6.16: Effect of cooling rate on PABA crystallization [196]: (a) process paths and (b) amount of solids versus time.

First, consider the process path under a very slow cooling rate, indicated as "Slow" in Figure 6.16a. The faster nucleation rate of the unstable form α relative to the stable form β initially leads to the crystallization of only form α, and the process path coincides with its solubility curve. Due to the instability of form α in this temperature range, and the slow cooling rate, all the crystallized form α have redissolved before the temperature reaches 5 °C, and is followed by nucleation and growth of form β crystals. Note that this is consistent with the so-called *Ostwald's rule of stages*, which states that the least stable polymorph would crystallize first, followed by recrystallization to the more thermodynamically stable forms [198, 199]. As form α transforms to β, the process path gradually moves away from the α-saturation curve towards the β-saturation curve. The final product consists of 100% form β.

Next, consider the process path under a fast cooling rate (marked as "Fast" in Figure 6.16a), which quickly passes through the saturation curve of the unstable form α into the region in which both forms are supersaturated. Since form α nucleates faster, it crystallizes first and provides seeds for subsequent growth. The fast cooling rate does not allow enough time for the stable form β to nucleate, and the process path stays close to the saturation curve of form α. By the time the temperature reaches 5 °C, no crystals of form β has formed. By quickly separating the crystals from the solution to prevent solvent-mediated transformation to the more stable form β, a product consisting of 100% form α can be obtained.

With a moderate cooling rate, some (but not all) of the crystallized form α transform to form β by the time the temperature reaches 5 °C. Thus, both forms are found in the final product. These results are consistent with the experimental observations of Gracin and Rasmusson [197]. Therefore, controlling the cooling rate can be an effective means to controlling the polymorphic form of the final product.

Example 6.6(I) – Liquid crystal intermediate

A liquid crystal material, A, was to be recrystallized in solvent S as part of its manufacturing process. Results of preliminary tests showed that crystals of different forms were obtained from crystallization under different conditions. The desirable form was a crystalline powder, which could be obtained with high yield from crystallization at a very low temperature (-20 °C or below). However, highly thixotropic slurry was formed under some crystallization conditions, causing the mixture to turn into a lump of solids, so that further agitation and cooling became impossible. In order to scale up the process, it was desirable to identify suitable crystallization conditions so that the desirable crystalline form could be consistently obtained.

Figure 6.17 shows a hypothesized SLE phase diagram for the binary system of A and S. The right edge of the diagram represents the liquid crystal material in pure state, which, in general, can exist as various phases or forms at different temperatures [200]. At very low temperatures, most liquid crystal materials form a conventional crystal, which is the crystalline form. As temperature is increased, a variety of phases can be observed, until, finally, an ordinary liquid, or isotropic phase is reached. The most common phases are smectic, where the molecules form well-defined layers that can slide over one another, and nematic, wherein the molecules have directional order but no positional order, so that they have similar fluidity to that of the isotropic phase. Some liquid crystal materials can also reach other phases such as columnar and helical microfilament, which are structurally and morphologically different from each other [201]. Based on this understanding, the smectic and crystalline forms can be considered as analogous to two polymorphic forms, with their own solubility curves in the given solvent. Two different sets of experimental solubility data, shown as triangles and squares on the phase diagram, also suggest the presence of two different solubility curves. As the results of the preliminary tests suggest that the crystalline form is stable at low temperatures, there should be a critical temperature below which the crystalline form becomes more stable than the smectic form. Furthermore, based on results from additional crystallization tests, the critical temperature is around 20 °C, as illustrated in Figure 6.17.

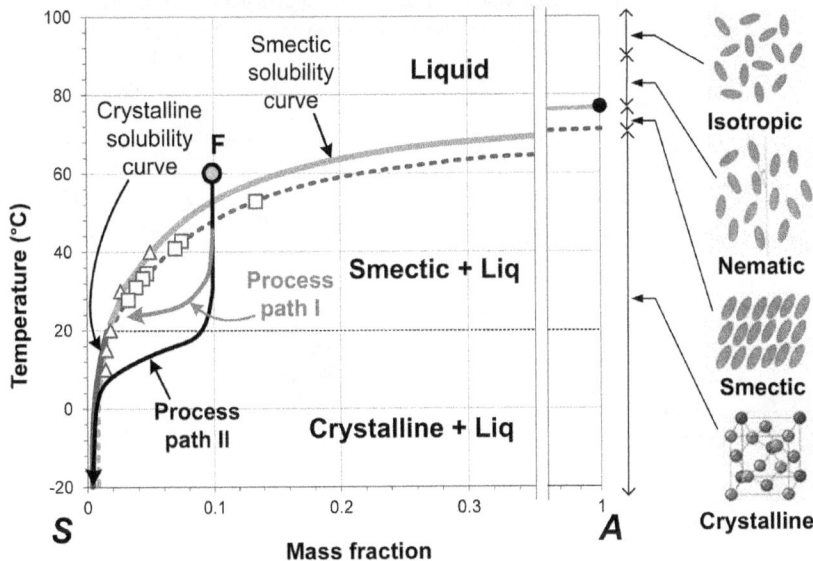

Figure 6.17: Conceptual binary SLE phase diagram of a liquid crystal material and a solvent.

To come up with a suitable process to obtain the desired crystalline form, two process paths were considered. Starting from a given feed in solution form (point F), cooling brings the process path through the crystallization region of the smectic form. If cooling is done slowly, significant crystallization of the smectic form, which is thermodynamically more stable at higher temperatures, would occur before the temperature reaches the critical temperature, as illustrated by process path I. As the thixotropic behavior of the slurry prevents further cooling, it is not possible to reach the crystallization region of the crystalline form. In contrast, when cooling is performed quickly to reach a temperature below the critical temperature before significant crystallization occurs (process path II), the crystalline form can be obtained at low temperatures. Essentially, the fast cooling is an attempt to bypass the smectic region, although in practice, some crystallization of the smectic form may be unavoidable. Once the mixture is successfully cooled below the crossover temperature, cooling can be continued at a slower rate, until the desired final temperature is reached. The slow cooling in this second step would also give enough time for the smectic form to be transformed into the crystalline form, which is more stable at low temperatures. This procedure was successfully verified in laboratory scale, but implementation at a larger scale was challenging due to the need for intense agitation during the initial cooling, in order to achieve a sufficiently high cooling rate.

Example 6.7(I) – Crystallization of the desired form

During scale-up of a generic drug preparation process that involved crystallization of compound A from solvent S, some of the production batches had exceptionally small particle size and high residual solvent levels after drying [202]. The problem was traced to the formation of solvate crystals, which lost the solvent during drying and were rapidly transformed to very small anhydrate crystals. Therefore, it was desirable to identify a set of operating conditions such as temperature, cooling rate, and solvent content, under which anhydrate crystals would be consistently obtained without the contamination of solvate.

To begin, experiments were performed to determine the solubility of the anhydrate as well as the solvate AS at different temperatures. Figure 6.18 shows an enlarged section of the binary SLE phase diagram involving compound A and solvent S, with the experimental data marked as diamonds (anhydrate) and squares (solvate). Note that although this is not a polymorphic system, the intersection

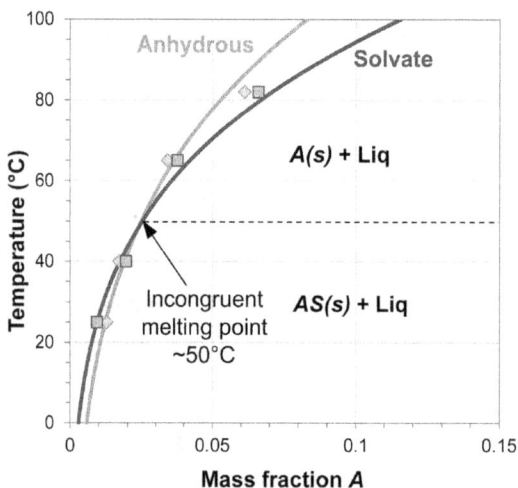

Figure 6.18: SLE phase diagram of drug compound A in solvent S.

between the two solubility curves resembles that of a polymorphic system with enantiotropic behavior, so that strategies for polymorphic system were similarly applied to this problem. The two solubility curves intersect at the incongruent melting point of AS, which was estimated to be about 50 °C (based on other evidence from existing production data). Above this temperature, the anhydrate is the more stable form that can be obtained at equilibrium conditions. On the other hand, operating at equilibrium below 50 °C would give solvate crystals as the product.

In order to design a kinetically controlled crystallization process, information such as metastable limits and relative crystallization rates of the two forms are needed. However, since detailed measurements would involve elaborate experiments, results from several previous trial runs were used to qualitatively approximate the metastable limits for both A and AS. In one of the trial runs, crude A was mixed with solvent S and heated at 92 °C, until A was completely dissolved. Some solvent was then removed by evaporation at the same temperature, and followed by cooling crystallization and filtration to isolate the solids. Crystals were sampled at various points during the process, and the solvate content in the crystals was determined by gravimetry. The results are summarized in Table 6.4.

Table 6.4: Solvate content in various samples taken during a trial run.

Sample	Description	Solvate in solids, %
1	End of evaporation, 92 °C	50
2	After cooling to 30 °C	11
3	After cooling to 22.5 °C	1
4	After stirring for 2 hrs at 22.5 °C	31
5	After stirring for 4 hrs at 22.5 °C	68

Figure 6.19a depicts the process paths of this trial run on the phase diagram. Starting from an unsaturated solution with a composition given by point 1, the process began with evaporation of

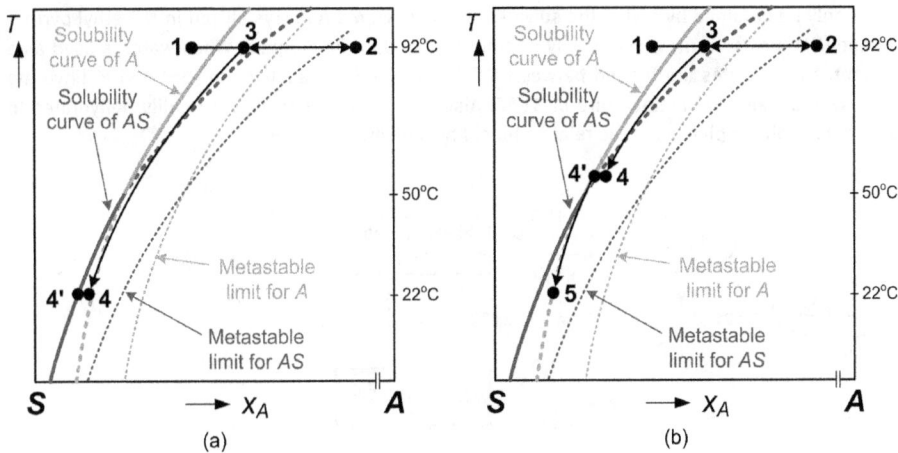

Figure 6.19: Illustration of composition changes during the drug crystallization process: (a) trial run and (b) recommended recipe.

about 50% of the solvent, causing the composition to move from point 1 to point 2, located deep inside the two-phase region. As is evident from the analysis results of sample 1, crystals of both *A* and *AS* appeared after such an operation, suggesting that point 2 must be located beyond the metastable limit of both forms. Crystallization of the solids shifted the liquid composition to point 3. During the cooling process, the solids became richer in the anhydrate, as indicated by the analysis results of samples 2 and 3. Therefore, point 3 must be located inside the metastable region of the anhydrate but outside the saturation region of *AS* (that is, between the solubility curves of *A* and *AS*), so that only the anhydrate could crystallize out during cooling while most of the already formed solvate crystals were redissolved. As a result, the solution composition stayed close to the solubility curve of *A*, reaching point 4 at the end of cooling. Finally, as the mixture was kept at 22.5 °C, the solution composition shifted to point 4', as the anhydrous form was transformed into the more stable solvate, as suggested by the analysis results of samples 4 and 5.

Based on an understanding of the phase diagram, a recipe was proposed to consistently obtain the anhydrate. The corresponding process path is shown in Figure 6.19b. The first part of the recipe follows the same procedure as the trial run, as the analysis results of sample 3 indicate that crystals consisting mostly of the anhydrous form can be obtained using this procedure. However, to ensure that only the anhydrate form is obtained, the end of cooling (point 4) should be kept slightly above the incongruent melting temperature (50 °C), and enough time should be allowed for the mixture to reach equilibrium (point 4'). This would cause all the solvate generated during evaporation to dissolve and recrystallize in the form of anhydrate. The recipe also calls for a second cooling step to point 5 to achieve a higher yield, followed by immediate filtration of the solids to prevent the anhydrate from converting to the solvate. The recipe was implemented during another trial run, during which a product with low solvate content was successfully obtained.

Example 6.8 – Process alternatives for crystallization of a drug
MK-A, a development compound at Merck, has four polymorphs *A*, *B*, *C*, and *E*, and is capable of forming hydrates and solvates [203]. Some of the stability relationships among the various polymorphs and hydrates are summarized in the diagram shown in Figure 6.20. Form *B* is monotropic with respect to both form *A* and form *C*, and it is comparatively less stable. Forms *A* and *C* are enantiotropic, with a crossover temperature of 21 °C. *A* is the stable form above this temperature. Form *E* can only be obtained by drying the solvate formed when MK-A is crystallized in *N*-methyl pyrrolidinone (NMP) as solvent, which is why it is not included in the diagram. When water is used as a solvent, two hydrates are formed between MK-A and water: a hemihydrate (HH) and a dihydrate (DH), with a transition temperature of 31 °C. Also, it is known that the DH readily dehydrates to anhydrous polymorph *A* when the relative humidity is below 60%.

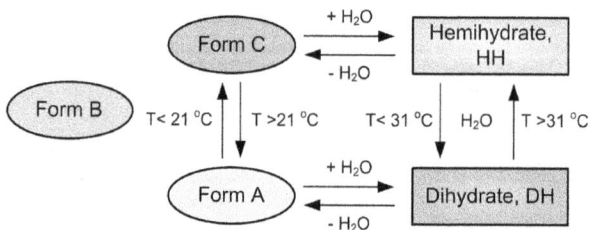

Figure 6.20: Stability relationship between various polymorphs and hydrates of MK-A.

In the last synthesis step of MK-A, a mixture of form C and HH solids (semipure) are obtained. It is desired to develop a strategy for synthesizing form A, the desired polymorphic form, from the semipure solid above the crossover temperature (21 °C). A second solvent is added to increase the degree of freedom of the system. Isopropyl acetate is chosen, since its azeotrope with water helps facilitate solvent removal by distillation. In addition, unlike water and NMP, isopropyl acetate does not form solvate with MK-A. Based on the stability information, a sketch of the ternary phase diagram as shown in Figure 6.21 can be developed. The solubility surfaces of A and C cross at 21 °C, while that of B lies below both surfaces. The solubility surfaces for the HH and DH extend from the MK-A/water side into the interior of the prism, but they do not reach the MK-A/isopropyl acetate side, because a hydrate cannot form without the presence of water. A polythermal projection showing the various crystallization compartments including those of thermodynamically stable polymorphs is shown on the triangular base.

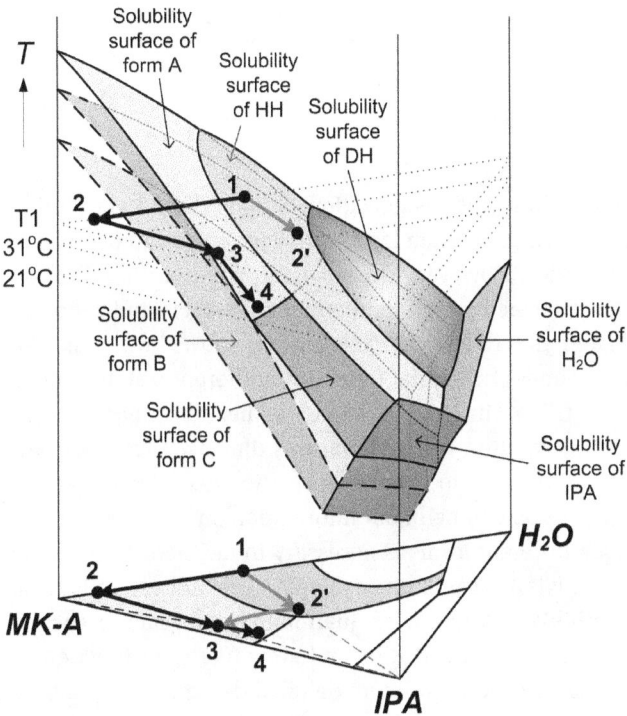

Figure 6.21: Process paths for crystallizing form A of MK-A [196].

Since the objective is to get form A, the process path has to include movements that would bring the composition into the compartment of form A. The starting point is a wet cake of form C and HH at temperature T_1 (indicated by point 1). There are two possible ways to get from point 1 to the desired region: either remove water and then add isopropyl acetate, or add isopropyl acetate first, and then remove water. In the first alternative, removing water by drying moves point 1 to point 2, which is close to the MK-A vertex but still contains a small amount of water. As the HH would give form C as it releases water (Figure 6.20), this solid product would be rich in form C. Since transformation to form *A* is unlikely under such a dry condition, isopropyl acetate is added to the system so that point 2 moves to point 3. The ratio of the added isopropyl acetate to the remaining water determines the location of point 3, which must stay within the compartment of form *A* to prevent the formation of hydrates. At point 3, suspended form *C* crystals undergo solvent-mediated transformation to form *A*, which is the most stable form at this temperature. With the resulting crystals of form *A* acting as seeds, cooling of this solution produces more form *A*, as represented by the movement from point 3 to point 4.

In the second alternative, the addition of isopropyl acetate to the wet cake moves point 1 to point 2' in the unsaturated region (above the saturation surfaces). Then, water is removed by distillation (as an azeotrope with isopropyl acetate), resulting in a composition inside the compartment of form *A*, such as point 3. Just like in the first alternative, solvent mediated transformation followed by cooling would produce pure form *A*.

This example illustrates that the knowledge of the SLE phase behavior, even if it is just a sketch, can help guide the selection of movements in the right direction. The phase diagram serves as a map, which makes it possible to locate the origin and destination, and then figure out the possible routes to reach the destination.

6.6 Summary

Phase diagram is a tool that allows the visualization of solid–liquid equilibrium phase behavior. Movements in phase diagram lead to conceptual design of crystallization-based separation processes. However, a component may not crystallize out even when the composition is located inside its thermodynamic crystallization region because of nucleation and growth kinetics. This chapter shows how phase diagrams illustrating metastable zones and crystal forms with different stability can be used to capture such effects. Other kinetic effects such as heat and mass transfer and reaction kinetics can be accounted for, with various dimensionless numbers. Also, techniques have been developed to overcome kinetic effects. For example, seeding with the desired polymorph can help, but information on the solubility and stability relationship between different forms is necessary to facilitate the development of the seeding process [204]. As seen in the examples, this understanding can be qualitative or semi-quantitative. While more quantitative information such as the exact width of the metastable zone, nucleation and growth rates as functions of supersaturation, and so on may be useful for more detailed design, conceptual design of the process can still proceed in their absence. The effects of crystallization kinetics on PSD, impurities, and so on are discussed next, in chapter 7.

Exercises

6.1. Maia and Giulietti [205] reported that the solubility of acetylsalicylic acid (aspirin) in ethanol in the temperature range of 276–337 K can be adequately represented by:

$$\log x = 27.769 - \frac{2500.906}{T} - 8.323 \log T \tag{6.9}$$

where x is solubility in mole fraction and T is temperature in K. The molecular weights of aspirin and ethanol are 180.16 and 46.07 g/mol, respectively.

a. If 150 g of a solution containing 40 wt% aspirin in ethanol is cooled down to 25 °C and 35 g of aspirin crystallizes out, what is the final supersaturation (in g aspirin/g ethanol)?

b. What is the final degree of supercooling?

6.2. An organic compound is to be crystallized from a solution containing an impurity. The growth rate of the crystals in the presence of the impurity is measured at the same degree of supersaturation, and compared to the growth rate in a clean solution (with no impurity). The results are summarized below.

Impurity concentration (%)	0	3	5	7	10
Growth rate (μm/min)	3.52	2.40	1.75	1.16	0.43

Determine the suitable parameter values to fit the data, using the correlation proposed by Kubota and Mullin (eq. (6.8)).

6.3. Preferential crystallization can be an attractive alternative for separating a mixture of diastereomeric salts, which features asymmetric phase behavior as shown in Figure 6.22, particularly if temperature swing does not significantly shift the double saturation point. Sketch the process flowsheet for separating $(S)\text{-}A \cdot (S)\text{-}B$ and $(R)\text{-}A \cdot (S)\text{-}B$ from their racemic mixture (M) using preferential crystallization, and draw the corresponding process paths on the phase diagram.

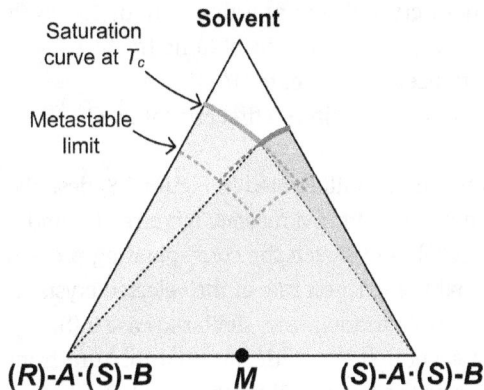

Figure 6.22: SLE phase diagram of a diastereomeric salt system in a solvent (problem 6.3).

6.4. Suppose that a continuous preferential crystallization process to separate D- and L-threonine is to be designed based on the results shown in Figure 6.6 (Example 6.2). Specifically, the crystallizers are to be operated at 33 °C, and the location of point 2 is assumed to correspond to the maximum yield of pure L-Thr suggested by the experimental data, as shown in Figure 6.23. It is further assumed that the preferential crystallization using seeds of D-Thr behaves in a symmetrical manner, so the location of point 4 is chosen correspondingly, as indicated in the diagram. In such a continuous operation, seeds are only introduced at the beginning of the operation (no more seed is introduced at steady state).

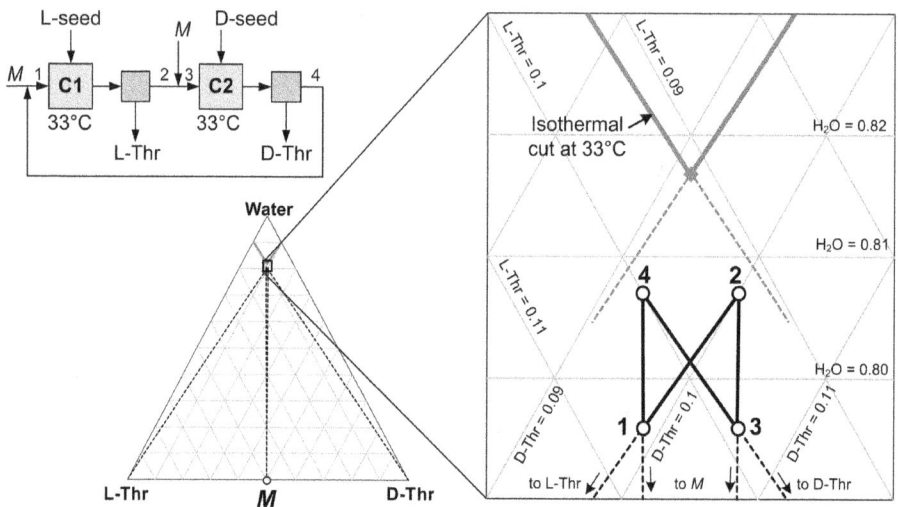

Figure 6.23: Preferential crystallization process to separate D- and L-threonine (problem 6.4).

a. If the racemic mixture M is fed to both crystallizers at a flow rate of 100 kg/h each, how many kg/h of L-Thr and D-Thr products will be obtained?
b. What is the flow rate of the first crystallizer feed (stream 1)?
c. What is the per-pass crystallization yield of L-Thr in the first crystallizer?

6.5. Based on the idea of asymmetric transformation illustrated in Figure 6.8, describe an ideal batch process to obtain pure S enantiomer from a racemic mixture of R and S with a very high recovery (theoretically 100%), and sketch the corresponding process path on an SLE phase diagram. Assume that the reaction rate at the selected crystallization temperature is much faster than the crystallization rate, similar to case iii(b).

Hint: The process should consist of several steps, with the final mixture from the last step having the same composition as the fresh feed for the next batch.

6.6. Component A has two polymorphic forms: α and β. The SLE phase diagram of a binary system involving A and a solvent S is given in Figure 6.24. Two process paths (I and II) are indicated on the diagram, representing different scenarios that can occur during cooling of a feed mixture (point 1).

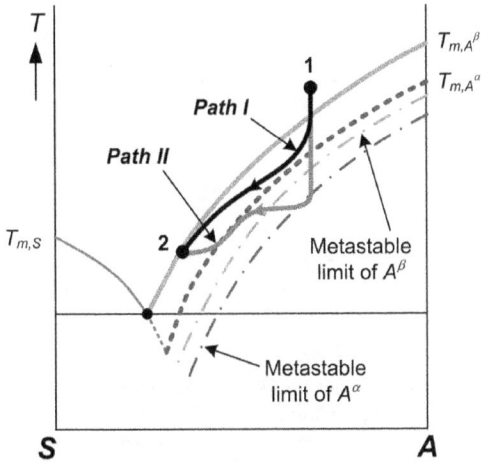

Figure 6.24: Two process paths in crystallization of component A (problem 6.6).

a. Describe the possible scenarios that would result in Path I and Path II, including which form(s) crystallizes out during cooling and whether the crystallization occurs spontaneously or seeding is needed.

b. What can be done to obtain form α as the final product?

6.7. Batra et al. [206] discussed the development of the crystallization process of treprostinil diethanolamine salt, which is a drug for lung disease. Two polymorphic forms were observed: form A (crystalline, melting point 103–104 °C) and form B (crystalline, melting point 106–108 °C). Both forms readily dissolve in water with solubility greater than 500 mg/mL at neutral pH. From the available information, there is no change of the relative stability of the two forms at any temperature, that is, one form is always more stable than the other.

a. From the melting point information, which is the more stable form, A or B?

b. Does the system exhibit monotropic or enantiotropic behavior?

c. Sketch the binary SLE phase diagram (temperature vs. composition) involving the two polymorphic forms and water.

6.8. The thermodynamic stability and transformation behavior of the polymorphs and solvates of a thiazole-derivative (BPT) in methanol/water solvent system have been reported by Kitamura [207]. There are four forms of BPT: two polymorphic forms (*A* and *C*), a monohydrate (BH), and a methanolate (solvate with methanol, *D*). Analysis of pure samples of the four forms by DSC and TGA gives the following results:

Form	DSC curve	TGA results
A	Single endothermic peak around 480 K	Weight loss above 480 K
C	Endothermic peak around 470 K, followed by a small exothermic peak and another endothermic peak around 480 K	Weight loss above 480 K (same as A)
BH	Endothermic peak around 328 K in addition to the peaks observed in C	Weight loss of 5.3% (consistent with a molar ratio of water: BPT=1:1) at 325 K, then same trend as A and C
D	Endothermic peak around 370 K in addition to the peaks observed in C	Weight loss of 9.2% (consistent with a molar ratio of MeOH : BPT=1:1) at 368 K, then same trend as A and C

Therefore, it is concluded that form *A* is the most stable form at higher temperatures. Furthermore, the solubility of the forms is determined by adding an excess amount of each form to a water-methanol mixture with various methanol volume fraction in the solvent (V_{MeOH}) and measuring the concentration at 323 K. The following observations have been made:
- Solubility increases significantly with V_{MeOH} (water is an anti-solvent)
- Form *C* has a lower solubility than form *A* over the entire range of solvent compositions studied, implying that form C is the most stable form at 323 K.
- At high methanol content (V_{MeOH}=0.95), all forms transformed to form *D*.
- At V_{MeOH} between 0.7 and 0.8, all forms eventually transformed to form *C*.
- At V_{MeOH}=0.5, forms *A* and *D* transformed to form *BH* (which is believed to be metastable), while form *C* did not undergo any transformation.

Based on the information:
a. Provide a sketch of the three-dimensional *T*-*x* diagram of BPT/water/methanol system, featuring the solubility surfaces of forms *A*, *BH*, *C*, and *D*.
   Hint: First think about how each side of the prism should look like, then fill in the interior.
b. Show a sketch of the isothermal cut at 323 K.
c. Starting from a solution of form *C* in methanol, synthesize a possible process to obtain crystals of form *A*.

7 Management of Particle Size Distribution and Impurities in Crystallization Processes

7.1 Crystallizer Model

Particle size distribution (PSD) and product purity are among the most important attributes of a solid product. After synthesizing a process to crystallize out the desired product, additional considerations must be taken to ensure that the solid product has the desirable PSD and meets the purity specifications. The PSD in a crystallizer is controlled by the nucleation and growth rates during crystallization, which in turn are affected by the supersaturation in the crystallizer. Impurity incorporation into the product crystals generally depends on the growth rate and is often affected by the PSD. Therefore, in order to manage the PSD and impurities in a crystallization process, it is imperative to consider supersaturation, nucleation rate, and growth rate in the crystallizer. Such an analysis can be performed quantitatively with the help of a crystallizer model.

7.1.1 Population balance

Population balance equations (PBE) are needed to track the PSD during the crystallization process so that the effect of various factors on the PSD can be quantified. In this approach, the PSD is mathematically defined in terms of *population density*, n [no/m^4], which is a continuous function of particle size. The integral of this function over a small size segment dL gives the number of particles of size between L and $L+dL$ per unit volume or mass of slurry, N [no/m^3]. A basic population balance for particles of size L in a well-mixed crystallizer can be written as

$$\begin{bmatrix} \text{accumulation of} \\ \text{crystals of size } L \\ \text{over time } \Delta t \end{bmatrix} = \begin{bmatrix} \text{no. of crystals} \\ \text{of size } L \\ \text{in inlet stream} \end{bmatrix} - \begin{bmatrix} \text{no. of crystals} \\ \text{of size } L \\ \text{in outlet stream} \end{bmatrix} + \begin{bmatrix} \text{no. of crystals} \\ \text{growing into} \\ \text{interval } \Delta L \end{bmatrix}$$

$$- \begin{bmatrix} \text{no. of crystals} \\ \text{growing out of} \\ \text{interval } \Delta L \end{bmatrix}$$

$$Vn|_{t+\Delta t}\Delta L - Vn|_t \Delta L = Q_{in}n_{in}\Delta L\Delta t - Q_{out}n_{out}\Delta L\Delta t + VGn|_L\Delta t - VGn|_{L+\Delta L}\Delta t \tag{7.1}$$

https://doi.org/10.1515/9781501519901-007

where G is the crystal growth rate [m/s], V is crystallizer volume [m³], and Q is volumetric flow rate [m³/s]. After dividing by $\Delta L \Delta t$ and taking the limit of both ΔL and Δt to zero, the equation can be cast in the form of a differential equation,

$$\frac{\partial(nV)}{\partial t} + V\frac{\partial(Gn)}{\partial L} = Q_{in}n_{in} - Q_{out}n_{out} \qquad (7.2)$$

There are two boundary conditions. The first one states that the PSD at $t = 0$ is equal to an initial PSD (which is equal to the PSD of the seeds or zero for unseeded crystallization),

$$n(L, 0) = n_0(L) \qquad (7.3)$$

The second boundary condition introduces the nucleation rate, B [no/m³.s], as the generation term for particles of zero size,

$$B = \frac{dN}{dt}\bigg|_{L=0} = \frac{dN}{dL}\bigg|_{L=0} \cdot \frac{dL}{dt} \qquad (7.4)$$

By definition, dN/dL is the population density (n) and dL/dt is the growth rate (G). Therefore, eq. (7.4) gives

$$n(0, t) = B/G \qquad (7.5)$$

Note that eq. (7.2) can also be written on the basis of slurry mass instead of volume, with V replaced by slurry mass M [kg], Q replaced by mass flow rate F [kg/s], and the units of n and B changed to no/m.kg and no/kg.s, respectively. Equation (7.2) can also be simplified into an equation for batch crystallization by setting the right-hand side to zero or into an equation for continuous crystallization at steady state by setting the accumulation term ($\partial(nV)/\partial t$) to zero.

The ideal case for a continuous crystallizer is the mixed suspension mixed product removal (MSMPR) crystallizer in which the slurry is assumed to be perfectly mixed and homogeneous throughout the crystallizer. For an MSMPR crystallizer with no solids in the inlet stream and a growth rate that is independent of crystal size, eq. (7.2) simplifies to

$$G\frac{dn}{dL} = \frac{Q_{out}}{V}n \qquad (7.6)$$

Using the boundary condition given in eq. (7.5), this equation can be integrated to give an analytical expression for $n(L)$,

$$n(L) = \frac{B}{G}\exp\left(-\frac{Q_{out}L}{GV}\right) \qquad (7.7)$$

This expression shows that the PSD depends on the residence time (V/Q_{out}) as well as supersaturation, which determines the values of B and G.

The MSMPR model is often inadequate to describe an actual crystallizer, which is often purposely designed not to be well-mixed in order to obtain a desirable PSD. Two of the most common mechanisms featured in industrial crystallizers are removal and dissolution of fine particles and selective withdrawal of larger crystals. To describe those crystallizers, a generic model that includes fines dissolution and product classification can be used instead [208]. The population balance for this generic crystallizer is written as

$$VG\frac{dn}{dL} + ZQ_pn + (R-1)Q_pn\left[1 - H(L-L_f)\right] = Q_fn_f + (Z-1)Q_pn[1 - H(L-L_c)] \quad (7.8)$$

where the Heaviside function H is defined as

$$H(L-x) = \begin{cases} 0 \text{ if } L < x \\ L \text{ if } L \geq x \end{cases} \quad (7.9)$$

The explanation for the notations and the correspondence between the model and a draft-tube baffle crystallizer, which is one of the most commonly used crystallizers in commercial processes, is shown in Figure 7.1.

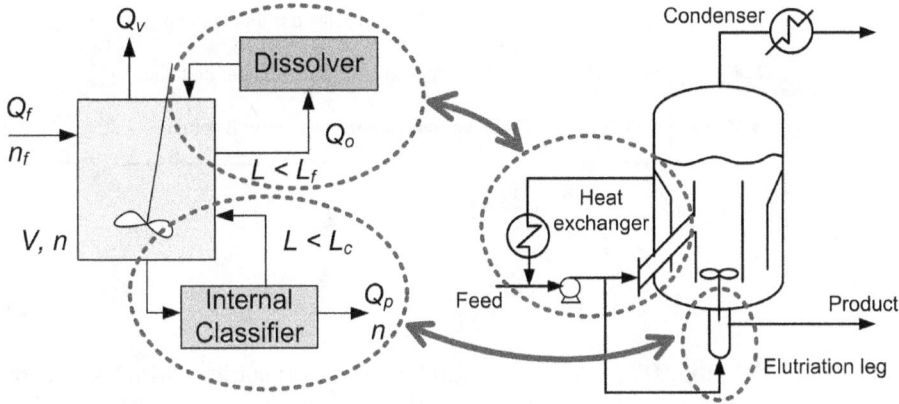

Figure 7.1: A generic crystallizer model and the correspondence to a draft tube baffle crystallizer in Figure 7.9b.

In this model, R and Z are proportional to the circulation flow rates through the heat exchanger and the elutriation leg, respectively, in the crystallizer. However, because classification of particles in the actual crystallizer is not as clear-cut as assumed in the model, R and Z should be viewed as parameters rather than actual proportionality constants. The cut sizes, L_f and L_c, are also parameters, the values of which depend on the crystallizer design. For example, L_c would depend on the diameter of the elutriation leg as well as the circulation flow rate.

7.1.2 Solutions of population balance equations

To calculate the supersaturation and PSD in the crystallizer, the population balance equations must be coupled with the mass balance of the liquid side. This can be conveniently done using the method of moments [209]. The i-th moment of a distribution is defined as

$$M_i = \int_0^\infty n(L)L^i dL \qquad (7.10)$$

The most commonly used moments and their physical significance are summarized in Table 7.1. The surface factor k_a and volume factor k_v are defined to allow the use of a characteristic length L to represent the particle size for irregular-shaped particles. For example,

Table 7.1: Physical significance of moments.

Moment	Formula	Significance
Zeroth	$M_0 = \int n(L)dL$	Total number of particles per unit volume
First	$M_1 = \int n(L)LdL$	Total length of particles per unit volume
Second	$k_a M_2 = k_a \int n(L)L^2 dL$	Total surface area of particles per unit volume
Third	$k_v M_3 = k_v \int n(L)L^3 dL$	Total volume of particles per unit volume

$$k_a = \frac{\text{particle surface area}}{L^2} \qquad (7.11)$$

$$k_v = \frac{\text{particle volume}}{L^3} \qquad (7.12)$$

Using these definitions, $k_a = 6$ and $k_v = 1$ for cube-shaped particles with the side of the cube as the characteristic length. For spherical particles with the diameter as the characteristic length, $k_a = \pi$ and $k_v = \pi/6$.

The link between the mass balances of the solid and liquid sides is made by equating the mass of solute leaving the liquid side and the mass of solids formed. For batch crystallizers,

$$V\frac{dc_A}{dt} = -V\frac{dM_T}{dt} \qquad (7.13)$$

where c_A is the solute concentration and M_T is the magma density, which is related to the third moment,

$$M_T = \rho_S k_v \int n(L)L^3 dL \qquad (7.14)$$

where ρ_S is the true density of the solids. The PSD can also be represented as a mass-based differential distribution, which is obtained by multiplying the population density with the mass of each particle,

$$W(L) = \rho_S k_v n(L)L^3 \qquad (7.15)$$

Using this representation, the magma density, which is the total mass of particles of all sizes per unit volume, is simply given by

$$M_T = \int W(L)dL \qquad (7.16)$$

7.2 Particle Size Distribution Management

There are various factors that control the PSD, the most dominant ones being nucleation and crystal growth, as implied by the basic PBE (eqs. (7.2), (7.3), and (7.5)). Primary nucleation and crystal growth depend on the supersaturation level in the solution, while secondary nucleation mainly depends on magma density (the mass of crystals per unit volume of slurry). There are other phenomena that can affect the PSD, such as agglomeration and breakage, *Ostwald ripening* (very small particles may dissolve and deposit on larger particles to minimize surface area), and *growth dispersion* (crystals of the same size grow at different rates because of different structures or different environments in the crystallizer). Additional terms can be added to the PBE to quantitatively account for those mechanisms [210, 211].

7.2.1 PSD manipulation in batch crystallization

A batch crystallizer usually takes a simple design – an agitated vessel equipped with a jacket or a coil and it is often a multipurpose unit used for different products. Additional accessories such as a forced circulation line or a draft tube may be added to assist mixing [212]. The key to a successful batch crystallization process often depends on the profile of supersaturation generation rate (either by cooling, evaporation, MSA or reactant addition, or combinations thereof). For example, in batch cooling crystallization, to go from the same initial to final temperatures in the same period of time, different temperature profiles can be applied, such as natural, linear, and stepwise linear (slower at the beginning and faster later), as illustrated in Figure 7.2a. The applied profile determines how the saturation condition (solubility) changes with time. Meanwhile, nucleation and growth rates determine how the

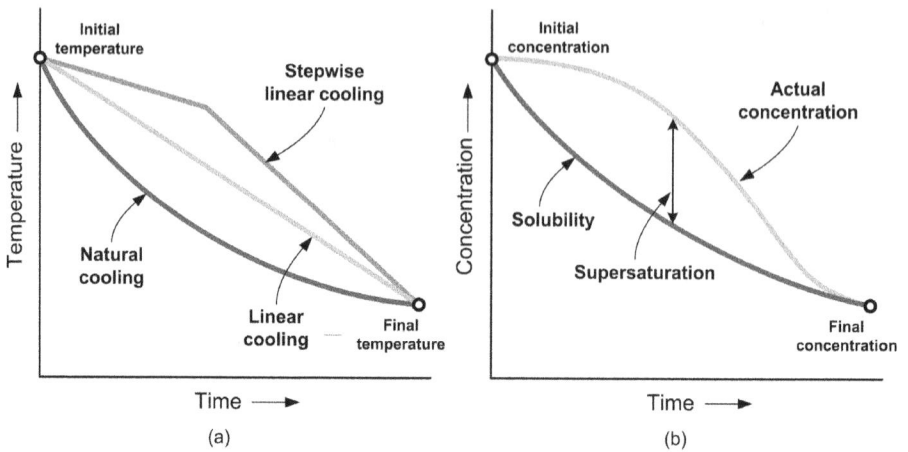

Figure 7.2: Temperature and concentration profiles in batch cooling crystallization: (a) three typical temperature profiles and (b) typical concentration profile for linear cooling.

actual solution concentration changes with time. A typical profile features a slow decrease in the beginning as nucleation begins to occur, followed by a faster drop as more crystals are formed, before a final slowdown as the solution concentration approaches equilibrium, as illustrated in Figure 7.2b. The difference between the actual solution concentration and solubility gives the supersaturation profile, which typically peaks near the beginning and later diminishes towards equilibrium. The supersaturation profile, in turn, affects the outcome of the process, such as PSD and impurity content. Besides supersaturation generation rate, the addition of seeds at the appropriate time can greatly influence the supersaturation profile in a batch crystallization operation.

In batch crystallization, the product PSD is mainly controlled by the supersaturation profile over the course of crystallization as well as by its spatial distribution in the crystallizer. Hence, there are two types of actions to improve product quality and avoid PSD-related problems. The first type includes operation-related issues such as choosing a suitable operating profile (the time variation of cooling, evaporation, or anti-solvent addition rate, or reactant addition rate) and proper seeding policy. The second type is equipment-related which involves the proper design of the crystallizer, such as maintaining good circulation, distributing anti-solvent or reactant over several inlet points to avoid high local supersaturation, and minimizing "cold spots" and areas of high shear in the vessel.

As mentioned, the profile of supersaturation generation rate (cooling profile in cooling crystallization, evaporation profile in evaporative crystallization, anti-solvent addition profile in drowning-out crystallization, or reactant addition profile) in reactive crystallization determines the solubility versus time relationship. The supersaturation profile can be controlled by selecting a suitable profile. Normally, excessive supersaturation at the beginning should be avoided as it can lead to excessive and uncontrollable nucleation.

Seeding can have a more pronounced effect on supersaturation profile compared to controlling the operating profile. This is because the presence of seed particles minimizes primary nucleation by providing surface area for growth and also provides a template for the final product PSD. Using the right amount of seed with the right size is very important. If the amount is not enough, primary nucleation may still occur, often resulting in a bimodal PSD, with one peak corresponding to the growing seed particles and another corresponding to the newly generated particles. Adding too much seed is of course undesirable from an economic point of view and it can also lead to final particles that are too small. If there is no nucleation, the relationship between the size and the mass of the seed and the desired size of the product is given by a simple equation based on mass balance [157]:

$$\frac{m_S}{m_P} = \left(\frac{L_S}{L_P}\right)^3 \tag{7.17}$$

where m_S is the mass of seed added, m_P is the mass of product crystals at maximum theoretical yield, L_S is the size of seed added, and L_P is the target size of product crystals. Mersmann [30] proposed a more detailed expression for the required seed mass in case of diffusion-controlled growth and constant cooling rate,

$$m_S = -\frac{2k_v}{k_a}\left(\frac{dc^*}{dT}\right)m_{solv}\frac{\dot{T}L_S}{G_{max}}\left(1 + \frac{G_{max}\Delta T_{max}}{2\dot{T}L_S}\right)^{-2} \tag{7.18}$$

where k_a and k_v are shape factors, m_{solv} is the solvent mass, $\dot{T}$ is the cooling rate, G_{max} is the maximum growth rate, and ΔT_{max} is the metastable zone width. Kubota et al. [213] empirically determined the critical seed loading to guarantee a unimodal distribution of final products over a wide range of supersaturation generation rates. Other aspects of seeding pertinent to the control of a crystallization process have been reviewed by Yu et al. [214].

A workflow to guide the decision making for seeding application in batch crystallization has been developed [215, 216]. At the center of the workflow are two causal tables which summarize the qualitative relationships between kinetic phenomena and product attributes (Table 7.2) and between seed qualities/seeding practice and kinetic phenomena (Table 7.3). The row headings represent causes and the column headings are effects. A plus sign (+) indicates that if the entity shown in the row heading is increased, the entity in the column heading is enhanced, while a minus sign (−) indicates the opposite relationship. For example, Table 7.2 suggests that increasing nucleation leads to product crystals with smaller mean size, wider or bimodal distribution, and increased impurity level. Since Table 7.3 indicates that increasing seed mean size enhances nucleation, one way to minimize nucleation would be to use seeds of smaller size. Similarly, since seed mass affects nucleation

Table 7.2: Qualitative relationships between kinetic phenomena and product attributes (after [215]).

Kinetic phenomenon	Product attribute				Batch reproducibility or scalability
	Mean size	Narrow & unimodal size distribution	Impurity level	Fraction of desired polymorph/enantiomer	
Nucleation	–	–	+	Of desired form +	–
				Of undesired form –	
Growth	+	0	+	Of desired form +	+
				Of undesired form –	
Dissolution	–	+	–	Of desired form –	–
				Of undesired form +	
Agglomeration	+	–	+	0	–
Attrition/ breakage	–	–	–	0	–

Table 7.3: Qualitative relationship between seed qualities/seeding practice and kinetic phenomena (after [215]).

Seed quality or seeding practice	Super-saturation	Kinetic phenomenon				
		Nucleation	Growth	Dissolution	Agglomeration	Attrition/ breakage
Mean size	+	+	+	–	–	+
Size distribution (width/standard deviation)	!	!	!	!	!	!
Mass	–	–	–	–	+/– [a]	+
Lattice stress/strain	+	+	–	+	+	0
Degree of supersaturation when seeds are introduced	+	+	+	–	+	0

[a] '+' when small seed crystals are used, '–' if large seed crystals are used.

in a negative way, introducing more seeds can be an effective way for reducing nucleation. Zero (0) indicates that there is no direct relationship, while an exclamation mark (!) indicates that the relationship cannot be determined unless more information is available. While the causal tables do not provide a quantitative prediction and may not be valid for a particular system, these guidelines can help minimize experimental efforts by pointing to the right direction.

Apart from controlling the supersaturation profile, physical means are sometimes used to affect the PSD during crystallization. For example, when the target is to obtain small crystals but it is not desirable to operate at high supersaturation, the application of ultrasound helps to break up the crystals without any adverse effects on the crystal quality [155].

Example 7.1(I) – Batch crystallization of an active pharmaceutical ingredient (API)

A pharmaceutical company was developing a process for manufacturing an API by a batch reaction, followed by crystallization of the API by cooling to 0 °C. The obtained product was highly pure but the crystals had undesirable product attributes, such as small mean size, bimodal PSD, and low bulk density. The company would like to fix the crystallization process to produce acceptable materials without any need for reprocessing.

To find out the best operating conditions to produce crystals with the target attributes, the crystallizer operation was linked to product attributes via population balance modeling. The model followed the change in crystal PSD and API concentration in the solution during the cooling process, accounting for nucleation and growth kinetics. Relevant thermodynamic information, including the solubility of the API in the solvent as a function of temperature, was also obtained to allow the calculation of the supersaturation profile by comparing the API concentration to its solubility. The values of the kinetic parameters were tuned to match the available PSD data from different runs. Although it was impossible to perfectly match the actual PSD because the simplified model did not account for complexities such as breakage, agglomeration, or ripening, the overall trends could be captured. Table 7.4 summarizes the modeling results for three runs with different cooling profiles. Plots of the cooling profile, calculated supersaturation profile, and the final PSD for the three runs are shown in Figure 7.3.

Table 7.4: Cooling profile and final PSD from three different runs.

Run	Cooling profile	Particle size distribution			
		Mean (μm)	d_{10} (μm)	d_{50} (μm)	d_{90} (μm)
1	77 °C → 0 °C at 1 °C/min	44.1	9.9	29.6	103.6
2	77 °C → 53 °C at 0.173 °C/min, 53 °C → 0 °C at 0.398 °C/min	66.4	13.0	40.6	168.0
3	77 °C → 60 °C at 0.19 °C/min, 60 °C → 0 °C at 0.51 °C/min	73.4	15.7	46.6	180.0

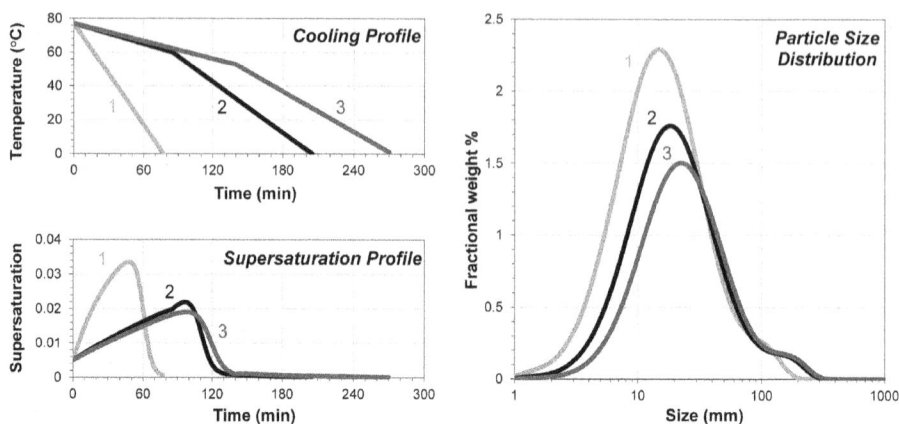

Figure 7.3: Cooling profile, supersaturation profile, and final PSD for three different runs.

The results suggest that slow growth leads to supersaturation build-up and nucleation is much faster than growth at high Δc, resulting in a wide PSD. At the fastest cooling rate (run 1), the supersaturation quickly builds up and reaches a large peak before significant crystallization occurs and relieves the supersaturation. As a result, the final PSD features a peak at 15 μm and the mean size is only 44.1 μm. A slower initial cooling rate (run 2) results in a lower maximum supersaturation and the mean size increases to 66.4 μm. The cooling rate is increased after the temperature of the mixture reaches 60 °C when some crystals have been formed to provide growth sites. But apparently the amount is not enough, so the faster cooling rate leads to a further increase in supersaturation beyond the transition point at about 90 min. In run 3, the initial cooling rate is almost the same but the increase in cooling rate occurs much later, and by that time the supersaturation has already passed its peak. The transition to the faster cooling rate still produces a small second peak but it is almost unnoticeable. This cooling rate results in a lower maximum supersaturation compared to run 2 and a further increase in mean size to 73.4 μm.

The overall trend is clear: when the maximum supersaturation decreases, bigger particles are produced. To minimize the supersaturation, it was suggested to use a slow initial cooling rate and delay the transition to the faster cooling rate so as to provide more time for crystal growth to consume the supersaturation. Once the supersaturation has passed its peak, a faster cooling rate can be used since it will not have much of an effect on the final PSD. Another idea was to introduce seeds to provide growth sites and lower the maximum supersaturation during the initial cooling period. The results from one of the trial runs indicated the effectiveness of seeding to obtain bigger particles. However, it is important to use seeds of the right amount and PSD to avoid generating a bimodal distribution.

Figure 7.4 shows the effect of seed amount on the final PSD based on simulation results. Seeds with a unimodal distribution and mean size of 10 μm are used and a cooling profile with a slightly faster rate compared to Run 2 is implemented. The results indicate that bigger particles with narrower distribution are obtained by introducing seeds. With more seeds, a larger area becomes available for crystal growth, allowing the supersaturation to be consumed quicker, thus lowering the maximum supersaturation achieved during crystallization. However, a larger amount of seeds also implies that the crystallized material is distributed over a larger number of crystals, resulting in a reduction in mean particle size of the product.

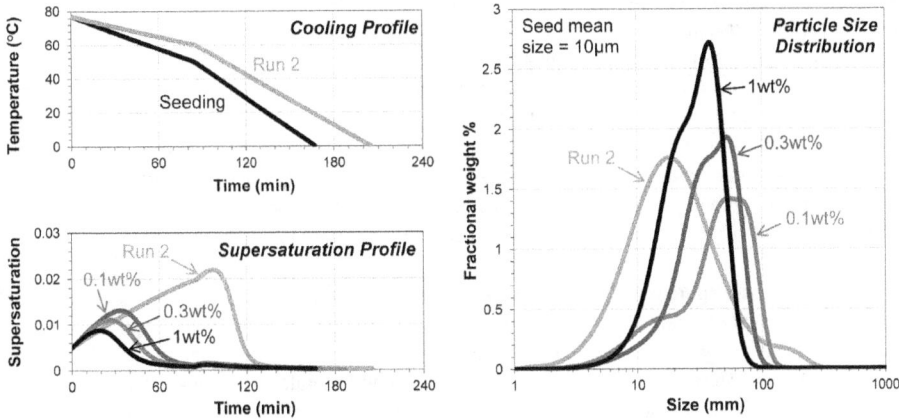

Figure 7.4: Effect of seed amount on the final PSD.

The effect of seed size is illustrated in Figure 7.5. With the same amount of seeds (1 wt%), smaller seeds (5 µm mean size) produce a more uniform PSD despite a smaller mean size of the product due to the larger number of particles. On the other hand, when using seeds of 20 µm mean size, a bimodal PSD is obtained. The first peak at larger particle size is attributable to the growing seed particles during the entire run, while the second peak at smaller particle size is caused by the newly formed particles during a spike in supersaturation when a faster cooling rate is applied. As shown in the zoomed section of the supersaturation profile, the maximum supersaturation achieved during the spike is higher for seeds with larger mean size.

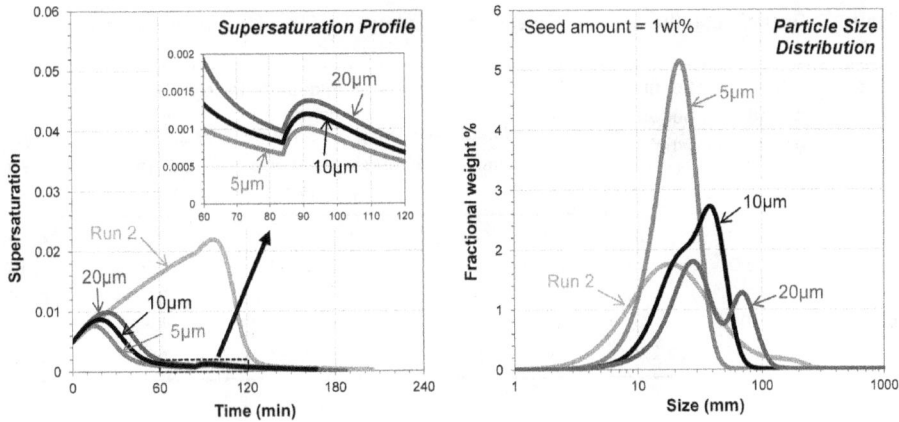

Figure 7.5: Effect of seed size on the final PSD.

Example 7.2(I) – Production of an organic acid by reactive crystallization

An organic acid R was produced from an anhydride P using a catalytic batch process which involves two liquid phase reactions promoted by two catalysts: $P + H_2O \rightarrow Q$ (hydration; catalyst A) and $Q \rightarrow R$ (isomerization; catalyst B). Since R is sparingly soluble in water, it crystallizes from the solution. The final product is in the form of a fine white powder, smaller than 40 mesh in size. Depending on the market demand, products of different sizes need to be made. The desired product is obtained by screening and the off-spec powder is sold as a lower grade product. Therefore, it is desirable to control the PSD of the crystallizer output such that the production of the desirable size is maximized.

In a typical batch, pure water and catalyst A are fed into the vessel. Next, solid P, catalyst B, and water are added at constant flow rates at 72 °C. After completing the addition process, the reaction mixture is cooled to 40 °C. To evaluate the effect of different operating procedures on the PSD, a model of the batch crystallizer consisting of both mass and population balances was developed. Based on an analysis of reaction kinetic data, it was concluded that dissolution of P and hydration reaction are very fast, so that Q is formed instantaneously once P is introduced. The rate of the isomerization reaction can be expressed as

$$-\frac{d[Q]}{dt} = k_r[Q][\text{cat} - B] \tag{7.19}$$

where k_r is the reaction rate constant which has an Arrhenius-type temperature dependence. The values of reaction kinetic parameters as well as those of crystallization kinetic parameters were estimated based on a comparison of the model results with available data.

Five cases as summarized in Table 7.5 were simulated using the model. The reactant addition rate is varied while keeping the cooling rate constant in cases 1, 2, and 3, while cases 4 and 5 focus on the effect of cooling rate with the same reactant addition rate. The temperature, liquid holdup, and calculated supersaturation profiles are shown in Figure 7.6.

Table 7.5: Operation profile, calculated yield, and calculated final PSD from three different runs.

Case	Addition period (min)	Cooling period (min)	Solid R yield (kg)	Calculated final particle size distribution				
				Mean (µm)	Passing 60 mesh (%)	Passing 80 mesh (%)	Passing 100 mesh (%)	Passing 150 mesh (%)
1	60	80	5298	120	100.00	99.98	95.02	24.85
2	90	80	5710	141	100.00	97.66	67.45	3.45
3	120	80	5976	159	100.00	84.00	34.38	0.28
4	90	40	5691	141	100.00	97.73	67.85	3.55
5	90	120	5743	141	100.00	97.52	66.75	3.29

It can be seen that supersaturation peaks very early due to the generation of R by reaction and then diminishes as R crystallizes out. Faster reactant addition leads to an earlier peak with higher maximum value as shown in the zoomed section showing the supersaturation profile in the first 20 mins. After the peak, the supersaturation approaches a steady value as the reactant addition rate balances the crystallization rate. During the cooling period, the supersaturation quickly drops in the absence of reaction. A much smaller degree of supersaturation is maintained as the generation rate by cooling

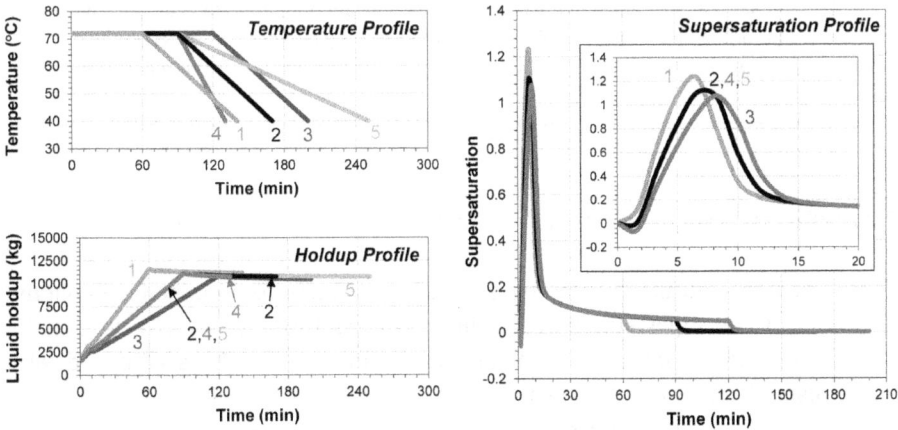

Figure 7.6: Temperature, liquid holdup, and supersaturation profiles during batch crystallization of organic acid.

balances the crystallization rate. The supersaturation profiles for cases 4 and 5 are very similar to that of case 2. The difference in cooling rate only causes a very small difference in the degree of supersaturation during the cooling period.

Figure 7.7 shows the comparison of the yield of solid R and the final PSD for the five cases. The PSD of cases 2, 4, and 5 are almost identical. The results are also summarized in Table 7.5. Faster reactant addition (case 1) leads to smaller particles due to the higher maximum supersaturation level. The solid yield at the end of the reaction is the lowest because the shorter batch time leads to a lower conversion from Q to R (higher Q concentration in the liquid). With the same addition time, the same mean particle size is obtained for cases 2, 4, and 5, showing that the cooling rate has no effect on PSD.

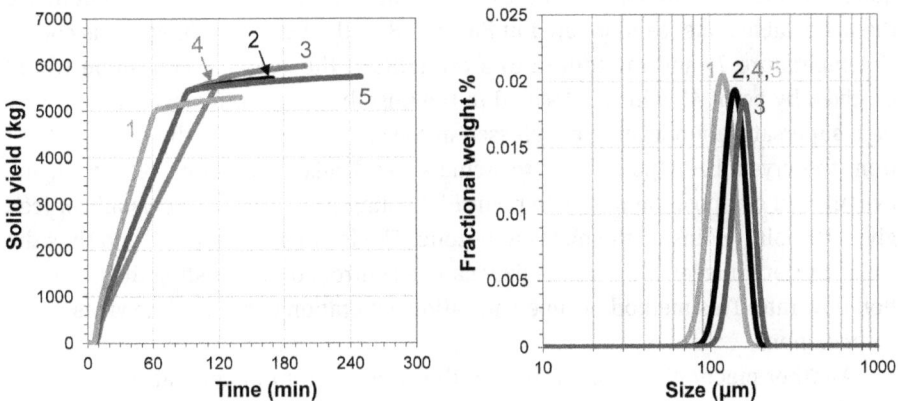

Figure 7.7: Comparison of solid yield and particle size distribution for the five cases.

From the results, it can be concluded that varying the reactant addition rate is an effective way to control the PSD of the final product, while changing the cooling rate has no impact on the PSD. The model is very useful in pointing to the right direction and understanding the reason for the observed results.

7.2.2 PSD targeting in continuous crystallization

Continuous crystallizers usually involve a special design to promote favorable hydrodynamics and supersaturation distribution inside the crystallizer. They normally include minimal or no mechanical agitation and are sometimes equipped with mechanisms for fines dissolution and classified product removal. A suitable crystallizer type can be selected by considering the appropriate mechanisms for generating supersaturation, controlling the supersaturation generation, and relieving the supersaturation [217]. The choice of supersaturation generation mechanism should take into consideration the SLE phase behavior of the system, such as how the solubility changes with temperature. When evaporation or adiabatic cooling is considered, the VLE behavior of the system must be considered as well. For example, using evaporation is not practical if a very low pressure is required to evaporate the solvent at the desired crystallizer temperature.

To get crystals with the desired quality, it is crucial to operate the continuous crystallizer at the desired supersaturation level, which normally corresponds to an operation within the metastable zone. However, at the same time a high recovery of the solute is desirable. In contrast to a batch crystallization process, in which the solution concentration gradually changes over time, there is only one point of operation in a continuous crystallization process. Since the metastable zone is usually quite narrow, recirculation is usually the only way to put this operation point inside the metastable zone, as illustrated in Figure 7.8. If the fresh feed (point F) is cooled to temperature T_c and introduced to a crystallizer, the overall composition would be given by point M, which is located in the unstable zone corresponding to a very high supersaturation level to be released in the crystallizer. But if the mother liquor from the crystallizer (point L) is recirculated and mixed with the fresh feed, the overall feed composition moves to point F'. Cooling to T_c leads to a crystallizer feed given by point M' inside the metastable zone. The location of point M' and thus the supersaturation level in the crystallizer can be controlled by adjusting the recirculation flow rate. This method of supersaturation generation control is known as *liquor recirculation*.

Another mechanism for controlling the supersaturation generation is *magma recirculation*, which is recirculating the entire slurry instead of just the mother liquor. From mass balance point of view, this accomplishes the same objectives as liquor recirculation, but from an operational point of view, there are two key differences between liquor and magma recirculation. The first is that liquor

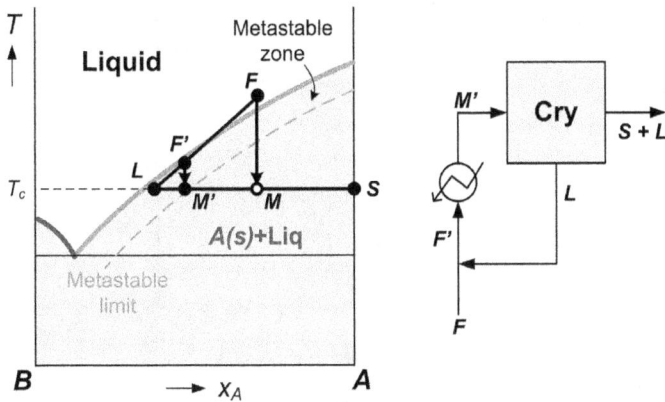

Figure 7.8: Liquor recirculation for controlling supersaturation generation in a continuous crystallizer.

recirculation creates a supersaturated liquid and brings it into contact with growing crystals whereas magma recirculation brings growing crystals into contact with a liquid that becomes supersaturated. Because of this, the PSD resulting from liquor and magma recirculation is usually different. The second difference is that liquor recirculation requires the ability to maintain fluidized classification in the active volume so that a clear liquid can be drawn from the crystallizer. This leads to two different mechanisms of relieving supersaturation in the crystallizer: classified suspension for liquor recirculation and mixed suspension for magma recirculation.

The most common types of industrial continuous crystallizers are illustrated in Figure 7.9. The simplest type is *forced circulation* (Figure 7.9a), in which the mixed suspension serves as the active volume for releasing supersaturation and magma recirculation is employed to control supersaturation generation. The heat exchanger can function as a cooler or a heater. It can be omitted if the mechanism for supersaturation generation is by adiabatic cooling. Conceptually, a forced circulation crystallizer approximates a MSMPR crystallizer, which is equivalent to the continuous stirred tank reactor in reaction engineering.

Figure 7.9b shows the *draft tube baffle* (DTB) crystallizer, which features a draft tube connecting the upper and lower bodies of the crystallizer. An agitator inside the draft tube (not shown) facilitates an upward flow inside the draft tube which induces downward flow outside the tube, thus creating circulation. On the top portion of the lower body, there is a calm region that serves as a settling zone from which a clear solution (and fine particles) is withdrawn and circulated through an outside recirculation loop. Fines are dissolved by passing the recirculation flow through a heater. There is also an elutriation leg at the bottom which can be designed with a suitable diameter to provide an upward flow that prevents smaller

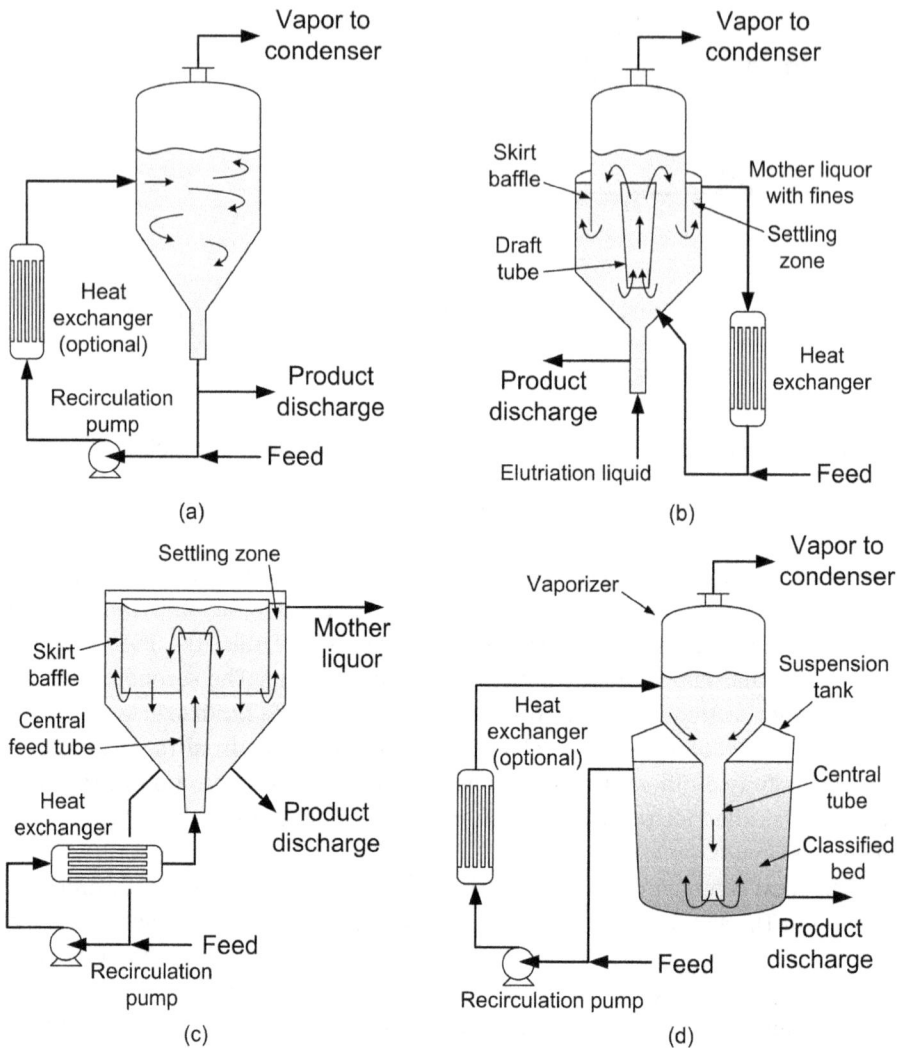

Figure 7.9: Common types of industrial continuous crystallizers: (a) forced circulation, (b) draft tube baffle (DTB), (c) surface-cooled, and (d) classified suspension (Oslo).

particles from moving downwards. As a result, only larger particles would be discharged as product. The elutriation liquid can be drawn from the circulation loop or from the mother liquor separated in the downstream solid-liquid separation unit. The mechanism for controlling supersaturation generation can be considered to be magma circulation although the recirculating magma contains only the fines. Note that fines removal and product classification mechanisms provide additional handles to control the PSD.

The *surface-cooled* type (Figure 7.9c), often called the Swenson–Walker crystallizer, also employs magma recirculation as the mechanism for controlling supersaturation generation. While DTB is normally used for evaporative or adiabatic cooling crystallization, the surface-cooled crystallizer is only used for cooling crystallization. The recirculation flow mixed with fresh feed is cooled in the heat exchanger and enters the crystallizer via a central feed tube. The skirt baffle creates a settling zone near the top of the crystallizer from which the mother liquor is withdrawn.

Figure 7.9d shows a *classified suspension* type, also known as Oslo crystallizer, which uses a liquor recirculation to control supersaturation generation. The solid particles are classified by fluidization in the suspension tank so that clear liquid can be withdrawn from the top and recirculated, while a product slurry containing mainly large particles is taken off near the bottom. If the mechanism for supersaturation generation is by cooling, the vaporizer (top part) is omitted and the recirculation stream is fed directly to the central tube. Other less common types of industrial crystallizers are also available in the market. For example, the scraped surface crystallizer [218], suitable for cooling crystallization from a very viscous liquid, takes the shape of a double pipe heat exchanger. A rotating helical screw inside the pipe continuously scrapes the crystals formed on the walls and pushes them to the outlet point.

In continuous crystallization, product PSD is affected by the supersaturation level and spatial distribution in the crystallizer. In principle, at steady-state condition the supersaturation level in the crystallizer is constant. Of course, in reality, a certain degree of fluctuation is unavoidable as it is virtually impossible to have a truly steady-state operation. Similar to batch crystallizers, actions to improve product quality and avoid PSD-related problems in continuous crystallizers can be classified into equipment-related and operation-related issues. The equipment-related issues include picking the right crystallizer type and configuration, maintaining good circulation, and minimizing "cold spots" and areas of high shear, while operation-related issues include controlling the supersaturation level and utilizing special configurations such as fines dissolution and/or product classification.

The outside circulation loop in a continuous crystallizer also functions as the primary means to introduce or remove heat from the crystallizer. The circulation stream passes through a heat exchanger, which cools (in the case of cooling crystallizer) or heats (in the case of evaporative crystallizer). Supersaturation control in a continuous crystallizer is achieved by controlling the temperature difference (ΔT) across the heat exchanger. Higher circulation through the heat exchanger means lower temperature decrease (or increase) across the exchanger, thus lower supersaturation. As a rule of thumb, the ΔT across the heat exchanger should be limited to 1–2 °C (for cooling) or 1–5 °C (for evaporative) [157]. To achieve this range, a high circulation rate through the heat exchanger is often necessary.

Another issue to consider is the possibility of PSD cycling or oscillation in continuous crystallizers [219]. Excessive fines removal may lead to insufficient crystal surface area to relieve supersaturation, such that a burst of nucleation occurs and the average particle size decreases. The fines are then gradually removed by classification, causing the average particle size to increase until a subsequent rise of supersaturation creates another burst. Controlling the supersaturation level is usually the key to avoid this undesirable behavior.

Example 7.3(I) – Improvement of ammonium sulfate crystallization process

An ammonium sulfate plant consisting of three DTB crystallizers operating in parallel were used to obtain crystals from an aqueous solution [220]. Each crystallizer was operated at a different temperature and vacuum level to allow for multiple-effect evaporation. Crystals from the three crystallizers which have similar PSDs were combined to give the product. The objective was to improve the yield of coarse, round-shaped crystals with a certain target size. Since undersized and oversized crystals were treated as a lower grade product and sold at a cheaper price, it was desirable to produce crystals with a narrow PSD. In addition, it was desirable to minimize PSD oscillations which lead to a decrease in the overall yield.

The DTB was modeled using the generic population balance model (eq. (7.8)) coupled with an overall mass balance around the crystallizers. For the given feed flow rate, ammonium sulfate concentration in the feed, and heat exchanger duty, the vapor flow rate was obtained from mass and energy balances. The crystallizer temperature could then be obtained as a function of the circulation flow rate through the heat exchanger via energy balance. The difference between the solubility of ammonium sulfate at the crystallizer temperature and the actual concentration calculated from the boiling point elevation of the solution gave the supersaturation. Kinetic models (eq. (6.4)–(6.7)) were used to calculate growth and nucleation rates so that the PSD could be calculated. The values of unknown model parameters were assumed such that the calculation results roughly matched actual plant operational data. Simulations using the model were then performed to study the sensitivities of the key operating variables on product PSD.

Figure 7.10a shows the effect of proportionally changing R in all three crystallizers by manipulating the circulation flow rate through the heat exchanger with constant cut size L_f. As the circulation flow rate through the heat exchanger (R) increases, the average particle size increases as more small particles are destroyed and larger particles grow faster. Figure 7.10b depicts the effect of changing Z in all three crystallizers. The cut size, L_c, automatically increases as Z increases, because the higher circulation flow rate through the elutriation leg having a constant diameter returns larger crystals back into the crystallizer. The average particle size decreases as Z increases due to the faster withdrawal of large particles and faster growth on small particles, but this change does not affect supersaturation. Figure 7.10c illustrates the effect of simultaneously changing R and Z by keeping the increase in R proportional to the decrease in Z. Note that the proportional change serves only as a case study as R and Z can be changed independently. Figure 7.10d indicates that fresh steam flow rate slightly affects the PSD. Increasing the fresh steam supply leads to more vaporization which, in turn, causes crystallization of more ammonium sulfate. Based on mass and energy balance calculations, the supersaturation in the crystallizer decreases, leading to an increase in mean particle size.

Based on the simulation results, it was decided to increase the recycle flows in the actual plant. It was found that the average particle size increased from 2.45 to 2.5 mm and the yield of particles larger than 10 mesh in size increased from 70% to about 75%. At the same time, the increase in the

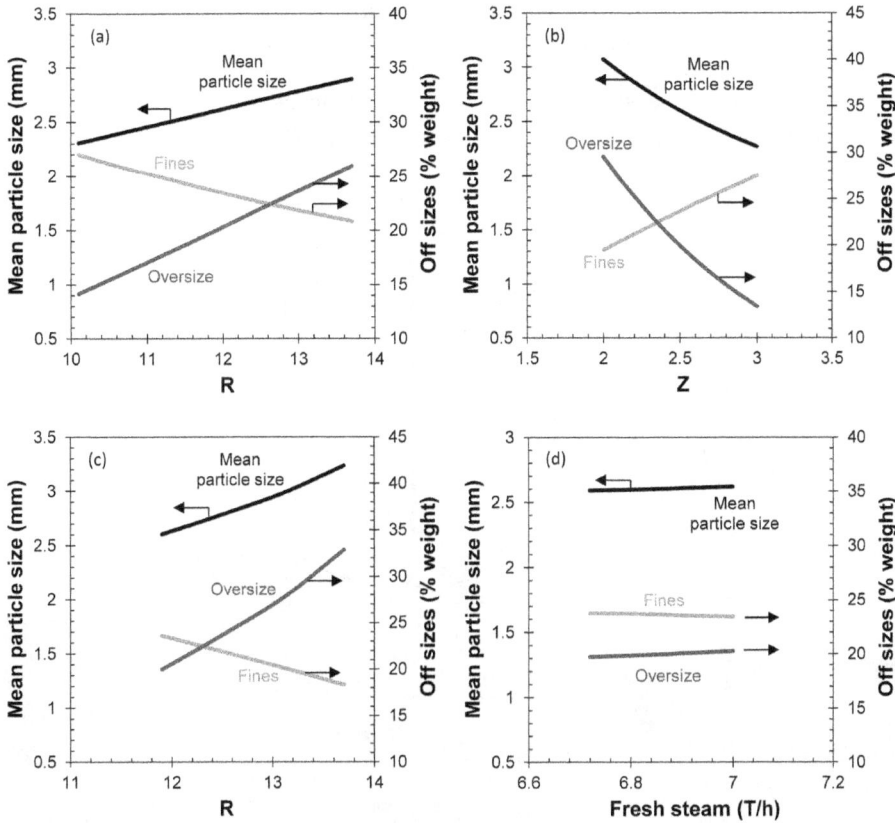

Figure 7.10: Effect of key operating variables on PSD: (a) circulation ratio R, (b) circulation ratio Z, (c) both R and Z, and (d) fresh steam flow rate.

recycle flow through the heat exchanger reduced the supersaturation, which eventually had an effect of calming down the PSD oscillation.

7.2.3 Scale-up consideration in PSD management

For both batch and continuous crystallizers, it is virtually impossible to achieve dynamic similarity in vessels of different sizes used in laboratory-scale experiments, pilot tests, and large-scale production. Heat transfer behavior is also different due to the different volume-to-area ratio despite geometric similarity. All these lead to the crystallizers' reputation of being difficult to scale up [221].

Large-scale crystallizers are often designed based on results obtained in a smaller-scale unit using scale-up rules. Vessel design normally follows geometric similarity rule,

$$\frac{D_1}{D_2} = \frac{W_1}{W_2} = \frac{T_1}{T_2} = \frac{Z_1}{Z_2} = \frac{B_1}{B_2} = \frac{C_1}{C_2} \tag{7.20}$$

where the various dimensions of the vessel are as indicated in Figure 7.11. Selection of agitation speed follows either equal tip speed or equal power per volume rule.

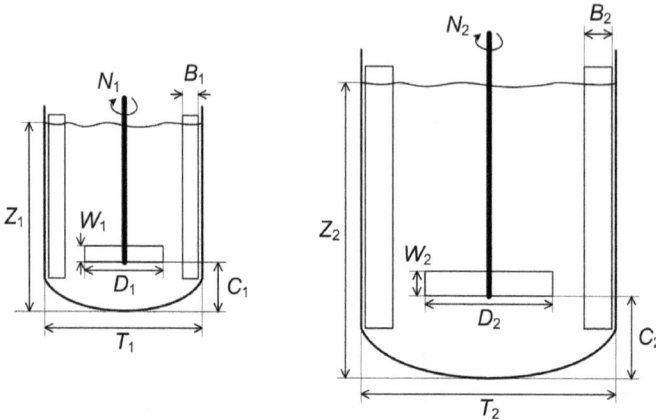

Figure 7.11: Scale up of an agitated vessel.

$$N_1 D_1 = N_2 D_2 \tag{7.21}$$

$$N_1 D_1^{2/3} = N_2 D_2^{2/3} \tag{7.22}$$

where N is the rotational speed of the impeller.

It is also possible to scale-up by keeping the mixing intensity constant, represented by torque per volume,

$$\frac{N_1^2 D_1^5}{V_1} = \frac{N_2^2 D_2^5}{V_2} \tag{7.23}$$

The appropriate choice of which scale-up rule to use depends on both the nature of the material being crystallized and the equipment configuration. The selection of impeller diameter has to account for the necessary tip velocity to produce the required flow to suspend the particles, but at the same time it must consider the required specific power intensity for mixing [221]. In general, the power input P is a function of Reynolds number as well as equipment geometry, [222]

$$N_P = f(N_{Re}, \text{Geometry}) \tag{7.24}$$

$$N_P = \frac{P}{\rho N^3 D^5} \tag{7.25}$$

$$N_{Re} = \frac{\rho N D^2}{\mu} \tag{7.26}$$

Therefore, once N and D are selected, the power input is fixed. Excessive power input per unit volume may cause too much crystal damage and secondary nucleation due to a high shear rate, especially when slurry density is low (2–5% solids content), as crystal-impeller contacts are the most important sources of nuclei [223]. Therefore, if the crystals are fragile, constant power per volume is more appropriate.

It is known that nucleation may not start at the same supersaturation level at different scales. For example, Yi and Myerson [224] observed that nucleation in the larger reactors occurred earlier or at a lower supersaturation level, irrespective of supersaturation generation methods (cooling or antisolvent addition). Because of this, larger crystals were obtained from the larger reactor even the same tip speed and supersaturation generation profile (cooling rate or antisolvent addition rate) was used in vessels of different sizes.

Because of such a complex relationship as well as the different level of homogeneity between the large-scale and small-scale crystallizers, it is usually impossible to get the same PSD upon scale-up. Therefore, the best that can be done is to try to balance the nucleation and growth rates to get a desirable PSD in the large-scale unit. Analysis of various mixing mechanisms and identification of the dominant mechanisms have been attempted to help pick the suitable conditions at different scales. For example, an experimental study of batch reactive crystallization by Torbacke and Rasmuson [225] concluded that the particle size can be correlated fairly well to the ratio of the feeding time to the homogenization time constant.

7.3 Impurity Management

Impurity can be present in the solid products of crystallization for various reasons. One possible reason is the SLE phase behavior. Co-crystallization of impurities can occur if the mother liquor composition is close to the boundaries of both the product and impurity crystallization regions. Solid solution behavior can also be responsible for the existence of impurities in the product crystals. But even if the mother liquor is well within the product crystallization region, impurities that are still dissolved in the mother liquor can be incorporated into the crystals during the crystallization process. Another possibility is that the impurity simply adsorbs on the crystals already formed. Just by looking at the impurity content in the solids, it is impossible to identify whether the presence of impurities is due to the thermodynamic behavior or non-equilibrium effects. However, further analysis can provide important clues. For example, if impurities are present because of liquid incorporation, then the molar ratio of impurity to the crystallization solvent in the crystals should be the same as that in the solution [226].

In general, two types of non-equilibrium impurities can be distinguished by the location of the impurity. The first type is *inclusion impurities* (sometimes referred to as *occlusion impurities*), which are caused by trapping of mother liquor or adsorption into the crystals during growth. Some less common types such as substitutional impurities and impurities incorporated at defect sites can also be considered as inclusion impurities for practical purposes. Inclusion impurities are relatively hard to remove since the crystals have to be partially dissolved or even completely dissolved and recrystallized to release the trapped impurities. The second type is *surface impurities*, which include impurities adsorbed on the crystal surface and imperfect removal of the mother liquor among the crystals. The latter can be removed by proper washing and deliquoring.

7.3.1 Control of inclusion impurities

The amount of inclusion impurities in the product can be controlled in two ways: by preventing it from happening in the first place (or at least keeping it to a minimum) and by removing it after the crystallization step. Minimizing the extent of inclusion during crystallization can be done by manipulating the crystallization conditions such as temperature, residence time, or recirculation rates in the crystallizer. Although the exact mechanism of inclusion has never been elucidated, theories based on experimental observations have been developed since the 1950s. These observations provide the general direction for manipulating the crystallization conditions. For example, faster growth rate leads to increased impurity inclusions [227]; hence, slow growth (which can be achieved at low supersaturation level) can minimize inclusions. The existence of a critical growth rate and a critical crystal size below which no inclusion is formed has also been reported [228, 229]. Higher concentration of solute in mother liquor leads to increased impurity inclusion. This suggests the use of a more dilute feedstock to minimize inclusion.

Sometimes inclusion of impurities simply cannot be avoided and removal after crystallization becomes the only option. The choice of a suitable removal method depends on how deep the impurities are located inside the particles. The impurity profile inside the particles can be determined by gradually dissolving large particles and analyzing a sample taken after each layer has been dissolved. If the profile turns out to be uniform, which indicates that inclusion has started soon after nucleation and continued throughout growth, recrystallization is the only option to remove the inclusion impurities. But if the impurities tend to concentrate on the outer layer of the particles, it means the impurities are either adsorbed onto the surface or inclusion happens only after the particle has reached a critical size. In this case, recrystallization may not be necessary as partial dissolution may be sufficient to remove most of the adsorbed or included impurities. It should also be kept in mind that partial dissolution occurs, to some extent, during washing of the crystals.

Analysis of the impurity content in particles of different sizes (can be classified by sieving) can provide insights into the impurity profile. Results of experimental studies reported in the literature also indicate that the impurity content may vary with particle size [230, 231]. A higher concentration of impurities in larger particles than in smaller particles can be an indication that the impurity inclusion occurs later during the growth process so that the impurities are concentrated in the outer side of the crystals. If this is the case, selective dissolution of large particles would be an effective way to remove inclusion impurities.

Models of impurity inclusion during crystallization can be incorporated in the overall model of the crystallization process to quantify the impact of inclusion impurities on both the process performance and product quality. Commonly used models are summarized in Table 7.6. Theoretical models relate the concentration of impurities in the crystals to that in the liquid (melt or solution) using the distribution coefficient,

Table 7.6: Summary of impurity inclusion models.

Theoretical models			
Burton–Prim–Slichter [232]	$K_{eff} = \dfrac{K_0}{K_0 + (1 - K_0)\exp(-\delta G/D)}$	(7.28)	K_0 = value of K at equilibrium G = crystal growth rate D = diffusivity of impurity in solvent δ = length of diffusion boundary layer
Sangwal-Pałcyńska [233]	$K_{eff} = K_0 + \dfrac{B(\Delta c)^{1+n_2}}{K_{imp}\, c_{imp,L}}$ $B = \dfrac{1}{19}\dfrac{\gamma_\ell}{kT}(1 - n_2)\exp\left(\dfrac{W_0}{kT}\right)$	(7.29) (7.30)	Δc = supersaturation K_{imp} = constant in Langmuir adsorption isotherm $c_{imp,L}$ = impurity concentration in liquid γ_ℓ = linear edge free energy k = Boltzmann constant T = temperature W_0 = activation energy for crystal growth n_2 = parameter
Empirical models			
Power law [234]	$c_{imp,S} = k_{imp}\left[d_p H(d_p - d_{p,ic})\right]^m [GH(G - G_{ic})]^n$	(7.31)	d_p = particle size $d_{p,ic}$ = critical particle size G = growth rate G_{ic} = critical growth rate H = Heaviside function k_{imp}, m, n = empirical parameters

$$K = \frac{c_{imp,S}}{c_{imp,L}} \qquad (7.27)$$

where the indices S and L refer to solid and liquid, respectively. The Burton–Prim–Slichter model (eq. (7.28)) is based on theoretical consideration of crystallization from the melt, which includes the effect of diffusion of impurity molecules from the liquid to the crystal surface. When the growth rate approaches zero, the value of K_{eff} approaches K_0, the equilibrium value. The Sangwal–Pałcyńska model (eq. (7.29)) describes the effective distribution coefficient as a function of supersaturation, assuming that the impurity concentration in the crystals is relatively small. The value of K_{eff} also approaches K_0 at equilibrium, at which point the supersaturation is zero. Alternatively, an empirical correlation such as a power-law expression can be developed based on experimental observations. For example, eq. (7.31) directly correlates the concentration of included impurities to particle size and growth rate. It accounts for the possibility of having no impurity inclusion when the particle size and growth rate are below certain critical values using the Heaviside function.

In practice, it is impossible to obtain the parameters of the theoretical models *a priori*. For this reason, some parameters are often lumped and determined from experimental data. For example, δ/D in eq. (7.28) can be considered as a single model parameter and determined from the experimental data of K_{eff} at different values of G. The values of B/K_{imp} and n_2 in eq. (7.29) can be obtained by plotting experimental values of K_{eff} against Δc.

Example 7.4 – Vitamin C crystallization

Ascorbic acid (AsA), also known as vitamin C, is typically produced by fermentation from D-glucose or sorbose, or by a catalytic reaction from 2-keto-L-gulonic acid [15]. The reactor effluent contains the product (AsA), solvent (50/50 wt% ethanol/water), as well as some intermediates and byproducts. AsA is recovered from this reactor effluent by cooling crystallization. Cheng et al. [231] studied the inclusion of two major impurities, oxalic acid (OxA) and 2-furaldehyde (2-Fur), during cooling crystallization of AsA.

A set of experiments were carried out with the same initial composition and temperature, but with different final crystallization temperatures and cooling rates. As indicated in Figure 7.12, the feed composition is located in the AsA region even at the lowest final crystallization temperature (8 °C). Therefore, no OxA can co-crystallize. The feed mixture contains the same amount of OxA and 2-Fur and the composition shown on the SLE phase diagram is the normalized composition without 2-Fur. Due to its very low melting point (−38 °C), 2-Fur cannot crystallize out at the crystallization temperature range. Therefore, any OxA or 2-Fur present inside the solid particles is incorporated by inclusion instead of co-crystallization.

In each experiment, the mixture is first heated until all solids are dissolved and then cooled to 2 °C below the dissolution temperature prior to adding some seeds. Cooling is continued at the desired cooling rate until the specified final crystallization temperature is reached and the mixture is kept overnight under constant stirring to achieve equilibrium. The solids are filtered out and washed thoroughly to remove surface impurities before being analyzed. The results are depicted in

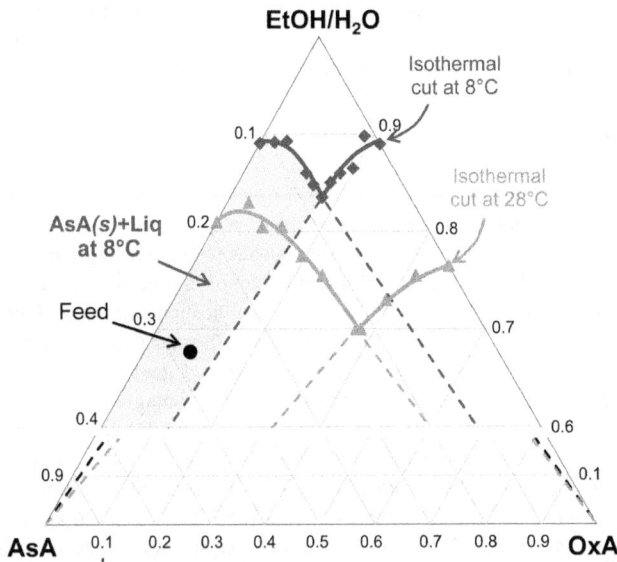

Figure 7.12: SLE phase diagram of ascorbic acid/oxalic acid/solvent system.

Figure 7.13, showing the OxA and 2-Fur content in the solids as a function of crystallization yield and cooling rate. In the 5 runs shown in Figure 7.13a, the cooling rate is fixed at 0.5 °C/min, while in the 6 runs shown in Figure 7.13b, the final crystallization temperature is fixed at 28.3 °C, giving an AsA crystallization yield of about 40%.

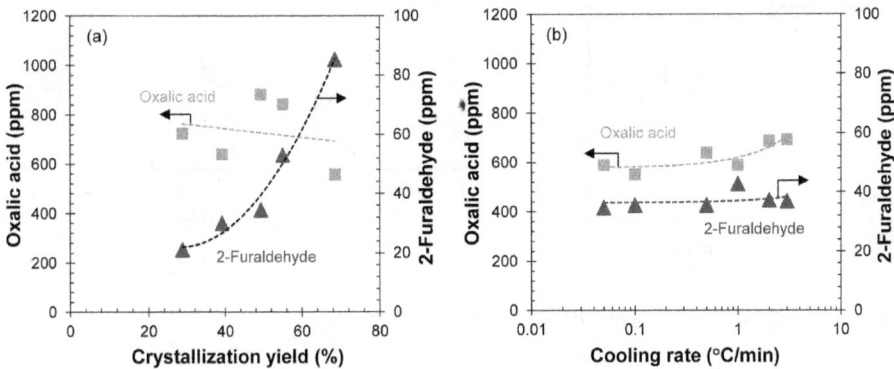

Figure 7.13: Inclusion concentration of oxalic acid and 2-furaldehyde in ascorbic acid crystallization process: (a) inclusion vs. crystallization yield and (b) inclusion versus cooling rate.

From the results, it can be observed that 2-Fur inclusion strongly depends on crystallization yield, while OxA does not show a clear trend. Faster cooling rate leads to higher inclusion of both OxA and 2-Fur, but the effect is very weak. The results suggest that 2-Fur is likely to be incorporated by adsorption during crystal growth so that the only way to minimize its inclusion is by lowering the crystallization yield. On the other hand, the OxA content is found to correlate well with the amount of fines [231], suggesting that it can be minimized by reducing the amount of fine particles in the product.

Example 7.5(I) – Yield improvement in monomer crystallization
A company is producing monomer *P* via a continuous reaction in aqueous phase, which produces a solution containing 70 wt% *P* and a small amount of impurities. The solution is sent to a crystalliza-tion section which includes a clarification tank, a crystallizer, and a purification column, as illus-trated in Figure 7.14. The feed enters the clarification tank at 30 °C. This tank is kept at the saturation temperature of 70 wt% solution (22 °C). The overflow from this tank goes to the crystal-lizer operated at 17 °C, which produces crystals of *P* and a mother liquor containing 40 wt% *P*. The mother liquor is sent to another unit for further processing. The crystals, which also contain some impurities, are sent back to the clarification tank, where they are put in contact with the incoming feed from the reactor as a means of washing. The crystals from the clarification tank are sent to a purification column. The mass flow rates of the components in different streams per 100 kg/h of feed solution are given in Figure 7.14. From mass balance, 50 kg/h of *P* crystals is produced. Note that the residual mother liquor in the crystals is assumed to be zero.

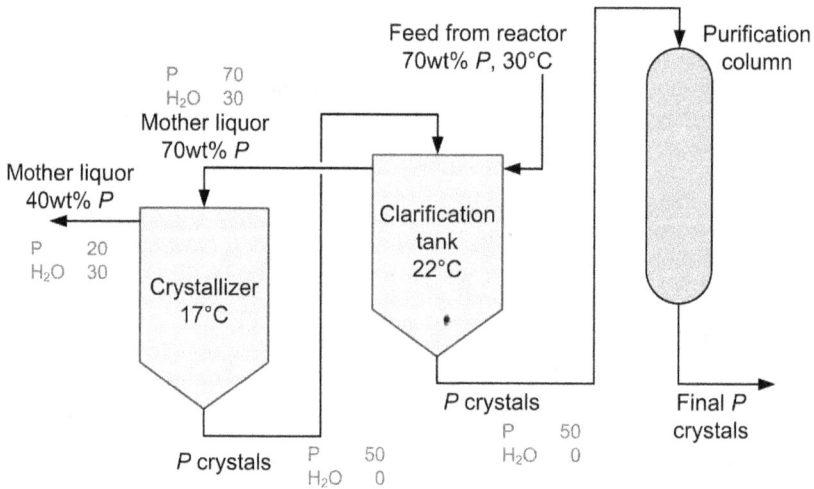

Figure 7.14: Schematic of monomer *P* crystallization process.

To increase the monomer production capacity, the company considered the idea of decreasing the crystallization temperature so as to increase the yield of crystals. Considering the temperature of the available coolant at the plant site, it was proposed to reduce the crystallization temperature to 5 °C, at which point the solubility of *P* is 20 wt%. By mass balance, 100 kg/h of feed solution would produce 62.5 kg/h of *P* crystals, corresponding to a 25% increase in yield. However, a plant test revealed that when the crystallizer temperature was decreased, the impurity content in the crystals became too high and could not be satisfactorily reduced to an acceptable level in the purification column.

This problem can be understood by realizing that producing 25% more crystals in a vessel of the same size would require a faster growth rate, which also corresponds to a higher supersaturation level in the crystallizer. This, in turn, leads to more inclusion of impurities in the product crystals. One alternative to resolve this problem would be to increase the residence time by using a bigger vessel or adding another vessel in parallel to decrease the supersaturation level and growth rate. But a simpler alternative is to install an additional crystallizer in series to the original crystallizer, as shown in Figure 7.15. The mother liquor from the original crystallizer containing 40 wt% P would go to the additional crystallizer operating at 5 °C, which should be designed with enough volume to produce 12.5 kg/h of P crystals at a sufficiently low growth rate such that the impurity content in the crystals is at an acceptable level. Crystallization experiments at 5 °C would be needed to determine the suitable growth rate and crystallizer volume.

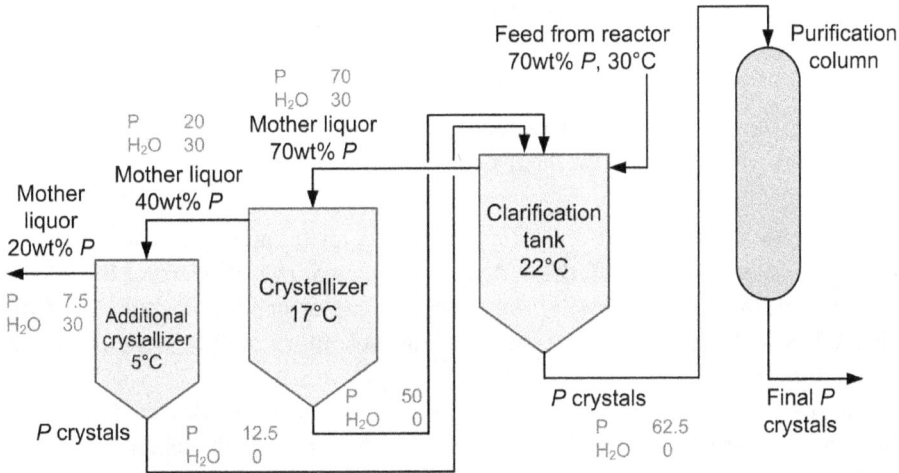

Figure 7.15: Additional crystallizer increases crystallization yield without increasing supersaturation level.

7.3.2 Removal of inclusion impurities by melt crystallization

Melt crystallization is a potentially suitable option for purification of a product with a relatively low melting point. As the name suggests, crystals are formed from the melt, so there is no need for a solvent. The operating temperature is higher compared to slurry crystallization and the viscosity of the melt is normally much higher than that of a solution. However, the equipment tends to be compact and since the final product is also collected as a melt, no additional solid-liquid separation unit is necessary. Figure 7.16 depicts two common types of melt crystallizer along with an illustration of the basic principle of melt crystallization [235, 236]. The first type is the static melt crystallizer, Figure 7.16a, which consists of a

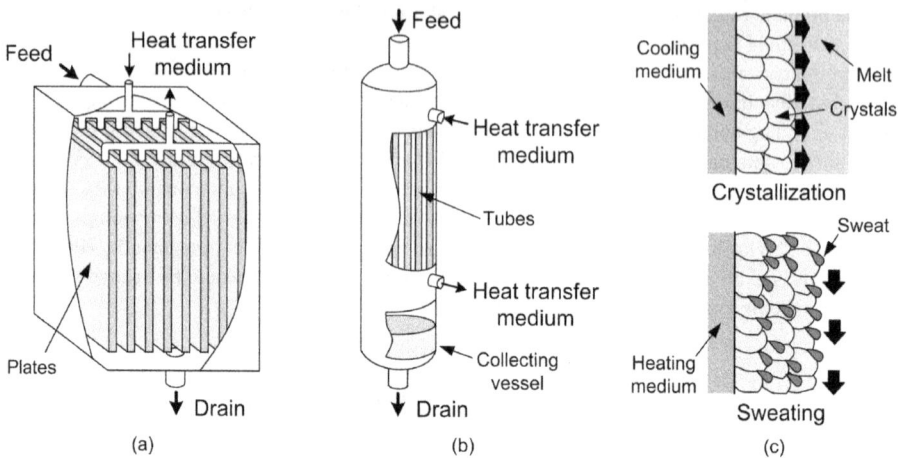

Figure 7.16: Melt crystallization: (a) static melt crystallizer, (b) falling film crystallizer, and (c) illustration of crystallization and sweating processes.

box filled with parallel plates. Cooling or heating medium flows inside the plates and crystallization occurs on the surface of the plates. The second type is the falling film crystallizer, Figure 7.16b, in which cooling or heating medium flows inside tubes and crystallization occurs as the melt flows outside the tubes as a falling film.

In the crystallization step, a feed that contains some impurities are introduced into the crystallizer and a cooling medium is used to decrease the temperature. In a static melt crystallizer, the feed fills the box until the plates are completely immersed and crystallization occurs on the surface of the plates as the melt remains static. At the end of this step, the remaining melt (mother liquor) is drained from the crystallizer. In a falling film crystallizer, the feed flows on the outside of the tubes and crystallization occurs on the surface of the tubes as the melt flows down as a falling film. Any melt that has not turned into solids by the time it reaches the end of the tube is collected at the bottom of the crystallizer and drained at the end of this step.

Impurity removal is achieved in the crystallization step as shown in the top figure of Figure 7.16c. The representation of the crystallization step on the SLE phase diagram is shown in Figure 7.17a. During crystallization, some impurities are included in the solids, resulting in a solid composition that lies inside the solid-liquid region even if the system exhibits simple eutectic behavior. The locus of actual solid compositions resembles the solidus curve in a solid solution-forming system (Figure 2.6), but the location depends on kinetics and mass transfer issues instead of thermodynamics. For example, a different cooling rate would result in a different inclusion level in the solids. In any case, the concentration of the impurities in the crystals is lower than that in the melt.

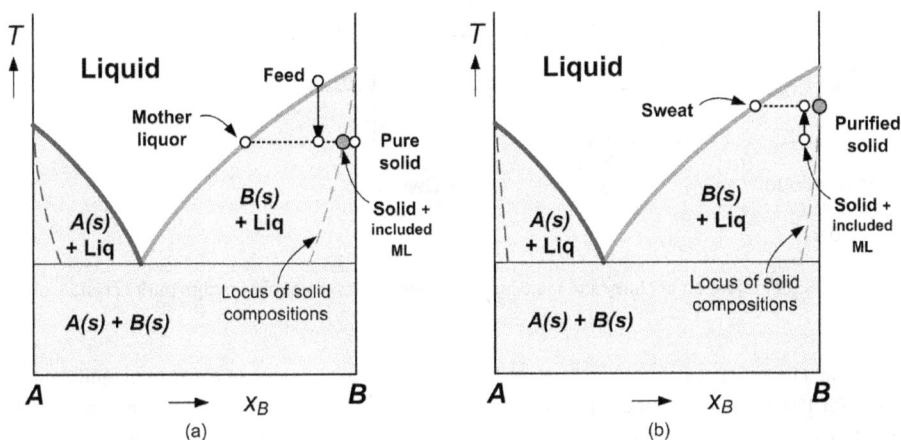

Figure 7.17: Representation of melt crystallization steps on SLE phase diagram: (a) crystallization and (b) sweating.

Additional impurity removal is achieved in the sweating step as shown in the bottom figure of Figure 7.16c. In the sweating step, a heating medium flows through the plates (in a static melt crystallizer) or tubes (in a falling film crystallizer), causing the crystals to partially melt. The formed liquid or sweat flows over the crystal layer to the bottom of the crystallizer. Figure 7.17b shows the representation of the sweating step on the SLE phase diagram. As the temperature is increased, the crystals release a mother liquor (sweat) having a higher impurity concentration, leaving behind a solid phase with lower impurity content. The locus of solid compositions has a shape similar to the one for crystallization step, but in general they do not coincide. The location of this solid composition locus depends on the heating rate during the sweating step. After sweating is completed, the temperature of the plates or tubes is further increased to completely melt the remaining crystals and obtain the final product as a liquid.

Similar to a solid solution system, melt crystallization process can be arranged as a multistage operation to reach the desired purity. Mother liquor and/or sweat from any stage can be recycled to another stage as an effort to increase the overall recovery.

Example 7.6(I) – Purification of a hydrocarbon product
One of the final steps of the hybrid process discussed in Example 4.10 is crystallization of chemical *D* from a mixture of *C* and *D*. In practice, a melt crystallization process as illustrated in Figure 7.18 can be used to obtain crystals of *D* with the desired purity. The feed containing 60 wt% *D* is first cooled down in the crystallization step. The mother liquor is drained, leaving crude crystals which contain *C* as inclusion impurities at the end of this step. These crystals are then slowly heated in the sweating step. The purified crystals are then completely melted and obtained as a liquid product. The collected sweat is sent to the next batch and mixed with a fresh feed.

To confirm the feasibility of this process, a laboratory-scale experiment was conducted. The corresponding steps are illustrated on the phase diagram in Figure 7.19. A mixture of *C* and *D* (Feed) is

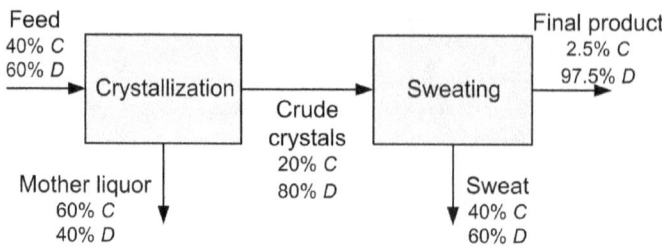

Figure 7.18: Combination of slurry and melt crystallization processes to obtain high purity crystals of *D*.

placed in a jacketed vessel, cooled from 45 °C to 20 °C at a cooling rate of 0.1 °C/min, and held at 20 °C for 120 min. The obtained crystals are then separated from the mother liquor by vacuum filtration and analyzed. As indicated in Figure 7.19a, these crystals are not pure, since they contain inclusions. The mother liquor composition is located close to the saturation curve of *D*, indicating that the mixture almost achieves equilibrium. A fraction of the crystals is then placed on a filter funnel equipped with a jacket to control the temperature. The funnel is heated in a stepwise manner to cause sweating and the temperature of the crystals is measured using a thermocouple. After each heating step, the sweat is collected as filtrate and the remaining crystals on the filter are sampled before being further heated in the next step. As a result of the sweating process, the crystals gradually become richer in *D*, as illustrated in Figure 7.19b.

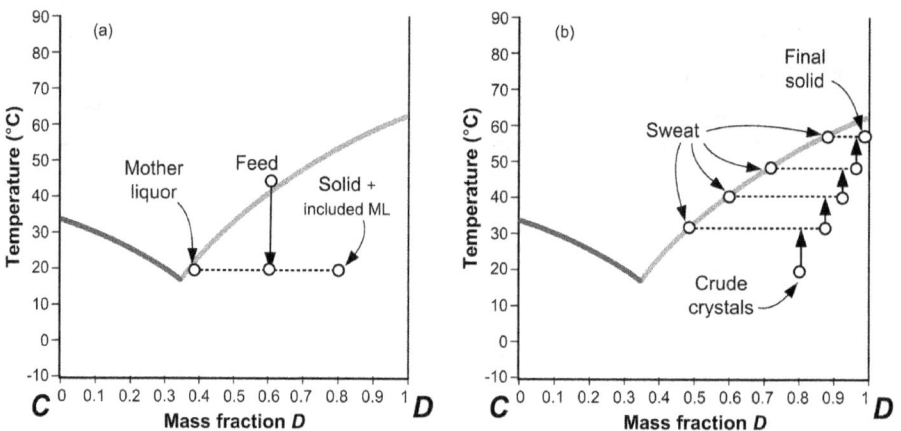

Figure 7.19: Representation of melt crystallization steps on SLE phase diagram: (a) crystallization and (b) sweating.

The analysis results of the crystals after each heating step is summarized in Table 7.7. The data suggest that the purity gradually increases as the temperature increases. The final crystals obtained after the fifth heating step has a purity of 97.5 wt%, which is considered satisfactory. While the required cooling and heating rate in the actual melt crystallization process still needs to be determined in a bench or pilot scale test, the results of the laboratory experiment confirm the technical feasibility of using crystallization to obtain product *D* with the desired purity.

Table 7.7: Crystal purity profile during sweating experiment.

Step	Temperature (°C)	D in solids (wt%)
0	20.0	81.47
1	36.5	86.90
2	39.2	no data
3	48.8	93.82
4	57.8	96.60
5	60.7	97.46

Example 7.7(I) – Purification of a monomer intermediate

Chemical A, which is an intermediate for producing a key monomer, is to be purified using a melt crystallization process. The main objective is to reduce the content of key impurities B and C to a very low level. The feed stream going to the melt crystallization step contains 99.42% A, 0.32% B, and 0.26% C, and the target is to obtain a final product containing less than 0.1 wt% C, with at least 60% overall recovery.

Table 7.8 summarizes the results of a plant test with 50,000 kg of feed using an existing melt crystallizer with somewhat optimized cooling and heating rates. The test was conducted in two stages as it was not possible to achieve the desired target purity and recovery at the same time using a single stage. The drained mother liquor and sweat were collected separately and their amount was measured. In the first stage, the sweat was collected in two fractions. The first fraction (Sweat 1) consists of the first 12,500 kg of collected sweat. All the sweat produced afterwards was collected separately as a second fraction (Sweat 2). The sweating process was ended when the total amount of drained materials (mother liquor + Sweat 1 + Sweat 2) reached 20,000 kg, corresponding to a 60% recovery. The test results clearly indicated that sweating was very effective in purging impurities, especially the initial fraction which had an even higher content of impurities than the mother liquor. As the product became gradually purer towards the end of the sweating step, the later sweat contained fewer impurities. Figure 7.20 illustrates this recycling concept along with the key mass balance numbers.

Table 7.8: Compositions from test results.

Stage 1	Composition (wt %)			Stage 2	Composition (wt %)		
	A	B	C		A	B	C
Feed	99.42	0.32	0.26	Feed	99.73	0.13	0.14
Mother liquor	98.84	0.69	0.47	Mother liquor	99.57	0.22	0.21
Sweat 1	98.64	0.79	0.57	Sweat 1	99.67	0.15	0.18
Sweat 2	99.36	0.35	0.29	Sweat 2	–	–	–
Product	99.79	0.09	0.12	Product	99.84	0.07	0.09

Since the product from Stage 1 contained more than 0.1 wt% C, it was processed further in a second stage. A total of 9,060 kg of mother liquor and sweat was collected in this stage,

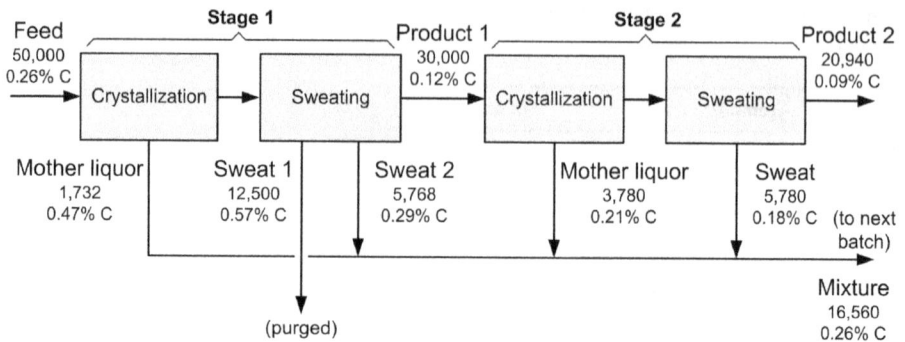

Figure 7.20: Two-stage melt crystallization process with recycle.

corresponding to a recovery of 69.8%. The final product met the desired specifications, but the overall recovery of the process would be only $(50,000-20,000 - 9,060)/50,000 = 41.9\%$. To increase the overall recovery, recycling of some fractions should be considered. Since Sweat 1 from the first stage was the most concentrated in impurities, it would be a good choice to use as a purge stream. Meanwhile, the combined composition of the other fractions is very similar to the fresh feed, so those fractions can be mixed with a fresh feed in the next batch. With the recycle, the fresh feed for the next batch is only $50,000-16,560 = 33,440$ kg, giving an effective overall yield of $20,940/33,440 = 62.6\%$, which is slightly above the target of 60%.

7.3.3 Crystallization downstream processing system

In addition to inclusion impurities, impurities can be adsorbed on the crystal surfaces or dissolved in the residual mother liquor trapped between the product crystals after solid-liquid separation. All of these impurities can be effectively managed by properly designing the downstream processing system, which includes filtration, washing, deliquoring, and drying, to ensure that the product crystals meet the impurity specifications [237]. A generic structure of the downstream processing system is depicted in Figure 7.21. The crystals produced in the crystallizer are separated from the mother liquor in a filter or centrifuge and the residual mother liquor on the crystals is removed in the deliquoring step before the crystals are sent to the dryer. A washing step prior to deliquoring is needed if the removal of residual mother liquor (that contains surface impurities) by deliquoring alone is insufficient. Washing is also necessary if the crystals are to be sent to another processing step in which a different solvent is used (the reason for this is not so much about impurities but rather about minimizing the amount of the original solvent going to the next step). Dissolution and recrystallization steps are necessary if the inclusion impurities exceed the purity specifications or if the product attributes such as the PSD do not meet the target. An adsorbent is often added during the dissolution process in

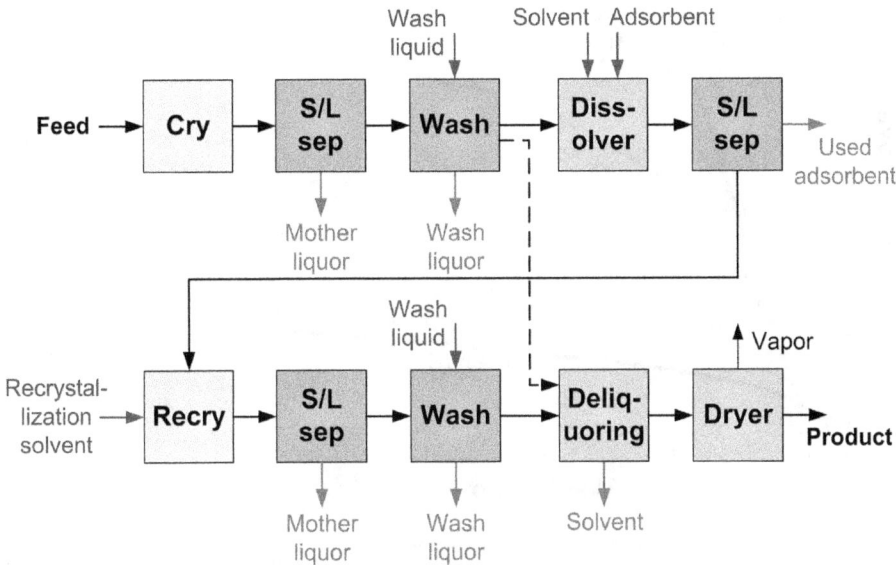

Figure 7.21: Generic structure of crystallization downstream processing system.

an effort to remove color-forming impurities, which are sometimes hard to remove by crystallization alone.

Washing can be done by two methods. In *displacement washing*, clean wash liquid is introduced to replace the dirty liquid that is being sucked out at the same time. This is usually performed in a filter such as a rotary vacuum filter (RVF) or bench filter, or a centrifuge. The wash liquid is normally sprinkled over the wet cake being filtered. In *reslurry washing*, the clean wash liquid is mixed with dirty residual liquid in a dilution tank and the resulting slurry is sent to a filter or centrifuge to separate the crystals. Significant dissolution may occur during reslurrying due to the presence of a relatively large amount of wash liquid. In general, reslurry washing that works by dilution is less efficient compared to displacement washing.

Washing performance can be quantified by the washing efficiency, represented by the fraction removal of solute from the wet cake as a function of the *wash ratio*, which is defined as the ratio of the amount of wash liquid to the initial amount of residual liquid in the cake. For displacement washing, the washing efficiency is a function of the amount of wash liquid used as well as cake and liquid properties. The typical relationship is shown in Figure 7.22. The parameter D_n is the dispersion number, which depends on cake and liquid properties such as particle size, porosity, and viscosity (eq. (7.34)). A higher value of D_n leads to a more effective washing. At very large values of D_n, the washing behavior approaches the ideal case of a perfect displacement, where the wash liquor displaces exactly the same amount of dirty residual liquid from the cake.

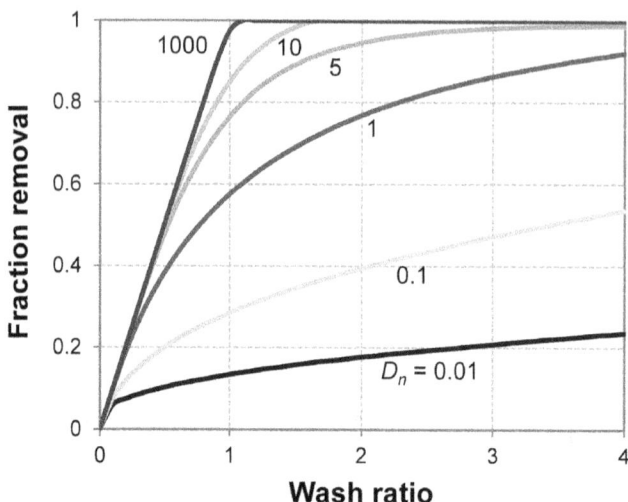

Figure 7.22: Typical displacement washing performance [238].

A higher washing ratio leads to better impurity removal, with an asymptotic approach towards complete removal at infinite wash ratio. Clearly, there is a tradeoff because while a higher washing ratio means more impurity removal, it also leads to higher loading to the recovery system that processes the wash liquor. Another tradeoff concerns the selection of washing liquid. The selection of a washing liquid in which the product has a higher solubility generally leads to better impurity removal due to partial dissolution effect, but obviously more product is lost through dissolution.

Although deliquoring appears as a separate block in Figure 7.21. It typically occurs at the same equipment as solid/liquid separation and washing. Conceptually, two types of deliquoring can be distinguished, as illustrated in Figure 7.23. *Compaction*, which involves rearrangement of solid particles in the cake as residual liquid leaves the cake, results in a decrease in cake porosity while the cake remains saturated; that is, the interstitial space among the particles is completely filled with liquid. *Desaturation*

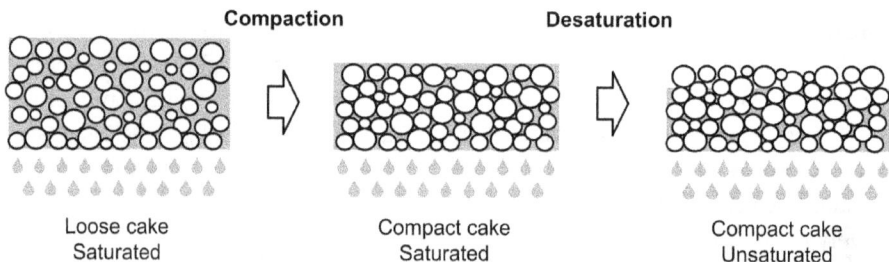

Figure 7.23: Two types of deliquoring.

involves partial draining of the residual liquid from the interstitial space as the cake porosity remains constant, leaving gaps that are subsequently filled with gas.

Models can be used to evaluate the performance of the solid/liquid separation, washing, and deliquoring units, so as to determine the final impurity content in the product. Selected models for batch operation and a continuous belt filter are summarized in Table 7.9. Similar expressions can be derived from models reported in the literature for other equipment types such as a rotary vacuum filter [239] or centrifugal filter [240]. By tying together those models with overall mass balance, the

Table 7.9: Summary of selected models for downstream processing units.

Filtration

Filtration time (batch filtration, constant pressure)	$t_f = \dfrac{\mu \alpha c V_f^2}{2 A_f^2 \Delta p} + \dfrac{\mu R_m V_f}{A_f \Delta p}$	(7.32)	t_f = filtration time μ = filtrate viscosity V_f = cumulative filtrate volume α = specific cake resistance c = mass of deposited solids per unit volume of filtrate A_f = filter area Δp = pressure drop R_m = filter medium resistance
Filtrate flow rate (continuous belt filter, constant pressure)	$q_F = \dfrac{vw}{\alpha c}\left(-R_m + \sqrt{R_m^2 + \dfrac{2\alpha c \Delta p L_F}{\mu v}}\right)$	(7.33)	q_F = filtrate flow rate v = linear speed of moving belt w = belt width L_F = length of belt section dedicated to filtration

Washing

Displacement washing performance [238]	$f = f(W, D_n)$	(7.34)	f = fraction removal W = wash ratio (mass of wash liquid to mass of residual liquid in wet cake) D_n = dispersion number
	$D_n = \dfrac{ux}{\varepsilon D_L}$	(7.35)	u = filtrate linear velocity x = cake thickness ε = cake porosity D_L = axial diffusivity
Length of belt section dedicated to washing (continuous belt filter)	$L_W = \dfrac{\mu q_w}{w \Delta p}\sqrt{R_m^2 + \dfrac{2\alpha c \Delta p L_F}{\mu v}}$	(7.36)	L_W = length of belt section dedicated to washing q_w = wash liquid flow rate

Table 7.9 (continued)

Deliquoring		
Required deliquoring time (compaction)	$t_D = -\displaystyle\int_{\varepsilon_0}^{\varepsilon_f} \dfrac{\mu\left[\rho_S(1-\varepsilon)\alpha x + R_m\right] x_0(1-\varepsilon_0)}{\Delta p(1-\varepsilon)^2}\, d\varepsilon$ (7.37)	t_D = deliquoring time ε_0 = cake porosity before compaction ε_f = cake porosity after compaction
Required deliquoring time (desaturation of vacuum-drained cake) [241]	$t_D = c_1 \dfrac{\mu\varepsilon(1-S_\infty)x^2}{k\Delta p}\left(\dfrac{1-S}{S-S_\infty}\right)^{c_2}$ (7.38) $S_\infty = 0.155\left(1+0.031 N_{Ca}^{-0.49}\right)$ (7.39) $N_{Ca} = \dfrac{d_p^2\varepsilon^3(\rho_L g x + \Delta p)}{(1-\varepsilon)^2 \sigma x}$ (7.40)	S = saturation S_∞ = irreducible saturation k = cake permeability c_1, c_2 = parameters N_{Ca} = capillary number d_p = particle size g = gravitational acceleration σ = surface tension
Length of belt section dedicated to deliquoring (continuous belt filter)	$L_D = t_D v$ (7.41)	L_D = length of belt section dedicated to deliquoring

impurity amount at different locations can be readily tracked, thus allowing for evaluation of the effect of equipment design parameters such as filter dimensions and fractions of filter area dedicated to filtration, washing, and deliquoring, as well as operating variables such as moving speed, pressure drop, wash ratio, and so on, on the impurity amount in the final product [231].

Again, *a priori* determination of the parameter values for these models is not easy. While typical values of some parameters are available in the literature [238], it is much more practical to use empirical correlations for the specific system in hand. For example, the dependence of washing performance on wash ratio can be experimentally determined instead of predicted with calculations based on the value of D_n, which is hard to estimate. It is also much easier to measure the required deliquoring time to achieve the desired cake wetness rather than determine the parameters for calculating N_{Ca}. The specific cake resistance α and filter medium resistance R_m can be obtained from a laboratory-scale batch filtration experiment.

Example 7.8(I) – Washing in precipitated calcium carbonate process
A process for producing precipitated calcium carbonate from waste materials such as steel con-
verter slag or fly ash using a recyclable solvent has been proposed [242, 243]. The key step in such
a process is the selective dissolution of calcium from the waste material using a solution of a proprietary salt LX. Here, L is the cation and X the anion. A simplified diagram of the process is shown
in Figure 7.24. The feed containing CaO is contacted with a concentrated solution of LX so that CaO
dissolves in water by forming $Ca(OH)_2$ which is then converted to CaX_2. The presence of LX enhances the solubility of calcium hydroxide because it reacts with OH^- to form a weak base LOH. Other
oxides in the feed are also converted to hydroxides but those hydroxides are only sparingly soluble
in water. The solution is separated from the residual solids by filtration and sent to the carbonation
unit, where CO_2 is introduced to precipitate the calcium as $CaCO_3$. The precipitated product is filtered and washed, and the mother liquor containing the recovered salt solution is concentrated
and recycled.

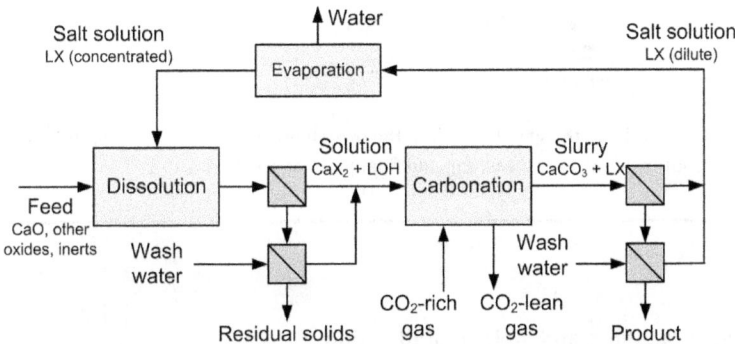

Figure 7.24: Simplified process diagram of mineral ion recovery process using salt solution.

Washing of the residual solids after filtration is crucial because the amount of the expensive proprietary salt going to the waste residual solids needs to be minimized. Obviously, extensive washing using a large amount of water would recover much of the salt, but this would lead to dilution of
the resulting solution, which means that more water would have to be evaporated from the recycle
stream. For this reason, a series of tests were performed in the laboratory to obtain the washing
performance. The filtered residual solids from the dissolution step were washed with different
amounts of water, and the salt concentration after each wash was measured to determine the fraction removal. A similar test was also performed for the final product washing, which minimizes the
amount of LX in the precipitated $CaCO_3$ product. The data (not shown) led to fraction removal versus wash ratio curves that fall somewhere in between the curves for $D_n = 1$ and $D_n = 5$ in
Figure 7.22.
   The information about washing performance is very useful in evaluating the tradeoff between
the incurred cost due to the loss of proprietary salt to the residual solids and final product and the
cost for evaporating water from the recovered solution. Figure 7.25 shows the calculated variable
cost for the entire process, taking into account utility consumption such as steam for the evaporator, water for washing and cooling tower makeup, and electricity for moving parts, as well as fresh

salt to make up for the losses. The results clearly show that a higher wash ratio leads to a decrease in the portion of variable cost attributed to salt makeup, while at the same time causes an increase in the steam cost for the evaporator. The tradeoff leads to an optimum wash ratio of about 2.5.

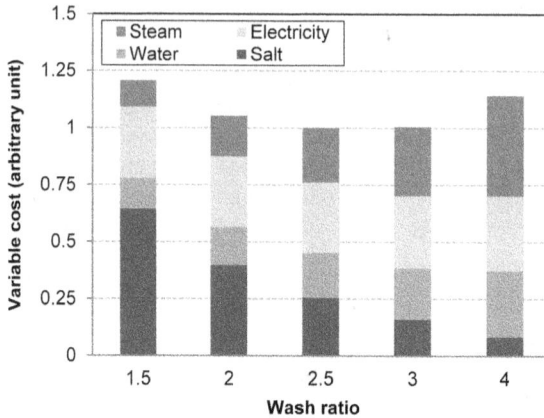

Figure 7.25: Effect of wash ratio on the variable cost of the overall process. For each case, the same wash ratio is used in both residual solids washing and final product washing.

7.3.4 Simultaneous washing and melt crystallization

Removal of surface impurities by washing is also applicable for crystallization from the melt. In addition to batch process discussed in the previous section, melt crystallization can also be conducted in a continuous process using a scraped surface crystallizer [244]. An example of commercial application of this technology is the Badger/Niro process for p-xylene crystallization [245]. High purity crystals are produced in suspension, allowing for a much larger surface area for growth compared to a static melt or falling film crystallizer. The slurry is then pumped into a wash column where the crystals are separated by gravity-buoyancy action and washed using a fraction of the melted final product. A schematic of the process is shown in Figure 7.26a. The wash column can be viewed as comprising of multiple stages with solids and liquid flowing in opposite directions from one stage to the next, as depicted in Figure 7.26b. In practice, a solid–liquid separator such as a belt filter or a centrifuge may be installed after the crystallizer to remove the bulk portion of the mother liquor so that only a relatively small amount of residual mother liquor enters the wash column along with the crystals.

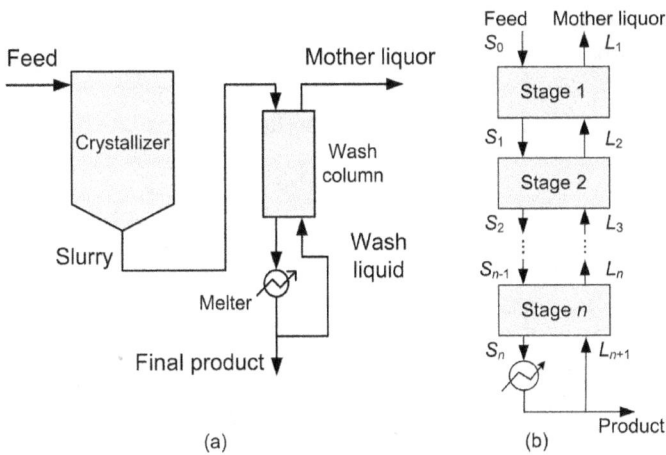

Figure 7.26: Continuous melt crystallization process with wash column: (a) process flow diagram and (b) stages in the wash column.

An alternative design that involves moving the crystals upwards by a pushing mechanism, while letting the wash liquid flow downwards by gravity is also commercially available. For example, the Kureha Crystal Purifier uses a screw conveyor to lift the crystals [246, 247]. In any case, the use of melted product as washing liquid leads to a temperature gradient in the column, with the product end (lower impurity content) having a higher temperature than the feed/mother liquor end (higher impurity content). Sweating occurs simultaneously with washing, which means both surface and inclusion impurities are removed during this process.

In the design and analysis of a wash column, it is possible to use an approach similar to the McCabe-Thiele method for binary distillation [248]. Analogous to the original method, the validity of this approach is subject to the *constant molar overflow* assumption, which requires that (1) the molar heats of crystallization of the feed components are equal; (2) for every mole of liquid crystallized, a mole of solid is melted; and (3) heats of mixing are negligible. Although not all of these conditions are satisfied in reality, such an approach would nonetheless be useful for a preliminary analysis, especially during conceptual design of the process. One key difference with the original McCabe-Thiele method is that this approach considers a pseudo-equilibrium between the solid and liquid phases – which depends on kinetics and mass transfer issues – rather than the true solid-liquid equilibrium as dictated by thermodynamics. That is, the operating line in the McCabe-Thiele diagram accounts for mass transfer effect, and so on. There is no need to distinguish between inclusion and surface impurities, as only the total content is considered in the pseudo-equilibrium composition.

As an example, consider a purification process of crystals of product B containing a certain amount of impurity A, which involves four pseudo-equilibrium stages in a wash column. Figure 7.27a shows a section of the binary SLE phase diagram involving A and B. Process points corresponding to the liquid and solid compositions at different stages are shown on the diagram.

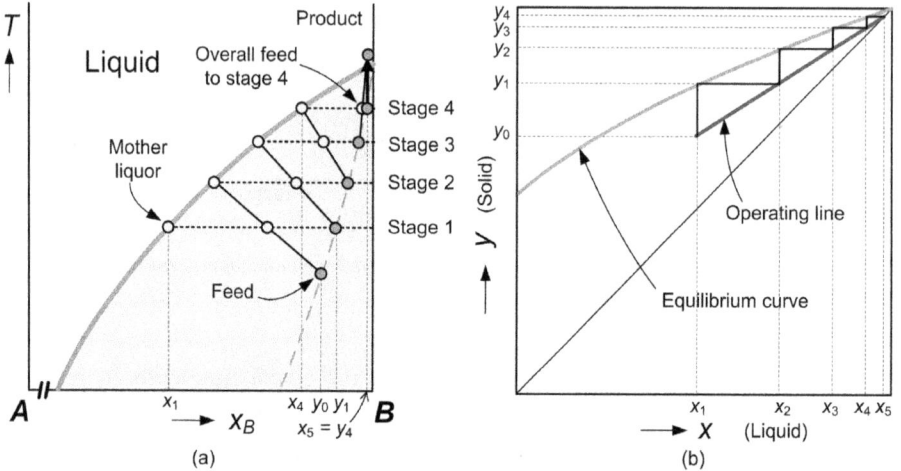

Figure 7.27: McCabe-Thiele approach in wash column analysis: (a) SLE phase diagram and (b) $x–y$ plot.

The feed with composition y_0 enters stage 1 together with the mother liquor from stage 2. The resulting solid and mother liquor compositions from stage 1 are y_1 and x_1, respectively. The solid from stage 1 then goes to stage 2 where it is mixed with the mother liquor from stage 3, as indicated in Figure 7.26b. In the last stage (stage 4), the solid from the previous stage (stage 3) is mixed with the reflux stream with composition x_5. The solid from stage 4 (composition y_4) is heated above its melting point to give the final product. Since the reflux stream is a portion of the melted product, $x_5 = y_4$.

Figure 7.27b is the equivalent of an x-y plot in binary distillation, in which the "equilibrium" solid composition (y) is plotted against the corresponding liquid composition (x) to give the equilibrium curve, while the liquid and solid compositions at the same location along the column are plotted against each other to give the operating line. This line starts at (x_1, y_0), which represents the compositions of the feed and mother liquor at one end of the column, and intersects the diagonal line at (x_5, y_4), which represents the compositions of the product and wash liquid. Note that x_5 is always equal to y_4 because the wash liquid is obtained by melting the product. The constant molal overflow assumption implies that the operating line is linear, so that it can be represented by

$$y = \frac{R}{R+1}x + \frac{1}{R+1}y_4 \qquad (7.42)$$

where R is the ratio of wash liquid to product, equivalent to the reflux ratio in a distillation column. The required number of stages to achieve the required product purity (y_4), which is determined by the number of steps that can be drawn between the operating line and the equilibrium curve, depends on the gradient of the operating line. Therefore, for a given column with a fixed number of stages, this approach can be used to determine the targeted reflux ratio to obtain the product of the desired purity from a given feed.

In practice, the real challenge is the determination of the equilibrium curve and the operating curve, as their dependence on kinetic and mass transfer instead of thermodynamics makes it impossible to predict these curves *a priori*. One possible approach in designing a new unit is to obtain data from a bench or pilot-scale unit having the same configuration as the desired full-scale unit. Alternatively, operation data from an actual unit can be used in a retrofit or process improvement exercise.

7.4 Summary

Chapters 2–6 provide the understanding and methods to recover a desired chemical compound from a multicomponent mixture. This chapter considers the PSD and purity of the compound, two of the most important product attributes in industrial crystallization. The required product specifications such as mean particle size and maximum impurity level can be met by taking advantage of the SLE phase behavior and the kinetic characteristics of the mixture under consideration. These include solubility, width of the metastable zone, nucleation and growth kinetics, reaction kinetics, and so on. The product quality also depends on how the crystallizer is designed and its operating conditions, which determine the transport processes faced by the crystals. Kinetics and transport models have been developed for the design of crystallization processes. In particular, the population balance model that counts the number of particles within a specified particle size range is indispensable. By capturing the effects of supersaturation on growth and nucleation rates, accounting for fines dissolution and product classification, and so on, it tracks the PSD of the crystals in a crystallizer. No matter how well-designed a crystallizer is, the presence of impurity in the product crystals is inevitable. These impurities can be removed downstream of the crystallizer using washing, deliquoring, partial redissolution, and melt crystallization.

Exercises

7.1. A sample of NaCl particles has a PSD that follows the Gaussian distribution, given by the normalized function,

$$f(L) = \frac{1}{\sigma\sqrt{2\pi}} \exp\left[-\frac{(L-L_m)^2}{2\sigma^2}\right] \qquad (7.43)$$

with $L_m = 1.5$ mm and $\sigma = 0.42$ mm. All particles have the shape of a rectangular prism, with sides having a length ratio of 1: 1: 2. The true density of NaCl is 1,300 kg/m^3.

a. Calculate the volume shape factor, k_v, for these NaCl particles, if the longest side is taken as the characteristic length L.

b. Derive an expression for calculating the zeroth moment, M_0, which is the total number of particles per kg of powder mass. (*Hint*: Start from an expression for the total volume of particles per kg powder. The units of population density $n(L)$ would be no/m·kg.)

c. Discretize the PSD using an equal-sized interval of 0.1 mm and ignoring particles larger than 4.0 mm in size. (Do this by tabulating the values of $f(L)$ for $L = 0.1$ mm, 0.2 mm, 0.3 mm, . . ., 4.0 mm.)

d. Calculate M_0 using the expression derived in part b, by numerical integration using the discretized PSD.

7.2. The PSD in a continuous MSMPR crystallizer with no solids in the inlet stream and a growth rate that is independent of crystal size is given by eq. (7.7),

$$n(L) = \frac{B}{G}\exp\left(-\frac{Q_{out}L}{GV}\right) \qquad (7.7)$$

a. Express this distribution as a mass-based differential PSD, $W(L)$.

b. Derive an expression for dominant crystal size L_d, which is defined as the size for which $W(L)$ reaches a maximum.

c. Derive an expression for the magma density, M_T, in terms of L_d.

7.3. A seeded batch crystallizer is used to crystallize pure A from 5,000 kg of solution containing 30 wt% A. Based on the solubility of A at the final crystallization temperature, it is expected that the mother liquor would contain 5 wt% A. The target size of the product crystals is 1 mm, while the size of the seed crystals is 50 μm. Assuming no nucleation and uniform crystal growth, how many kg of seed crystals should be added?

7.4. An MSMPR cooling crystallizer is used to crystallize adipic acid (AA) from its solution. The feed is a saturated solution of AA in water at 70 °C. The crystallizer is to be operated at 50 °C. It is desired to produce 4,600 kg/h adipic acid crystals with a dominant size (L_d) of 150 μm. The following data are available:

Density of adipic acid crystals, kg/m^3	1360
Density of final solution, kg/m^3	1300
Crystal shape factor	0.5
Solubility of AA in water, kg AA/kg water	$c^* = 0.00575\,T - 0.1775$
	(T = temperature in °C)
Empirical correlation for crystallization kinetics	$B = 5.36 \times 10^{31}\,G^{3.5}\,M_T^{0.4}$
	(B in no/m^3.s, G in m/s, and M_T in kg/m^3)

a. Calculate the feed flow rate in kg/h.
b. Calculate the magma density (M_T) in kg/m^3.
c. Using the result from part b, calculate the required growth rate, G, in m/s.
d. Determine the required crystallizer volume in m^3.
e. If a larger crystallizer is used for the same flow rate, feed temperature, and crystallizer temperature (that is, the residence time becomes longer), how would the dominant size change?

7.5. A batch crystallization process is to be scaled up from a pilot-scale unit that can accommodate 100 L of slurry to a commercial unit with a capacity of 10 m^3 of slurry. The slurry density is 1150 kg/m^3 and the viscosity is 2 cP. The pilot-scale vessel has a tank diameter of $T_1 = 0.5$ m and impeller diameter $D_1 = 0.25$ m, and is operated at an agitation speed of $N_1 = 50$ rpm. The commercial-scale vessel is designed to be geometrically similar to the pilot-scale vessel and it is decided that the constant torque per volume scale-up rule, given by

$$\frac{N_1^2 D_1^5}{V_1} = \frac{N_2^2 D_2^5}{V_2} \qquad (7.23)$$

would be the most appropriate for this system.
a. Show that for geometrically similar vessels, this scale-up rule would give a constant tip speed.
b. Determine the tank diameter, impeller diameter, and agitation speed of the commercial crystallizer.
c. If the relationship between power number N_P and Reynolds number N_{Re} for the selected agitator type is as shown in Figure 7.28, compare the power requirement per unit volume for the pilot and commercial units.

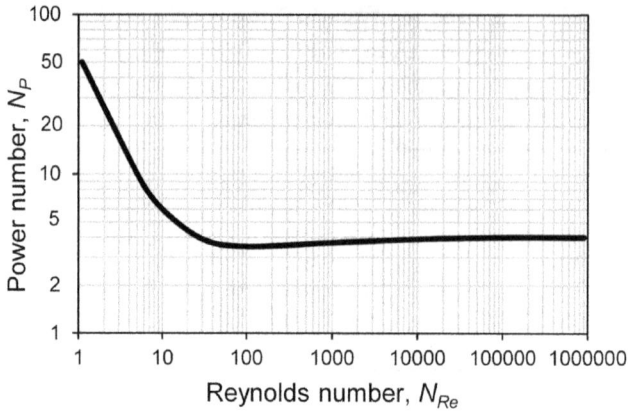

Figure 7.28: Power number as a function of Reynolds number (problem 7.5).

7.6. The Darcy's law for liquid flow through a porous medium can be expressed as

$$\frac{Q}{A} = \frac{k\,\Delta p}{\mu\,x} \tag{7.44}$$

where Q = volumetric flow rate, A = flow cross-sectional area, k = permeability, Δp = pressure drop across the porous medium, μ = liquid viscosity, and x = thickness of the porous medium. Applying this equation for the flow through a porous cake during filtration, Q can be replaced by dV/dt (V = filtrate volume, t = time), and the filter area A_f can be substituted for A. Accounting for the additional resistance to flow caused by the presence of filter medium, eq. (7.44) can be rewritten as

$$\frac{1}{A_f}\frac{dV}{dt} = \frac{\Delta p}{\mu(x/k + R_m)} \tag{7.45}$$

where R_m = filter medium resistance. Furthermore, the following two entities are defined:

$$\alpha = \frac{1}{\rho_S(1-\varepsilon)k} \tag{7.46}$$

$$c = \frac{\rho_S(1-\varepsilon)A_f x}{V} \tag{7.47}$$

where α = specific cake resistance, c = mass of deposited solids (cake) per unit volume of filtrate, ρ_S = solid density, and ε = cake porosity. Show that eq. (7.32) (Table 7.9) can be obtained by integrating eq. (7.45) under constant pressure drop.

7.7. A continuous belt filter is to be used to process 500 kg/h of slurry containing 30% (by mass) NaCl. The cake leaving the unit still contains 0.15 kg water per kg dry cake. The filtrate is a clear solution with a density of 1,050 kg/m³ and viscosity of 1 cP. The filter is operated at a vacuum level of 35,000 N/m² and a speed of 0.05 m/s. The filter belt has a width of 0.2 m.

a. Determine the filtrate volumetric flow rate in m³/h.
b. Determine the specific cake resistance α and filter medium resistance R_m from the batch filtration data given below. The batch filtration test is done at the same vacuum level and using the same slurry as the continuous unit. The filter area is 100 cm².

Hint: Use eq. (7.32) to obtain α and R_m from the slope and intercept of a plot of t_f/V_f against V_f.

Time (s)	5	10	20	30	40	60	90	120
Filtrate volume (mL)	34	50	72	88	101	124	152	175

c. Determine the required length of the belt section dedicated to filtration in m².
d. The specific cake resistance can be linked to the PSD. For example, a relationship based on the Kozeny-Carman permeability suggests that α is inversely proportional to d_p^2 (d_p is the particle size) [238]. Based on this relationship, how is the required filter length expected to change if the particle size were smaller, all other variables being the same?

7.8. A crystallizer is used to obtain AsA from a solution consisting of AsA, OxA, and water. The slurry from the crystallizer is to be processed in a belt filter, which also performs the washing and deliquoring function, followed by a dryer. The solid contains 5 ppm of OxA as inclusion and the cake after filtration contains 0.15 kg liquid/kg dry solid. The residual liquid contains 30 wt% AsA and 200 ppm OxA. A washing liquid containing 30 wt% AsA (saturated in AsA) and 10 ppm OxA is available. The following data is obtained from batch washing experiments using this washing liquid. (Note: a fraction removal of 1 would give 10 ppm OxA in the residual liquid.)

Wash ratio, W	OxA concentration (ppm)	Fraction removal, f
0	200	0
0.26	131	0.363
0.53	88	0.589
0.7	59	0.742

(continued)

Wash ratio, W	OxA concentration (ppm)	Fraction removal, f
0.88	50	0.789
1.06	45	0.816
1.23	34	0.874
1.54	30	0.895
2.2	21	0.942
3.01	16	0.968

The final wetness after deliquoring is 0.1 kg liquid/kg dry solid. It can be assumed that AsA and OxA are non-volatile and water is perfectly removed during drying.

a. If the fraction removal by washing is to be correlated to the wash ratio using an exponential function $f = f_{max} [1-\exp(-bW)]$, determine the appropriate values of f_{max} and b from the data.

b. If the final solid product should contain no more than 10 ppm OxA, how many kg of washing liquid is required per kg of dry solid in the slurry?

8 Determination of Solid–Liquid Equilibrium Phase Behavior and Crystallization Kinetics

8.1 Strategy for SLE Phase Behavior Determination

This chapter focuses on how to establish reasonably accurate SLE phase diagrams that are suitable for the synthesis and design of crystallization processes. With the exception of a few conventional systems that have been studied in detail, the SLE phase diagram is seldom available at the beginning of an industrial development project. Modeling and experimental activities are often needed to construct the relevant diagrams to be used in process synthesis. The key is to strike a balance between the need for accuracy and the required development effort.

The iterative development of the SLE phase diagram lies at the center of the workflow for synthesizing a crystallization process, as shown in Figure 1.3. Figure 8.1 provides a more detailed picture of the activities involved in the determination of the SLE behavior. Starting with the separation objectives, the first action is to identify relevant components in the system to be handled (point 1 in the figure). The analysis result of the potential feed stream often reveals a long list of components, which would require a complex, high-dimensional phase diagram to represent the phase behavior involving all of them. Therefore, the system should be simplified whenever possible, such as by considering only key components or lumping minor impurities into one component. Sometimes, the minor impurities can be lumped with the solvent, because they are unlikely to be crystallized out. Once the system has been narrowed down to three or four components, establishing the SLE phase diagram is much easier.

The recommended approach is to use basic information to proceed with modeling first (point 2), so as to obtain a better idea of the SLE behavior. Some information such as melting point, heat of fusion, solubility data, and so on may be available from the literature or from initial investigation by chemists. If no information is available, preliminary experiments can be performed to collect the basic information (point 3a). Alternatively, missing information can be estimated (point 3b). Relevant phase diagrams, based on which the process can be synthesized, are constructed (point 4). Focusing on selected regions as guided by process synthesis, additional experiments such as solubility measurement, crystallization tests, and so on can be performed to improve accuracy (point 5). The results are used to improve the model in the next iteration. At the end, if the separation objectives cannot be achieved using the model involving the selected components, the activities can be repeated with the inclusion of more components in the system (point 6).

https://doi.org/10.1515/9781501519901-008

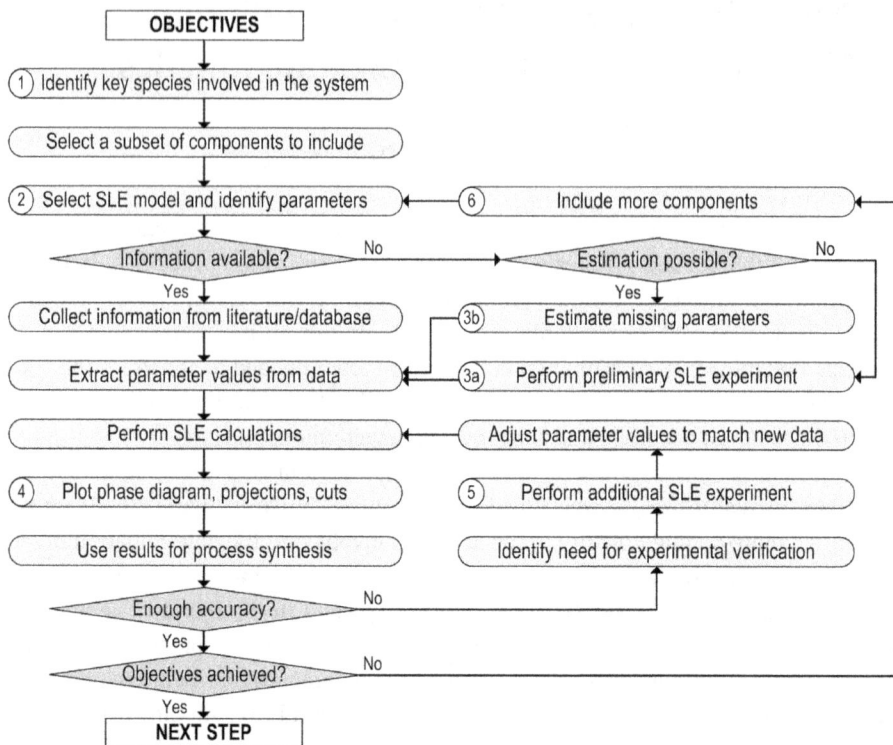

Figure 8.1: Flowchart of activities in the determination of SLE phase behavior.

8.2 Modeling of SLE Phase Behavior

Modeling can be viewed as an effort to make sense and provide structure to a set of data. Scattered data without a model is just a bunch of numbers, which have limited applications, as no understanding has been extracted. On the other hand, a model without data validation is just a random equation, which may have a scientific meaning but is not practically useful. For this reason, modeling and experimental efforts should always proceed hand-in-hand in conceptual design. In order for the model to be meaningful, it should be built with minimum complexity, as long as the objective of the exercise is met, and should include parameters that can be obtained or validated by experiments, in some way.

Levenspiel [249] illustrated the use of models of different complexities in various areas of chemical engineering throughout the twentieth century, and at the end, suggested to "always start by trying the simplest model and then only add complexity to the extent needed." In the same spirit, a three-step practical approach can be proposed. First, build the simplest model as it reflects the lowest cost exercise. Second, perform a sensitivity analysis to determine the effects of the critical parameters, so that the accuracy requirements for those parameters can be assessed, and the model can be tested by comparing the results with expectations from engineering knowledge. Finally, analyze the situation based on the model results and decide on the next move, such as adding more complexities, performing experiments to obtain better parameter values, and so on. The development in this section follows this philosophical approach, beginning with the model for describing SLE based on first principles.

8.2.1 Solubility equation and solubility product equation

There are two key questions in modeling phase equilibrium, regardless of the phases involved (solid–liquid, vapor–liquid, or liquid–liquid equilibrium): What are the criteria for phase equilibrium? What are the compositions of all phases that are in equilibrium with each other? The answer to the first question is provided by the second law of thermodynamics, which states that for any spontaneous process, the entropy change is positive [250]. Since no variation can be spontaneous at the stable equilibrium condition, the entropy of a multiphase system must reach a maximum with respect to its independent variables:

$$dS(U, v, x_1, ..., x_{C-1}) \geq 0 \tag{8.1}$$

where S is specific entropy, U is specific internal energy, v is specific volume, x is mole fraction, and C is the number of components in the system. Correspondingly, the specific Gibbs free energy (G) reaches a minimum with respect to its independent variables, which are temperature, pressure, and composition, as illustrated in Figure 8.2.

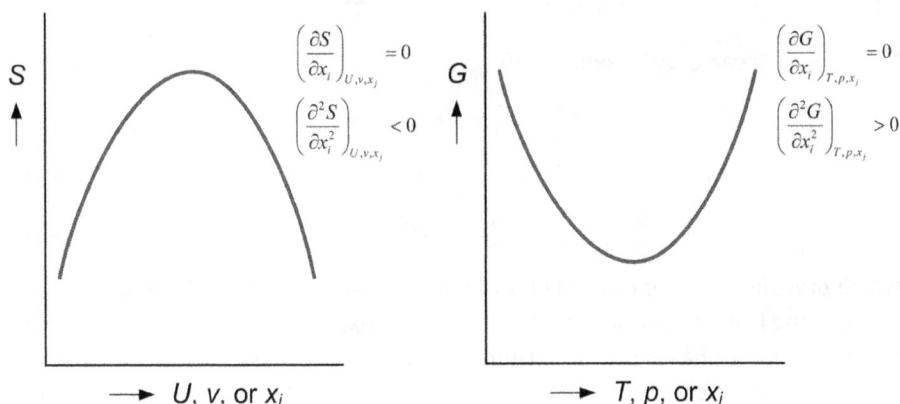

Figure 8.2: Criteria for a stable equilibrium condition.

For a closed, simple system, the entropy representation of the fundamental equation obtained from the combined first and second laws of thermodynamics gives

$$d\underline{S} = \frac{1}{T} d\underline{U} + \frac{p}{T} d\underline{V} - \Sigma \frac{\mu_i}{T} dn_i \qquad (8.2)$$

where the underbar signifies extensive properties, which depend on the size of the system. T, p, n_i, and μ_i are temperature, pressure, number of moles of component i, and chemical potential of component i, respectively. For a multiphase isolated system with no reaction, eq. (8.2) can be written as

$$d\underline{S} = \sum_{j=1}^{P} \frac{1}{T^{(j)}} d\underline{U}^{(j)} + \sum_{j=1}^{P} \frac{p^{(j)}}{T^{(j)}} d\underline{V}^{(j)} - \sum_{j=1}^{P} \sum_{i=1}^{C} \frac{\mu_i^{(j)}}{T^{(j)}} dn_i^{(j)} = 0 \qquad (8.3)$$

where $j=1, 2, \ldots, P$ in the superscripts indicates the multiple phases in the system. P is the total number of phases in the system. As the total energy, volume, and amount of each component across all phases must be constant in the isolated system:

$$\sum_{j=1}^{P} d\underline{U}^{(j)} = 0 \qquad (8.4)$$

$$\sum_{j=1}^{P} d\underline{V}^{(j)} = 0 \qquad (8.5)$$

$$\sum_{j=1}^{P} dn_i^{(j)} = 0 \qquad (8.6)$$

Subtracting a linear combination of eqs. (8.4)–(8.6) from eq. (8.3) gives

$$\sum_{j=2}^{P} \left(\frac{1}{T^{(j)}} - \frac{1}{T^{(1)}} \right) d\underline{U}^{(j)} + \sum_{j=2}^{P} \left(\frac{p^{(j)}}{T^{(j)}} - \frac{p^{(1)}}{T^{(1)}} \right) d\underline{V}^{(j)} - \sum_{j=2}^{P} \sum_{i=1}^{C} \left(\frac{\mu_i^{(j)}}{T^{(j)}} - \frac{\mu_i^{(1)}}{T^{(1)}} \right) dn_i^{(j)} = 0 \quad (8.7)$$

Equation (8.7) can only be satisfied if:

$$T^{(1)} = T^{(2)} = \ldots = T^{(P)} \qquad (8.8)$$

$$p^{(1)} = p^{(2)} = \ldots = p^{(P)} \qquad (8.9)$$

$$\mu_i^{(1)} = \mu_i^{(2)} = \ldots = \mu_i^{(P)} \, i = 1, 2, \ldots, C \qquad (8.10)$$

which provide the general criteria for multiple phases to be in equilibrium.

Starting from the phase equilibrium criteria, expressions relating the composition of different phases in equilibrium can be derived, thus providing the answer to the second question. For SLE of molecular systems at a given temperature and pressure, the chemical potential of any component i in the liquid phase has to be equal to the chemical potential of that component in the solid phase:

$$\mu_i^L = \mu_i^S \quad i = 1, 2, ..., C \tag{8.11}$$

The chemical potential of a component in a mixture is related to its fugacity:

$$\mu_i = \mu_i^0 + RT \ln \hat{f}_i \tag{8.12}$$

where μ_i^0 is the chemical potential at a reference state (ideal gas at temperature T and pressure of 1 bar) and $\hat{f}_i$ is the fugacity of component i in the mixture. Therefore, eq. (8.11) can be cast in terms of fugacity as

$$\hat{f}_i^L = \hat{f}_i^S \quad i = 1, 2, ..., C \tag{8.13}$$

The fugacity in the liquid phase can be related to the molar composition by introducing the activity coefficient, γ,

$$\gamma_i = \frac{a_i}{x_i} = \frac{\hat{f}_i^L}{f_i^L x_i} \tag{8.14}$$

where a_i is the activity of component i in the liquid phase, x_i is the mole fraction of i in liquid phase, and f_i^L is the fugacity of pure component i in liquid phase, which has been taken as the reference state to define activity. Since the solid phase is pure, $\hat{f}_i^S$ can simply be replaced by f_i^S. Therefore, eq. (8.13) can be written in terms of composition as

$$\frac{f_i^S}{f_i^L} = \gamma_i x_i \quad i = 1, 2, ..., C \tag{8.15}$$

The pure component fugacity ratio on the left-hand side can be related to measurable pure component thermodynamic properties such as heat of fusion and melting point, as demonstrated in Table 8.1.

Table 8.1: Derivation of the expression for pure component fugacity ratio.

Step		Equation
Definition of fugacity of pure component i	$dG_i = RTd \ln f_i$	(8.16)
Integration at constant temperature from $p \approx 0$ to p, and recognizing that $f_i \approx p$ at $p \approx 0$	$G_i - G_i^{id} = RT(\ln f_i - \ln p)$	(8.17)
Rearranging and taking derivative at constant p	$\left(\dfrac{\partial \ln f_i}{\partial T}\right)_p = \left[\dfrac{\partial(G_i/RT)}{\partial T}\right]_p - \left[\dfrac{\partial(G_i^{id}/RT)}{\partial T}\right]_p$	(8.18)

Table 8.1 (continued)

Step		Equation
Applying Gibbs-Helmholtz relationship to obtain the derivative of G_i with respect to T	$$\left(\frac{\partial \ln f_i}{\partial T}\right)_p = -\frac{H_i}{RT^2} + \frac{H_i^{id}}{RT^2}$$	(8.19)
Integrating from the melting point $T_{m,i}$ to T	$$\frac{f_i(T, p_{m,i})}{f_i(T_{m,i}, p_{m,i})} = \exp \int_{T_{m,i}}^{T} \frac{-H_i + H_i^{id}}{RT^2} dT$$	(8.20)
Derivative of $\ln f_i$ with respect to p, from the definition of fugacity	$$\left(\frac{\partial \ln f_i}{\partial p}\right)_T = \frac{V_i}{RT}$$	(8.21)
Integrating from melting pressure $p_{m,i}$ to p	$$\frac{f_i(T, p)}{f_i(T, p_{m,i})} = \exp \int_{p_{m,i}}^{p} \frac{V_i}{RT} dp$$	(8.22)
Equality of pure component fugacity in solid and liquid phases at melting point	$$f_i^S(T_{m,i}, p_{m,i}) = f_i^L(T_{m,i}, p_{m,i})$$	(8.23)
Combining eqs. (8.20), (8.22), and (8.23)	$$\frac{f_i^S(T,p)}{f_i^L(T,p)} = \exp\left[\int_{T_{m,i}}^{T} \frac{-H_i^S + H_i^L}{RT^2} dT + \int_{p_{m,i}}^{p} \frac{V_i^S - V_i^L}{RT} dp\right]$$	(8.24)
Expressing the temperature dependence of enthalpy and heat capacity	$$H_i = H_i(T_{m,i}) + \int_{T_{m,i}}^{T} C_{p,i} dT$$	(8.25)
	$$C_{p,i} = C_{p,i}(T_{m,i}) + \int_{T_{m,i}}^{T} \left(\frac{\partial C_{p,i}}{\partial T}\right)_p dT$$	(8.26)
Substituting eqs. (8.25) and (8.26) into the first integral term on the right-hand side of eq. (8.24)	$$\int_{T_{m,i}}^{T} \frac{-H_i^S + H_i^L}{RT^2} dT = [H_i^L(T_{m,i}) - H_i^S(T_{m,i})] \int_{T_{m,i}}^{T} \frac{dT}{RT^2}$$ $$+ [C_{p,i}^L(T_{m,i}) - C_{p,i}^S(T_{m,i})] \int_{T_{m,i}}^{T} \frac{T - T_{m,i}}{RT^2} dT$$ $$+ \int_{T_{m,i}}^{T} \frac{1}{RT^2} \int_{T_{m,i}}^{T} \int_{T_{m,i}}^{T} \left(\frac{\partial (C_{p,i}^L - C_{p,i}^S)}{\partial T}\right)_p dTdTdT$$ $$= \frac{\Delta H_{f,i}}{R}\left(\frac{1}{T_{m,i}} - \frac{1}{T}\right) + \frac{\Delta C_{p,i}}{R}\left(\ln \frac{T}{T_{m,i}} - \frac{T - T_{m,i}}{T}\right) + I$$	(8.27)
Neglecting $\Delta C_{p,i}$ and the triple integral I, and assuming that $V_i^S \approx V_i^L$ in eq. (8.24)	$$\frac{f_i^S(T,p)}{f_i^L(T,p)} = \exp\left[\frac{\Delta H_{f,i}}{R}\left(\frac{1}{T_{m,i}} - \frac{1}{T}\right)\right]$$	(8.28)

Substituting the final expression into eq. (8.15) gives the solubility equation:

$$x_i = \frac{1}{\gamma_i} \exp\left[\frac{\Delta H_{f,i}}{R}\left(\frac{1}{T_{m,i}} - \frac{1}{T}\right)\right] \tag{8.29}$$

where $\Delta H_{f,i}$ and $T_{m,i}$ are the heat of fusion and melting point, respectively, of component i. The values of heat of fusion and melting point for many components can be found in the literature [178] or in databases such as NIST, DIPPR, and Dortmund Data Bank. For an ideal solution for which the activity coefficient is unity, eq. (8.29) gives the ideal solubility of component, i. Note that slightly different expressions have been reported in the literature, mainly because of the different assumptions in obtaining the final equation. For example, Yang et al. [251] reported an equation for ideal solubility that includes additional terms on the right-hand side, because they accounted for the dependence of heat capacity on temperature in eq. (8.27), which is neglected in the derivation of eq. (8.29).

A similar criterion can be obtained for an electrolyte system, but the starting point is equating the chemical potential of the ions in liquid phase to that of the salt in the solid phase. Consider an electrolyte E that dissociates in solution into cation M and anion N:

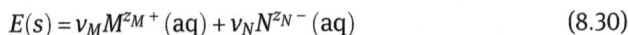

$$E(s) = v_M M^{z_M +}(\text{aq}) + v_N N^{z_N -}(\text{aq}) \tag{8.30}$$

where v is the stoichiometric coefficient and z is the ionic charge. The equilibrium criterion can be written as

$$v_M \mu_M^L + v_N \mu_N^L = \mu_E^S \tag{8.31}$$

By relating the chemical potential to activity via the activity coefficient, the criterion can be expressed in terms of the composition in liquid phase as

$$\left(\gamma_M^* m_M\right)^{1/v_N}\left(\gamma_N^* m_N\right)^{1/v_M} = K_{SP,E}(T) \tag{8.32}$$

$$K_{SP,E}(T) = \exp\left[-\frac{1}{RT}\left(\frac{\mu_M^*}{v_M} + \frac{\mu_N^*}{v_N} - \frac{\mu_{E(s)}^*}{v_M v_N}\right)\right] \tag{8.33}$$

where m is the molality and K_{SP} is the solubility product constant, which is related to the free energy of formation (or chemical potential at unit molality, μ^*) of the pure solid and the ions. The values of K_{SP} at various temperatures are available in the literature [252, 253, 254], or can be calculated from the standard state Gibbs free energy of formation, which are reported in various thermodynamic tables [255]. This expression is often referred to as the *solubility product equation*. For electrolyte systems, it is more common to use the infinite dilution standard state for activity coefficients, so that γ^* approaches unity (ideal condition) as the molality approaches zero. This makes sense as the strong ionic interactions, which are the main source of nonideality in electrolyte systems, are absent in very dilute solutions. In contrast, the pure component standard state ($y_i \to 1$ as $x_i \to 1$) is more appropriate for molecular systems.

8.2.2 Activity coefficient models

The activity coefficient can be physically interpreted as a measure of deviation from the ideal condition or standard state. In an ideal mixture, the interaction forces among any kind of molecules or species in the mixture are equal. In a real mixture, those interaction forces vary, based on the molecular size, shape, orientation, polarity, electronic charge, and so on. Thus, the interaction forces between two molecules of the same kind are not equal to those between molecules of different kinds, giving rise to nonideal behavior. Activity coefficients are related to the excess Gibbs free energy, ΔG^{EX} as:

$$\ln \gamma_i = \frac{1}{RT} \left[\frac{\partial \left(n \Delta G^{EX} \right)}{\partial n_i} \right]_{T, P, n_j} \tag{8.34}$$

Since $\ln y_i$ is a partial molar property, it must obey the Gibbs-Duhem relationship:

$$\sum_i x_i d \ln \gamma_i = 0 \tag{8.35}$$

which is automatically obeyed when derived from ΔG^{EX}. For this reason, activity coefficient models in the literature are normally expressed as ΔG^{EX}. Some of the most commonly used activity coefficient models for molecular and electrolyte systems are summarized in Tables 8.2 and 8.3, respectively.

In choosing the suitable activity coefficient model, one needs to consider the components involved as well as practicality. Assuming an ideal solution may be good enough in certain occasions, such as a molecular system consisting of similar molecules or an extremely dilute solution of electrolytes. Using a model with no adjustable parameter, such as regular solution model for mixtures of nonpolar molecules or Debye–Hückel for electrolytes can be a good first approximation for nonideal systems. Non-random two-liquid (NRTL) and universal quasichemical (UNIQUAC) models are generally the most suitable to fit experimental data.

8.2.3 Equation of state models

Equations of state (EOS) are routinely used for describing the relationship between temperature, pressure, and volume in the vapor phase. As the fugacity coefficient can be derived from such a relationship, it is a standard practice to use EOS for the vapor phase in VLE calculations. The application of EOS to the liquid phase generally does not give good results because liquid is rather incompressible. The standard

Table 8.2: Commonly used activity coefficient models for molecular systems [250].

Model	Equation		Features
Two-suffix Margules	$\Delta G^{EX} = \sum_i \sum_{j \neq i} A_{ij} x_i x_j$	(8.36)	Derived from lattice theory Completely symmetric for binary system One interaction parameter per binary pair Not complex enough for most mixtures
van Laar	$\Delta G^{EX} = \dfrac{\sum_i \sum_j x_i x_j \left[\frac{a_i b_j}{b_i} - a_i^{1/2} a_j^{1/2} \right]}{\sum_i x_i b_i}$	(8.37)	Derived by conducting isothermal mixing in the ideal-gas state, using van der Waals (vdW) equation of state for evaluating property changes in expansion/compression Two interaction parameters per binary pair (instead of using a, b from vdW)
Hildebrant-Scatchard (regular solution)	$\Delta G^{EX} = \sum_i x_i v_i (\delta_i - \bar{\delta})^2$ $\delta_i = \left(\dfrac{\Delta U_{v,i}}{v_i} \right)^{1/2}$ $\bar{\delta} = \sum_j \phi_j \delta_j; \ \phi_i = \dfrac{x_i v_i}{\sum_j x_j v_j}$	(8.38) (8.39) (8.40)	Derived from regular solution theory (entropy change of mixing and volume change during mixing are zero) No adjustable parameter, need molar volume v_i and solubility parameter δ_i Always gives positive deviation ($\gamma_i > 1$) Good for nonpolar solvents, but incorrect when strong associative interactions are present (such as for polar solvents)
Flory-Huggins	$\dfrac{\Delta G^{EX}}{RT} = \sum_i x_i \ln \dfrac{\phi_i}{x_i} + \sum_i \sum_j \chi_{ij} \phi_i \phi_j$	(8.41)	Derived from lattice theory, accounting for different molecular sizes Two interaction parameters per binary pair Good for polymer solutions
Wilson	$\dfrac{\Delta G^{EX}}{RT} = -\sum_i x_i \ln \left(\sum_j x_j \Lambda_{ij} \right)$ $\Lambda_{ij} = \dfrac{v_j}{v_i} \exp \left(-\dfrac{a_{ij}}{RT} \right)$	(8.42) (8.43)	Generally suitable for VLE calculation Unsuitable for LLE calculation since it cannot predict liquid-liquid phase split Stability criterion is always met Two interaction parameters per binary pair
Non-random two-liquid (NRTL)	$\dfrac{\Delta G^{EX}}{RT} = \sum_i x_i \dfrac{\sum_j x_j \tau_{ji} G_{ji}}{\sum_k x_k G_{ki}}$ $\tau_{ij} = \dfrac{a_{ij} + b_{ij} T}{RT} \ (a_{ij}, b_{ij} = 0 \text{ if } i = j)$ $G_{ij} = \exp(-\alpha_{ij} \tau_{ij})$	(8.44) (8.45) (8.46)	Theoretically weak, but very powerful for fitting experimental data Can be used for highly nonideal systems with liquid-liquid phase split Three interaction parameters per binary pair, but the nonrandomness parameter α_{ij} is typically fixed between 0.2 and 0.4

Table 8.2 (continued)

Model	Equation	Features
Universal quasichemical (UNIQUAC)	$\dfrac{\Delta G^{EX}}{RT} = -\sum_i x_i \ln\dfrac{\phi_i}{x_i} + \dfrac{z}{2}\sum_i q_i x_i \ln\dfrac{\theta_i}{\phi_i}$ $\qquad - \sum_i q_i x_i \ln\left(\sum_j \theta_j \tau_{ji}\right)$ (8.47) $\phi_i = \dfrac{x_i r_i}{\sum_j x_j r_j}\,;\ \theta_i = \dfrac{x_i q_i}{\sum_j x_j q_j}$ (8.48) $\tau_{ji} = \exp\left(-\dfrac{\Delta u_{ji}}{RT}\right)$ (8.49)	Combine both the effect of size/shape (combinatorial contribution) and differing intermolecular energies (residual contribution) Involve 2 pure component parameters (q_i, r_i) and 2 interaction parameters per binary pair $(\Delta u_{ij}, \Delta u_{ji})$ Parameters can be estimated using UNIFAC molecular group contribution method

Table 8.3: Commonly used activity coefficient models for electrolyte systems [250].

Model	Equation	Features
Debye–Hückel	$\ln \gamma_i^* = -Az_i^2 I^{1/2}$ (8.50) $A = \dfrac{(2{,}000\rho_S)^{1/2}}{8\pi N_A}\left(\dfrac{F^2}{\varepsilon_0 D_S RT}\right)^{3/2}$ (8.51) $I = \dfrac{1}{2}\sum_i z_i^2 m_i$ (8.52)	Derived from theory of electrostatics No adjustable parameter, but need solvent properties (density ρ_S, dielectric constant D_S) The value depends on ionic strength I Only good for very dilute solutions Explanation of other symbols: F=Faraday constant, ε_0=vacuum permittivity, z=ionic charge, m=molality, N_A=Avogadro number
Extended Debye–Hückel	$\ln \gamma_i^* = -\dfrac{Az_i^2 I^{1/2}}{1+bI^{1/2}}$ (8.53) $b = \left(\dfrac{2{,}000F^3}{\varepsilon_0 D_S RT}\right)^{1/2} r_0$ (8.54)	Less simplifying assumptions, leading to additional constant b No adjustable parameter, need the same solvent properties as Debye–Hückel as well as ionic radius r_0
Pitzer ion interaction	$\dfrac{\Delta G^{EX}}{W_S RT} = f(I) + \sum_{i,j}\lambda_{ij}(I)m_i m_j$ $\qquad + \sum_{i,j,k}\mu_{ijk}(I)m_i m_j m_k$ (8.55) $f(I)$=term representing Debye–Hückel limiting law	Derived from McMillan-Mayer osmotic virial equation, matching Debye–Hückel limiting law Contains binary and ternary interaction parameters that depends on ionic strength W_S=weight of solvent (water)
UNIQUAC-Debye–Hückel	$\dfrac{\Delta G^{EX}}{RT} = \left(\dfrac{\Delta G^{EX}}{RT}\right)_{DH} + \left(\dfrac{\Delta G^{EX}}{RT}\right)_C + \left(\dfrac{\Delta G^{EX}}{RT}\right)_R$ (8.56) Debye–Hückel term: same as extended D-H Combinatorial and residual terms: same as UNIQUAC	Combine Debye–Hückel term with combinatorial and residual contributions Involve 2 pure species parameters (q_i, r_i) and 2 interaction parameters per binary pair $(\Delta u_{ij}, \Delta u_{ji})$

practice is to use activity coefficient models for the liquid phase. However, in modeling solid–supercritical fluid equilibrium, the supercritical fluid phase is best modeled using an EOS. Commonly used EOS are summarized in Table 8.4. The Peng–Robinson and Soave–Redlich–Kwong equations are cubic type EOS, developed as improvements to the ideal gas EOS [250]. They are widely used in industry for low- and medium-pressure systems due to their simplicity and accuracy [256]. Parameters a, b, and α in these equations are functions of pure component critical properties. Adjustable binary interaction parameters appear in the mixing rules used to calculate the values of a, b, and α for mixtures. The perturbed-chain statistical association fluid theory (PC-SAFT) EOS has its foundation in statistical mechanics. For nonassociating systems, the model requires only three component-specific parameters related to molecular size and shape, to model bulk properties and phase equilibria. The complete equations can be found in the original publication [257]. The values of those parameters for common components are available in the database of various process simulators. For other components, these parameters are regressed from pure-component physical property data such as liquid density and vapor pressure. The model also includes a

Table 8.4: Commonly used equations of state for modeling solid–supercritical fluid equilibrium.

Model	Equation		Features
Peng-Robinson (PR)	$p = \dfrac{RT}{v-b} - \dfrac{a}{v(v+b)+b(v-b)}$	(8.57)	Good for vapor–liquid equilibrium calculations
	$a_m = \sum_i \sum_j x_i x_j \left(1-\delta_{ij}\right)\left(a_i a_j\right)^{1/2}$	(8.58)	No adjustable parameter for pure components; all parameters are functions of critical properties T_c and p_c
	$b_m = \sum_i x_i b_i$	(8.59)	Mixing rules for mixtures contain one adjustable binary interaction parameter δ_{ij}
Soave-Redlich-Kwong (SRK)	$p = \dfrac{RT}{v-b} - \dfrac{a\alpha}{v(v+b)}$	(8.60)	Good for vapor-liquid equilibrium calculations
	$(a\alpha)_m = \sum_i \sum_j x_i x_j \left(1-k_{ij}\right)\left(\alpha_i a_i \alpha_j a_j\right)^{1/2}$	(8.61)	No adjustable parameter, all parameters are functions of critical temperature T_c and critical pressure p_c
	$b_m = \sum_i x_i b_i$	(8.62)	Mixing rules for mixtures contain one adjustable binary interaction parameter k_{ij}
PC-SAFT	$Z = \dfrac{pV}{RT} = Z^{id} + Z^{hc} + Z^{disp}$	(8.63)	Developed from advanced molecular theory
	Ideal gas contribution (Z^{id})=1 Hard chain contribution (Z^{hc}) and perturbation contribution (Z^{disp}) are complex functions of molecular size and shape [257]		Good when effect of pressure is important such as in solid-supercritical fluid equilibrium Mixing rules for mixtures contain one adjustable binary interaction parameter k_{ij}

binary interaction parameter, k_{ij}, whose value can be adjusted to match mixture properties. PC-SAFT model is known for its excellent predictive capabilities and good precision for correlating mixture properties [256].

8.3 Prediction of Solid–Liquid Equilibrium Phase Behavior

For timeliness, cost, and the need for some preliminary data, even if the results are not very accurate compared to experimental data, it is always desirable to begin with the prediction of SLE phase behavior. Physical properties such as heat of fusion and melting point can be estimated using group contribution methods, while activity coefficients can be obtained in several ways. One way is to use a group contribution method such as the UNIFAC or NRTL-SAC (segment activity coefficient) model. Another way is to use a model with no adjustable parameter, such as the regular solution model, with required physical properties estimated by group contribution methods. Finally, quantum chemistry models can be used to compute interaction potentials and obtain the activity coefficients using statistical thermodynamics.

8.3.1 Group contribution methods

The basic notion of the group contribution method is that some simple aspects of the structures of chemical components are always the same, although the molecules themselves are different. Therefore, the property of a compound can be estimated based on the contributions of various groups in the molecular structure. The contributions of each group are regressed from a large database of compounds with known properties. While early methods only consider simple first-order groups such as the carbon chain, alkyl, hydroxide, amine, and so on, later developments include second- and third-order groups, such as aromatic ring, the location of double bond, and so on, which enable the distinction among isomers and similar molecules. The group contribution method has been widely used for predicting boiling points, vapor pressure, and so on, with reasonable accuracy. Unfortunately, they are often inaccurate for estimating the melting point and heat of fusion of complex solids.

One of the widely used group contribution methods is the Joback method [258]. Due to its simplicity, this method cannot distinguish between some isomers that have different physical properties in solid phase. A more complex method has been developed by Marrero and Gani [259], in which the molecular structure of a compound is considered to be a collection of three types of groups. The first-order groups are intended to describe a wide variety of organic compounds, while the second- and third-order groups provide more structural information about molecular fragments of compounds whose description is insufficient through the first-order groups. A property f of a molecule X can be described as the sum of group contributions in the general function:

$$f(X) = \sum_i N_i C_i + \sum_j M_j D_j + \sum_k O_k E_k \tag{8.64}$$

where C_i, D_j, and E_k are the contributions of a first-, second-, and third- order group, respectively, and N_i, M_j, and O_k are the number of occurrences of each group in the molecule. An important rule is that the first-order groups must be chosen in such a way that the entire molecule is described by first-order groups, with no overlap. If the same fragment is related to more than one group, the heavier group must be chosen to represent it instead of the lighter groups. The second- and third-order groups are allowed to overlap when they have atoms in common. However, one group should not overlap completely with another group, as this would lead to a redundant description of the same molecular fragment.

The Hildebrand-Scatchard or regular solution activity coefficient model (eq. (8.38)) allows the calculation of activity coefficients from the values of internal energy of vaporization (ΔU_v) and molar volume at saturation. The value of ΔU_v can be approximated by the enthalpy of vaporization (ΔH_v), which, in turn, can be estimated using group contribution methods. The molar volume at saturation can be calculated using the Rackett equation [250]:

$$V_{\text{liq(sat)}} = \frac{RT_c}{p_c} Z_{RA}^{\left[1+ \left(1-T_b/T_c\right)^{2/7}\right]} \tag{8.65}$$

$$Z_{RA} = 0.29056 - 0.08775\omega \tag{8.66}$$

$$\omega = \frac{3}{7} \left[\left(\frac{T_b/T_c}{1 - T_b/T_c}\right) \log\left(\frac{p_c}{1.013}\right) \right] - 1 \tag{8.67}$$

where R is the universal gas constant, T_b is the boiling point, T_c is the critical temperature, and p_c is the critical pressure. The values of T_b, T_c, and p_c can also be estimated using group contribution methods. Therefore, using the regular solution model, it is possible to use group contribution method to estimate the SLE phase behavior. However, it should always be kept in mind that this model is based on assumptions that are valid for nonpolar compounds only.

The UNIFAC (UNIQUAC functional-group activity coefficient) method [260] combines the functional group concept with a model for activity coefficients, based on an extension of the UNIQUAC theory. The method treats a nonelectrolyte liquid mixture as a solution of structural functional groups from which the molecules are formed, rather than as a solution of the molecules themselves. The activity coefficient is expressed as a combination of combinatorial and residual terms:

$$\ln \gamma_i = (\ln \gamma_i)_C + (\ln \gamma_i)_R \tag{8.68}$$

$$\left(\ln \gamma_i\right)_C = \ln\frac{\phi_{ic}}{x_i} + \frac{z}{2}\sum_i q_i x_i \ln\frac{\theta_i}{\phi_i} + L_i - \frac{\phi_i}{x_i}\sum_j x_j L_j \tag{8.69}$$

$$\left(\ln \gamma_i\right)_R = \sum_k v_k^{(i)}\left[\ln \Gamma_k - \ln \Gamma_k^{(i)}\right] \tag{8.70}$$

$$L_i = \frac{z}{2}(r_i - q_i) - (r_i - 1); \; z = 10 \tag{8.71}$$

The functions ϕ_i and θ_i are as defined in eq. (8.48), while r_i and q_i are expressed as the sum of functional group contributions:

$$r_i = \sum_k v_k R_k; \; q_i = \sum_k v_k Q_k \tag{8.72}$$

where r_k and q_k are the size parameters of functional group, k, which occurs v_k times in the molecule. The function, $\ln \Gamma_k$, in the residual contribution is defined as:

$$\ln \Gamma_k = Q_k\left(1 - \ln\sum_m \Theta_m \Psi_{mk} - \sum_n \frac{\Theta_m \Psi_{km}}{\Theta_n \Psi_{nm}}\right) \tag{8.73}$$

$$\Theta_m = \frac{Q_m X_m}{\sum_n Q_n X_n}, \; \Psi_{mn} = \exp\left(-\frac{a_{mn}}{T}\right) \tag{8.74}$$

where X_m denotes the mole fraction of group m in the mixture, a_{mn} is the interaction parameter between group m and group n, and T is temperature.

As UNIFAC functional group information is readily available for many molecules in public databanks and commercial simulators, the model can be used to predict activity coefficients for solvents and solutes, as long as these molecules are structured as a set of UNIFAC functional groups for which the group-group binary interaction parameters have been determined. For small linear organic molecules, UNIFAC yields acceptable qualitative or semiquantitative predictions. However, for complex molecules with rigid molecular configurations, UNIFAC predictions are often, at best, qualitative [256]. It should also be noted that since this model was originally developed for VLE, the validity range of the parameters may be unsuitable for SLE.

Instead of defining a molecule in terms of structural functional groups, the NRTL-SAC model [261] suggests that four conceptual segments should suffice to account for all major distinct molecular surface interaction characteristics: hydrophobic (X), polar attractive (Y^-), polar repulsive (Y^+), and hydrophilic (Z). The activity coefficients for all molecules in solution are determined based on inherent segment-segment binary interaction parameters among the four conceptual segments, which are established from phase equilibrium data of representative reference molecules that exhibit such molecular surface interaction characteristics, as well as the conceptual segment numbers. The interaction parameters are pure component constants, and are incorporated in many process simulators. The NRTL-SAC model has no adjustable binary interaction parameters, but the pure-component parameters for a new molecule that is not available in the database can be identified from regression of experimental phase equilibrium data. Therefore, it is often considered as a hybrid model that is both correlative and predictive.

Example 8.1 – Prediction of melting point and heat of fusion
The melting point and heat of fusion of three common drug molecules are predicted using Marrero and Gani's method. The identification of functional groups in the molecules is illustrated in Figure 8.3. The solid circles indicate first-order groups, whereas the dotted circles are second-order groups. No third-order group can be identified in these molecules. Table 8.5 summarizes the groups and their occurrences in each molecule, along with the calculated melting point and heat of fusion. The group contribution values used for the calculation are available in the original publication [259].

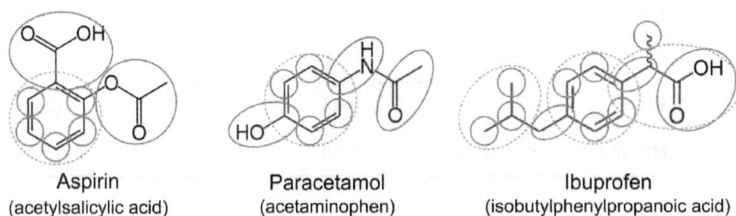

Aspirin	Paracetamol	Ibuprofen
(acetylsalicylic acid)	(acetaminophen)	(isobutylphenylpropanoic acid)

Figure 8.3: Identification of functional groups in three drug molecules.

Table 8.5: Estimation of melting point and heat of fusion by Gani's group contribution method.

		Aspirin	Paracetamol	Ibuprofen
First-order groups	CH_3	0	0	3
	CH_2	0	0	0
	CH	0	0	1
	aCH	4	4	4
	aC (nonfused)	1	0	0
	$aC-CH_2$	0	0	1
	aC-CH	0	0	1
	aC-OH	0	1	0
	COOH	0	0	1
	aC-COOH	1	0	0
	CH_3CO	0	1	0
	CH_3COO	1	0	0
	aC-NH	0	1	0

Table 8.5 (continued)

		Aspirin	Paracetamol	Ibuprofen
Second-order groups	$(CH_3)_2CH$	0	0	1
	aC-CH$_n$-COOH (n in 1..2)	0	0	1
	AROMRINGs^{1}s^2	1	0	0
	AROMRINGs^{1}s^4	0	1	1
Melting point T_m	Total group contribution	17.2181	12.9178	12.5160
	Estimated value, °C	146.5	104.1	99.5
	Literature value, °C	135–143	168.7	75–78
Heat of fusion ΔH_{fus}	Total group contribution	28.350	26.149	21.063
	Estimated value, kJ/mol	25.544	23.343	18.257
	Literature value, kJ/mol	29.8	27.84	25.7–27.94

The estimated value of aspirin melting point is higher than the average value of 135 °C listed in the NIST database. However, the experimental data reported over the past century has a relatively large variation, with some reported data as high as 143 °C [262]. The estimated heat of fusion of aspirin is about 15% lower than the reported literature value of 29.8 kJ/mol [263]. Taken together, the prediction results for aspirin are still reasonable, despite not being very accurate. The prediction results are worse for the other two molecules, especially for ibuprofen, which has a more complex structure. As discussed in Example 5.3, ibuprofen can exist as R and S enantiomers, and the reported literature data usually corresponds to a racemic mixture of the two, which has a higher melting point compared to the pure enantiomers. Since Marrero and Gani's method does not account for the different three-dimensional arrangement between enantiomers, the estimated value is presumed to be valid for the racemic mixture. It can be seen that the results from this model are not accurate for ibuprofen, as the predicted melting point (99.5 °C) is much higher than the literature value (75–78 °C). The estimated value for the heat of fusion is also quite far off. However, better values could be obtained using the latest version of commercial software programs implementing Marrero and Gani's method, such as ICAS [264], perhaps with updated group contribution values that are slightly different compared to those in the original publication.

Example 8.2 – Regular solution model
The solubility of aspirin in four organic solvents, namely ethanol, acetone, 2-propanol, and propylene glycol, are calculated using the regular solution model. The internal energy of vaporization and molar volume for each component are calculated from enthalpy of vaporization and saturated liquid molar density, respectively. These entities are calculated using DIPPR correlations as follows:

$$\Delta H_{v,i} = A\left(1 - \frac{T}{T_c}\right)^B \tag{8.75}$$

$$\rho_L^* = A/B^{\left[1 + (1 - T/T_c)^D\right]} \tag{8.76}$$

The parameters are summarized in Table 8.6, along with the melting point and heat of fusion values used for the SLE calculations. The calculated solubility curves are compared with the ideal solubility curve (which is the same, regardless of the solvent), as well as experimental data reported in the literature [205], in Figure 8.4. The results indicate that the predicted solubility using regular solution

Table 8.6: Values of parameters for SLE calculation using regular solution model.

Component	Melting point (K)	Heat of fusion (J/kmol)	Enthalpy of vaporization (J/kmol)		Saturated liquid molar density (kmol/m^3)			Critical temp, T_c (K)
			A	B	A	B	D	
Aspirin	408.15	2.98×10^7	1.048×10^8	0.3833	0.5141	0.2412	0.2857	765.0
Ethanol	159.05	4.93×10^6	5.579×10^7	0.3125	1.6288	0.2747	0.2318	514.0
Acetone	178.45	5.77×10^6	4.215×10^7	0.3397	1.2332	0.2589	0.2913	508.2
2-Propanol	185.28	5.41×10^6	6.308×10^7	0.3921	1.1898	0.2665	0.2399	508.3
Propylene glycol	213.15	7.57×10^6	8.070×10^7	0.2950	1.0923	0.2611	0.2046	626.0

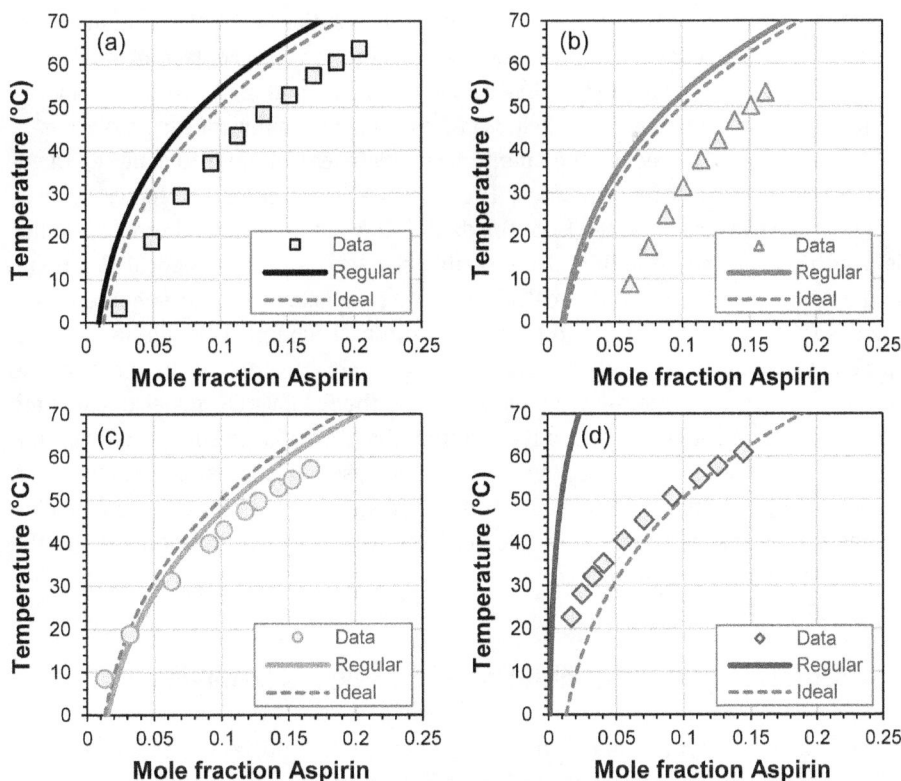

Figure 8.4: Calculated solubility of aspirin in organic solvents using regular solution model: (a) in ethanol, (b) in acetone, (c) in 2-propanol, and (d) in propylene glycol.

model can sometimes be closer to reality compared to ideal solubility, as in the case of 2-propanol (Figure 8.4c). However, in other cases, the regular solution model gives worse results compared to the ideal solution assumption, as is evident from the plots for the other solvents. Therefore, although this model can be useful to provide an initial estimate of solubility in an organic system, it may not give accurate results. Fitting of experimental data is recommended, whenever possible.

8.3.2 Quantum chemistry methods

Quantum chemistry models can be used to calculate solubility via the activity coefficient. An example is the conductor-like screening model for realistic solvation (COSMO-RS) theory, which is based on the so-called continuum solvation model [265]. In contrast to group contribution methods in which molecules are described as a collection of fragments with the consequence of losing any information about intramolecular interactions, the continuum solvation model approximates the electrostatic interaction between the solute and the solvent by classical dielectric theory. This allows for a self-consistent treatment of molecules in solution, even at quantum chemical level. The model describes the interactions in a fluid as local contact interactions of molecular surfaces, quantified by the values of screening charge densities σ and σ'. Molecules can be represented by their σ-profiles, which is unique for each molecule.

For any "new" molecule, the quantum chemistry calculations to compute the σ-profiles have to be done only once. The σ-profile for common molecules such as solvents can even be obtained from the database. Then, statistical thermodynamics is used to solve the ensemble of interacting surfaces, so that chemical potentials and activity coefficients can be calculated. The results are often reasonable for organic solvents, although time-consuming quantum mechanical calculations are required. To avoid lengthy calculations of many points spread over the entire phase diagram, a reasonable strategy would be to use the COSMO-RS model to calculate solubility data for a few points only, and treat the estimation results as experimental data. These simulated data points can then be used to regress the parameters in an empirical activity coefficient model such as NRTL, and eventually to calculate the entire SLE phase diagram. In this way, the required time and effort to get the phase diagram can be minimized.

8.4 Experimental Determination of SLE Phase Behavior

The role of experimental data in the development of SLE phase diagrams is twofold. First, experiments are often needed for the determination of model parameters, including physical properties (heat of fusion and melting point, if not already known), as well as the parameters of the activity coefficient model. Second, experiments can be used to validate model predictions, such as the location of a thermodynamic boundary or the yield at a given crystallization condition. As a general rule, the accuracy of the model is proportional to the amount of experimental data gathered. During the conceptual design stage, conducting a thorough experimental investigation is not necessary, as there is no need to have a highly accurate model. It is more important to obtain the overall image of the SLE behavior to get started with process synthesis. Additional experiments can be performed later, when the region of the phase diagram relevant to the separation objectives has been identified.

8.4.1 Thermal method for obtaining SLE data

The most common techniques of obtaining melting point and heat of fusion data are differential scanning calorimetry (DSC) and differential thermal analysis (DTA). A DSC machine measures the difference in heat flow to two separate chambers that are kept at the same temperature during a heating or cooling program. One chamber contains a standard, normally a metal piece that would not undergo a phase change, and the other contains the sample. When a phase change or reaction occurs to the sample, the heat flow to the sample chamber is automatically adjusted to compensate the heat effect and keep the same temperature as the standard chamber. This leads to a peak being displayed in the heat flow versus temperature chart. In DTA, the temperature difference between the sample and the reference is measured instead. Modern equipment usually combines this function into thermogravimetric analysis (TGA) systems, which simultaneously record the weight gain or loss of the sample. In addition to pure component phase transition data, the solidus and liquidus curves for mixtures are simultaneously measured in the same run using DSC, so that polythermal SLE data can be obtained [266]. From this information, it is also possible to get an initial idea on the type of phase behavior (simple eutectic, polymorphism, solid solution, and so on).

Figure 8.5 illustrates how a DSC curve is interpreted. If a pure solid is being heated, as indicated by the vertical movement from *a* to *b* to *c* on the phase diagram in Figure 8.5a, the resulting DSC curve would feature a single peak as shown in Figure 8.5b. The location of this peak corresponds to the melting point, while the area of the peak corresponds to the heat of fusion. The DSC equipment can usually be set to display these values on the chart for easy reading. If the solid sample is a mixture instead of a pure component, as represented by the path from *d* to *g* in Figure 8.5a, the corresponding DSC curve is depicted in Figure 8.5c. The first peak appears as the temperature reaches the eutectic temperature at point e, because at this point all B in the mixture turns into liquid, leaving only A in the solid phase, as the heating continues toward f. Between e and f, A continues to melt until the last piece of solid disappears at point f. This appears as a less sharp, second peak in the DSC curve. From a series of DSC curves taken at various compositions, it is possible to construct the binary SLE phase diagram.

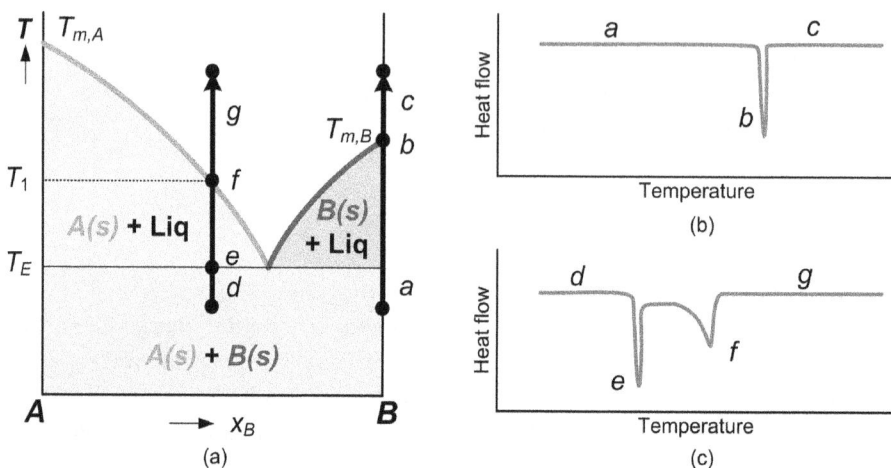

Figure 8.5: Interpretation of DSC results: (a) representation on the SLE phase diagram, (b) DSC curve of a pure component sample, and (c) DSC curve of a binary mixture sample.

Example 8.3 – SLE phase diagram of dimethylterephthalate/benzoic acid binary system
The determination of the binary SLE phase diagram of dimethylterephthalate (DMT) and benzoic acid (BA) by DSC has been demonstrated by Mettler-Toledo [267]. Samples containing mixtures of DMT and BA in different ratios are prepared and analyzed by DSC. The resulting DSC curves are depicted in Figure 8.6. The BA content is shown in mole percent. It is clear that, regardless of the BA content in the sample, a peak at 97 °C is always detected, corresponding to the eutectic point. The curves for 4% BA, 7% BA, and 13% BA feature a second peak, which becomes smaller as the BA content increases. The second peak is less clear in the curves for 48% BA and 89% BA, and virtually undetected in the curve for 62% BA. The curves for pure BA and pure DMT, which exhibit only one peak each, as expected, are also shown in the figure.

By plotting the heat of fusion (peak area) of the eutectic peak against composition, a eutectic composition at BA concentration of about 72 mol% is obtained by linear extrapolation, as demonstrated in Figure 8.7a. This method is appreciably more accurate than extrapolating the nonlinear liquidus (solubility curves) on the phase diagram. The eutectic point is then used, together with other points obtained from the onset temperature of the second peak, to reconstruct the entire solubility curves of DMT and BA, as shown in Figure 8.7b.

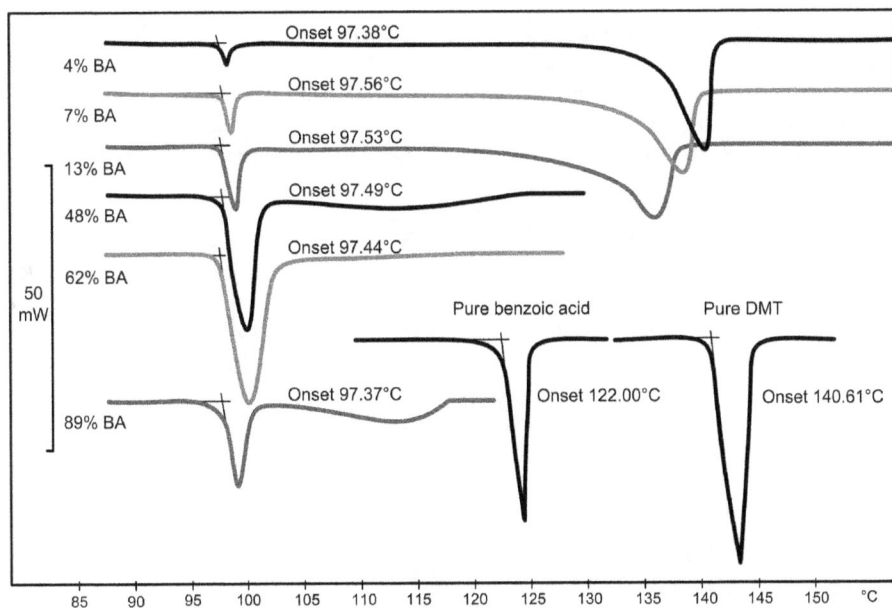

Figure 8.6: DSC curves of different mixtures of DMT and benzoic acid (data from [267]).

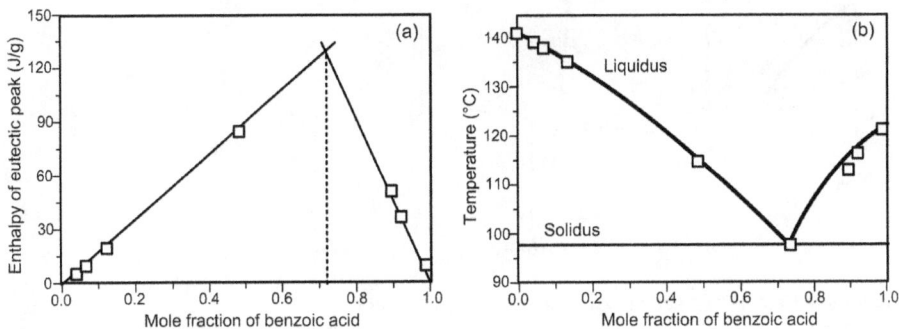

Figure 8.7: Determination of eutectic composition and SLE phase diagram of DMT/benzoic acid system (data from [267]).

Example 8.4 – Determination of ternary SLE phase diagram by DSC

Ozawa and Matsuoka [268] demonstrated the determination of the ternary SLE phase diagram of o-, m-, and p-nitroaniline (NA) system using DSC. Figure 8.8 illustrates the process path when a solid mixture containing the three components (point F) is heated in the DSC equipment. When the temperature reaches the ternary eutectic temperature at point 1, a liquid phase with the ternary eutectic composition begins to form. The melting process at this temperature continues until o-NA melts completely, leaving behind a binary solid mixture of p-NA and m-NA (point 1S). Afterward, the liquid composition moves along the p-NA/m-NA binary eutectic trough as the temperature rises and the solid is enriched in p-NA. At point 2, all m-NA has melted away, leaving behind pure p-NA in the solid phase (point 2S). Upon further heating, the process path turns away from the eutectic trough, as p-NA continues to melt. The solid disappears completely at point 3, which is located on the solubility surface of p-NA.

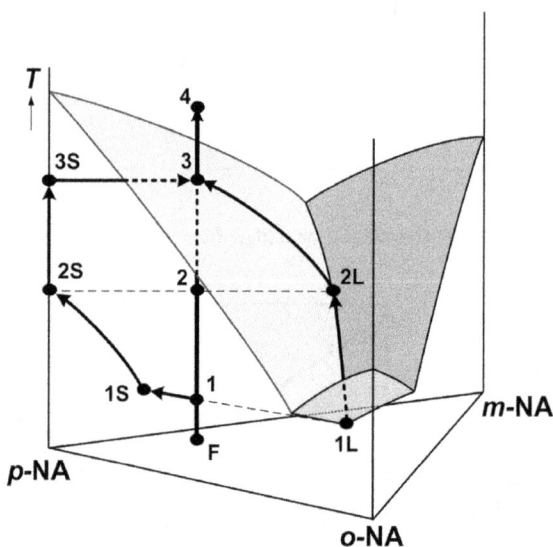

Figure 8.8: Heating of a solid mixture of nitroaniline isomers.

The resulting DSC curve typically looks like the schematic drawing shown in Figure 8.9. The first peak corresponds to the melting of the ternary eutectic (point 1 in Figure 8.8), the second one is attributed to the gradual melting of the binary eutectic mixtures (from 1 to 2), and the last one arises from the melting of the pure component (from 2 to 3). The peak areas (ΔH_{eut}, ΔH_{eub}, and $\Delta H'$) account for the heat of melting of solids from 1 to 1S, 1S to 2S, and 2S to 3S, respectively. Note that there is a significant overlap between the second and third peaks. One way to obtain their individual areas is by taking a cut at the valley between the two peaks, resulting in the shaded area as an estimate for ΔH_{eub} and the unshaded area for $\Delta H'$. The temperature of points 1, 2, and 3 can be approximated by the onset temperature T_o and peak temperatures $T_{p,2}$ and $T_{p,3}$, respectively.

In order to construct the ternary phase diagram, the three binary pairs of o-NA/m-NA, o-NA/p-NA, and m-NA/p-NA were studied first to obtain the binary eutectic compositions and temperatures.

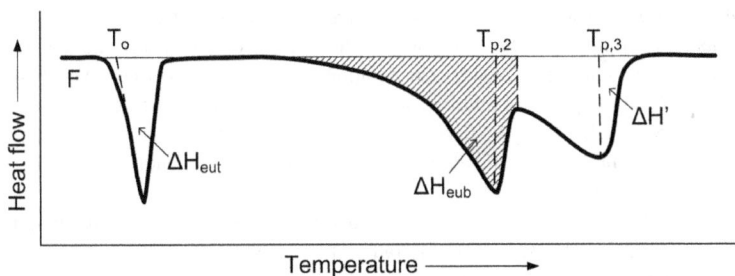

Figure 8.9: Schematic drawing of a DSC curve for a ternary eutectic mixture (reprinted from [268] with permission from Elsevier).

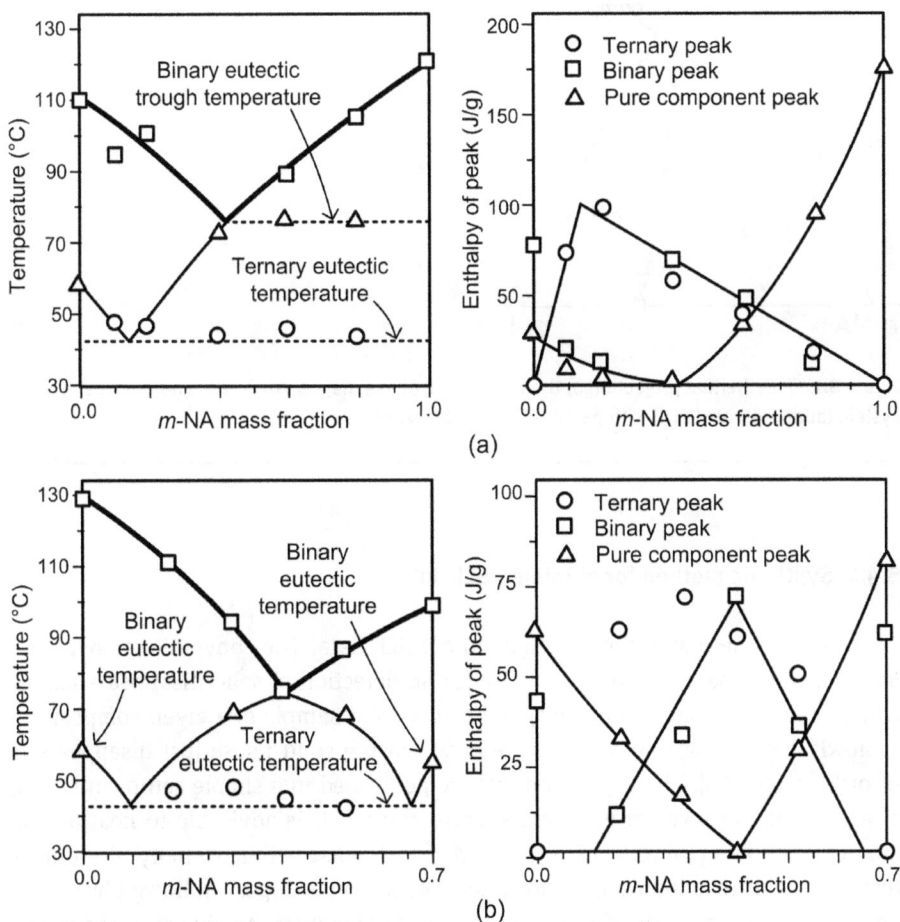

Figure 8.10: Polythermal cuts of the ternary SLE phase diagram and corresponding DSC peak enthalpies (reprinted from [268] with permission from Elsevier): (a) constant ratio cut at o-NA: p-NA = 1, (b) constant mass fraction cut at o-NA = 0.3.

Ternary mixtures along various polythermal cuts were then analyzed to obtain points on the solubility curves as well as to determine the location of the eutectic troughs. Figure 8.10 shows the plot of DSC peak temperatures and enthalpies against composition along two different cuts, which allows the determination of the temperature and composition of points along the eutectic trough. The cut at constant ratio of o-NA to p-NA (Figure 8.10a) is similar to the cut described in Figure 2.12c, while the cut at constant composition of o-NA (Figure 8.10b) is similar to the one in Figure 2.12b.

A total of five constant ratio cuts and six constant composition cuts were studied to determine all three binary eutectic troughs. The results were plotted on the polythermal projection shown in Figure 8.11. Excellent agreement with literature data was reported, demonstrating the effectiveness of using DSC for the determination of ternary SLE phase diagrams.

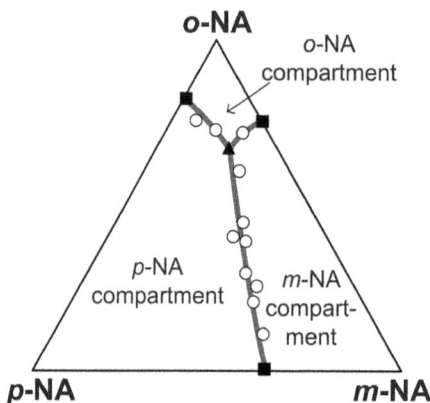

Figure 8.11: Polythermal projection of the ternary SLE phase diagram of o-, m-, and p-nitroaniline system (adapted from [268] with permission from Elsevier).

8.4.2 Synthetic method for obtaining SLE data

Besides DSC, there are other methods to get SLE data. The conventional synthetic method, which has several variants, relies on detection of solid disappearance (or reappearance). In the *step-warming method*, a solid sample of a given composition is slowly heated and the temperature at which the solid phase just disappears is recorded [269, 270]. The experiment can be performed in a simple setup consisting of a jacketed glass vessel with temperature control. It is advisable to heat up the mixture until everything turns into liquid first, to ensure homogeneity of the mixture. The mixture is then cooled down until some solids appear, followed by slowly heating it back up to identify the dissolution temperature. As shown in the phase diagram of Figure 8.12a, a point on the saturation curve (point 2) is obtained when the solid completely disappears. This method is recommended for generating a polythermal phase diagram, in which the saturation varieties are determined by

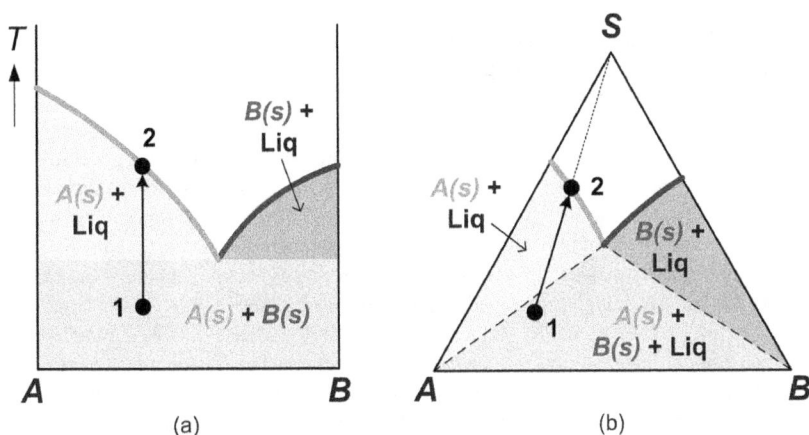

Figure 8.12: Synthetic methods for obtaining SLE data: (a) step-warming and (b) solvent addition.

repeating the experiment using different compositions of the solid sample. In the *solvent addition method*, the temperature is kept constant, while a solvent is added to a solid mixture until the solid just disappears [271]. A simple setup consisting of a jacketed vessel kept at a constant temperature is suitable for performing the experiment. Additional solvent is introduced using a syringe. By tracking the amount of solvent that has been added, a point on the saturation curve (such as point 2 in Figure 8.12b) can be located. This method is suitable for the determination of an isothermal phase diagram. In both methods, ample time should be allowed to ensure that equilibrium has been achieved.

Visual observation is the simplest way to detect the phase transition point, but it can be impractical, since it involves staring at the mixture for an extended period of time in order not to miss the moment at which the mixture experiences a change in visual appearance. Other possibilities include measuring opacity or conductivity, or relying on light scattering (Tyndall effect) and measuring the light transmittance to detect the presence of solid particles. Platforms for parallel processing and automation of data acquisition are available in the market, with varying capabilities and prices. Examples include the Crystal16® system [272], which allows the processing of 16 samples at the same time using the step-warming method, and the PolyBLOCK system [273], which can be equipped with an automatic solvent injection system, making it suitable for solvent addition method.

The principles of thermal and synthetic methods are combined in a method called discontinuous isoperibolic thermal analysis (DITA), which measures enthalpy change during the dissolution of a solid mixture using a highly accurate thermometer [274]. As solvent is added to a solid mixture such as point 1 in Figure 8.12b, slope disruptions in the plot of enthalpy change versus solvent amount are identified to locate region boundaries. As a result, both point 2 and the boundary between the A(s)+Liq and A(s)+B(s)+Liq regions can be obtained.

Example 8.5 – Solubility measurement using polythermal method

Chan [275] measured the solubility of adipic acid in water using the polythermal method. Six samples with measured quantities of adipic acid and water, with different compositions, were placed in jacketed glass vessels and subjected to stirring along with cooling and heating, following the same temperature profile. Solid appearance and disappearance were detected using a turbidity probe. The electronic signal from the probe was translated to light transmittance. A high value indicates a clear solution, while a low value suggests cloudiness due to the presence of solids.

Figure 8.13a shows the profile of transmittance and mixture temperature during the heating step for a sample containing 10.01 wt% adipic acid. As solids were present at the beginning, the transmittance started from a low value, and stayed nearly constant as temperature was ramped up. When the temperature reached about 50 °C, there was an abrupt increase in transmittance, followed by a more gradual increase until a maximum value was reached at about 120 mins. The transmittance stayed pretty much constant afterward, and the maximum value indicated that all solids have dissolved. By plotting transmittance against temperature, as shown in Figure 8.13b, the determination of the dissolution point is relatively straightforward. Taking the temperature at which the transmittance leveled up to the maximum value, the dissolution temperature was determined to be 52.8 °C.

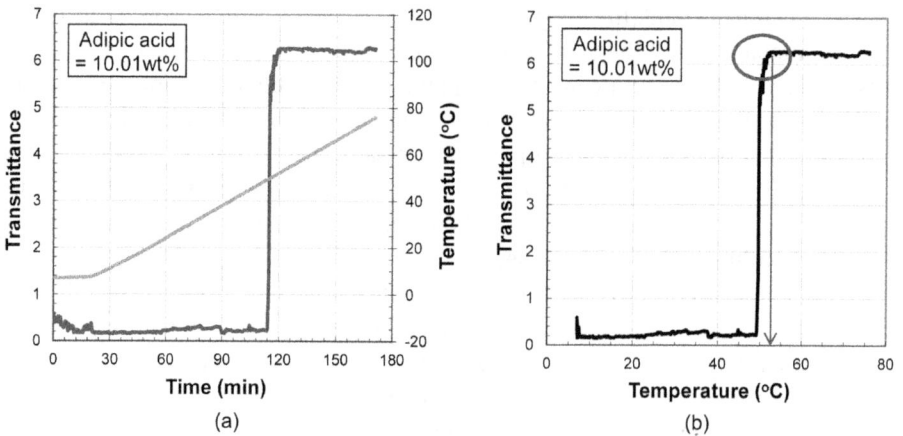

Figure 8.13: Dissolution temperature measurement of 10 wt% adipic acid solution in water: (a) transmittance and temperature profiles and (b) transmittance vs. temperature plot.

The dissolution temperatures of other samples with different adipic acid content were also determined. The results are plotted in Figure 8.14, alongside literature data from other studies [276, 277, 278], which were obtained using various methods. In particular, the data of Lin et al. were also obtained using the solid disappearance method, but the presence of solids was detected by visual observation [278].

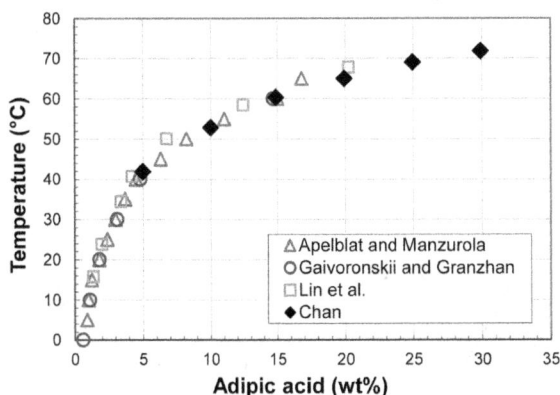

Figure 8.14: Dissolution temperature data of adipic acid/water mixtures.

8.4.3 Analytical method for obtaining SLE data

The most direct method for measuring SLE data is the analytical method, in which an excess amount of solid mixture is partially dissolved in a solvent at a constant temperature, and the liquid phase concentration is measured [279, 280]. If necessary, the solid phase can be extracted and analyzed as well. The experiment can be done using the same setup as the solvent addition method. To collect isothermal data for a ternary system containing A, B, and S as shown in Figure 8.15, start with a mixture of solid A and solid B (point F), and then add an amount of S while keeping the mixture at the desired temperature. If both A and B are still present in the solid phase after the addition of solvent, then the overall mixture is still in the two-solid region, as indicated by point 1. The liquid composition represents the double saturation point (point 1L), and the solid is a mixture of A and B (point 1S). If a relatively large amount of solvent is added, one of the components may completely dissolve, as in the case of point 2. The analysis of the solid phase would indicate that the solid only contains A (point 2S), and the liquid composition would correspond to a point on the saturation curve of A (point 2L).

In reality, the sampled solid particles are always wet with adhering mother liquor, so the actual composition would be at points 1S' and 2S', instead of 1S and 2S. Therefore, the true solid composition should be determined indirectly, by extrapolating the tie-line connecting the liquid and wet solid compositions. This so-called *Schreinemakers' wet residue method* [281] can be done graphically as indicated in Figure 8.15, but a mathematical extrapolation procedure [282] would be more practical for multicomponent systems.

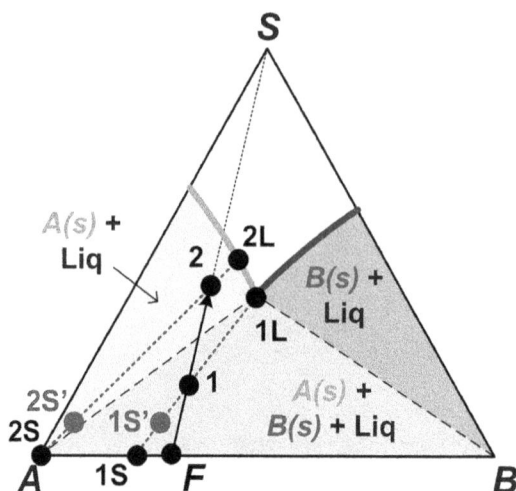

Figure 8.15: Analytical method for obtaining SLE data.

While the analytical method is straightforward, it is sometimes undesirable since sample preparation and analysis using HPLC or GC often require a significant amount of effort and cost. A simpler alternative, which is particularly applicable to measure the solubility of a single solute in a volatile solvent, is the *gravimetry method*. The solvent is kept in contact with an excess amount of solute solids at a constant temperature to produce a saturated solution. After equilibrium is achieved, a sample of the saturated solution is taken and weighed. The solvent is then completely removed by drying the weighed sample in an oven. The solubility can be determined by comparing the sample weight before and after drying.

8.4.4 Fitting of model parameters

Experimental SLE data can be used to calculate the activity coefficient using eq. (8.29), so that the parameters in the activity coefficient model can be fitted. In this connection, it is important to consider the design of experiments or data collection strategy. For molecular systems, the best strategy is to focus on the binary pairs first, because most activity coefficient models contain only binary interaction parameters. Ternary or higher interaction parameters are very rare. Therefore, the binary data would be sufficient to get a first estimate of all parameters in the model. In case getting binary data is impractical, such as between two solvent components with low melting points, the binary interaction parameters between those solvent components can be backed out from the solubility data in solvent mixtures.

To ensure that statistically meaningful parameter values can be obtained, the number of data points obtained has to be larger than (or at least equal to) the number of parameters in the model. For example, consider using the NRTL model (eqs. (8.44)–(8.47) for a ternary system shown in Figure 8.16. The model has three binary interaction parameters for each pair of components. Since the nonrandomness parameter, α_{ij}, is normally set to a constant value, there will be four parameters to be determined for each binary pair: a_{12}, a_{21}, b_{12}, and b_{21} (note that since the parameters are not symmetric, a_{12} is not equal to a_{21}). With three binary pairs, the total number of parameters is 12, which means at least 12 data points are needed. However, as a starting point, it is possible to neglect temperature dependence by assuming all b_{ij} to be zero, thereby reducing the number of adjustable parameters to 6. Therefore, at least 6 data points (2 data points for each binary pair) are needed. Data on the AB side can be obtained from DSC experiments, while those on the AS and BS sides are obtainable from the solubility data of A in S and B in S, respectively.

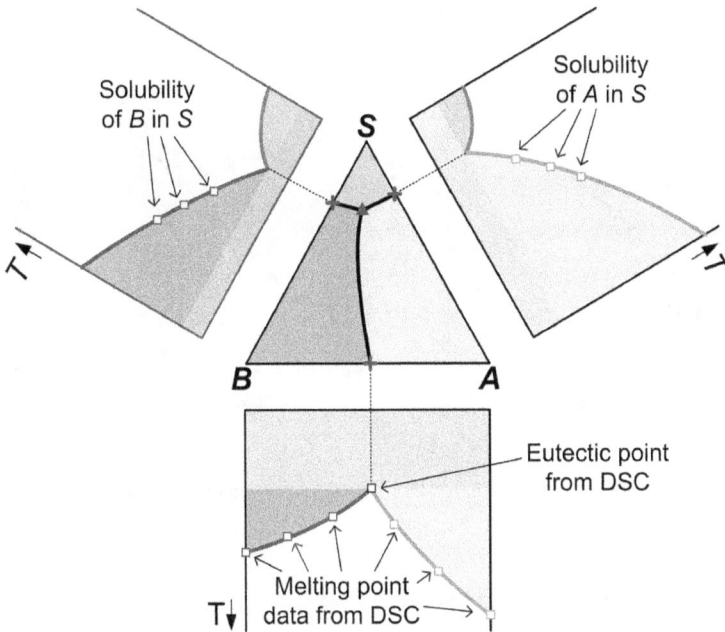

Figure 8.16: Strategy for determination of ternary SLE phase diagram.

Once the data have been collected, they can be used to fit the model parameters in the activity coefficient model by minimizing the difference between calculation results using the model and experimental data [283]. A convenient objective function can be chosen, such as:

$$\min F = \sum_k w_k \left(\gamma_i^{calc} - \gamma_i^{expt} \right)_k^2 \tag{8.77}$$

where w_k is the weighting factor for data point k. Other functions based on other comparable quantities such as mole fraction or temperature can also be used.

Example 8.6 – Solubility of aspirin in ethanol
Aspirin (acetylsalicylic acid) is a widely used anti-inflammatory drug, which is typically produced by esterification reaction of salicylic acid followed by crystallization. Owing to its industrial significance, solubility data of aspirin in several organic solvents, including ethanol, have been reported in the literature [205]. The data are shown in the T-x diagram of aspirin/ethanol binary system in Figure 8.17a. The open squares represent the experimental data points, and the solid curve is the calculated solubility curve assuming ideal behavior. The values of melting point and heat of fusion are listed in Table 8.7.

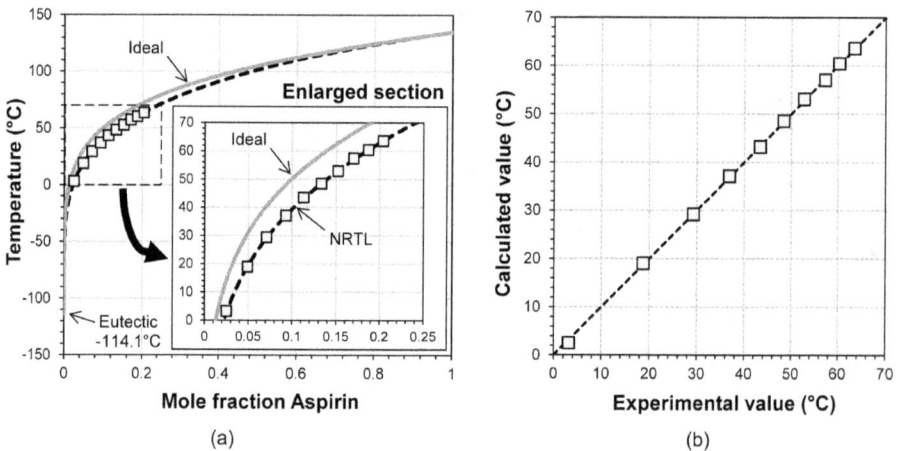

Figure 8.17: Aspirin solubility in ethanol: (a) binary SLE phase diagram and (b) dissolution temperature parity plot.

Table 8.7: Values of melting point and heat of fusion for SLE calculation.

Component	Melting point (K)	Heat of fusion (J/kmol)
Aspirin	408.15	2.98×10^7
Ethanol	159.05	4.931×10^6

It is clear from the comparison that the ideal solubility curve does not match the data. This can be fixed by using the NRTL activity coefficient model, with parameters obtained by regression of the experimental data. Only a_{ij} parameters are regressed, with b_{ij} and α_{ij} set to their default values of 0 and 0.3, respectively. Table 8.8 summarizes the regressed parameter values, which lead to an excellent match between the calculated and experimental dissolution temperature, as suggested by the parity plot in Figure 8.17b. The calculated solubility curve using these parameters is plotted as the dashed curve in Figure 8.17a.

Table 8.8: NRTL model parameters obtained from regression of experimental data.

Component *i*	Component *j*	a_{ij}	b_{ij}	α_{ij}
Aspirin	Ethanol	−1,121,460	0	0.3
Ethanol	Aspirin	7,622.52	0	0.3

Example 8.7 – Solubility of diamantane in common solvents

Diamondoids are saturated hydrocarbons consisting of fused cyclohexane rings in the form of a cage-like structure. They occur naturally in petroleum and have been used as indicators of natural oil cracking [284]. Diamondoids and their derivatives possess attractive optical and electronic properties for use in various nanotechnology devices and materials. It is expected that crystallization-based separation processes can be used for recovering diamondoids from petroleum. Chan et al. [285] reported the solubility data of several diamondoids and their derivatives in various organic solvents as measured by the solid disappearance method.

Figure 8.18a shows the reported solubility of diamantane ($C_{14}H_{20}$) in four organic solvents as a function of temperature, plotted in the format of a binary $T-x$ diagram. The data have been fit using the NRTL activity coefficient model, with melting point and heat of fusion values listed in Table 8.9 and NRTL parameter values summarized in Table 8.10. A parity plot showing the comparison between calculated and experimental solubility values is given in Figure 8.18b.

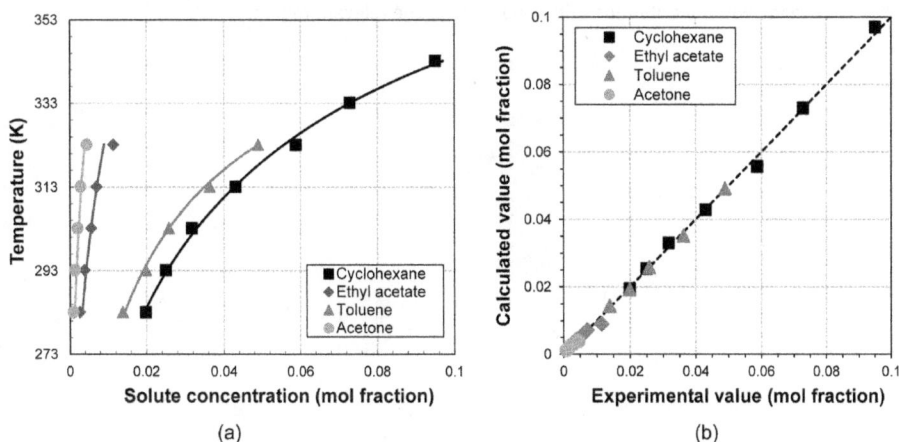

Figure 8.18: Solubility of diamantane in selected solvents: (a) SLE phase diagram, (b) parity plot.

Table 8.9: Values of melting point and heat of fusion for SLE calculation.

Component	Melting point (K)	Heat of fusion (J/kmol)
Diamantane	509.65	9.205×10^6
Cyclohexane	279.69	2.740×10^6
Ethyl acetate	189.60	1.048×10^7
Toluene	178.18	6.636×10^6
Acetone	178.45	5.774×10^6

Table 8.10: NRTL model parameters obtained from regression of experimental data.

Component i	Component j	a_{ij}	b_{ij}	α_{ij}
Diamantane	Cyclohexane	−1,307,000	0	0.3
Cyclohexane	Diamantane	7,349,040	0	0.3
Diamantane	Ethyl acetate	−576,947	0	0.05
Ethyl acetate	Diamantane	10,061,500	0	0.05
Diamantane	Toluene	−2,380,740	0	0.3
Toluene	Diamantane	9,959,170	0	0.3
Diamantane	Acetone	660,993	0	0.02
Acetone	Diamantane	10,969,800	0	0.02

A reasonable fit can be obtained for all data sets without regressing b_{ij}, although lower values of α_{ij} have to be used for ethyl acetate and acetone, in which the solubility of diamantane is very low. While these parameters give a good fit in the vicinity of the experimental data, they may lead to dubious results when they are used to calculate the SLE phase diagram over the entire composition range. For example, the parameters for diamantane/ethyl acetate binary system lead to a discontinuity in the solubility curve, as shown in Figure 8.19. While such a discontinuity may indicate the presence of a liquid-liquid phase split (solid-liquid-liquid equilibrium behavior), no conclusion should be made without obtaining additional data at the relevant region.

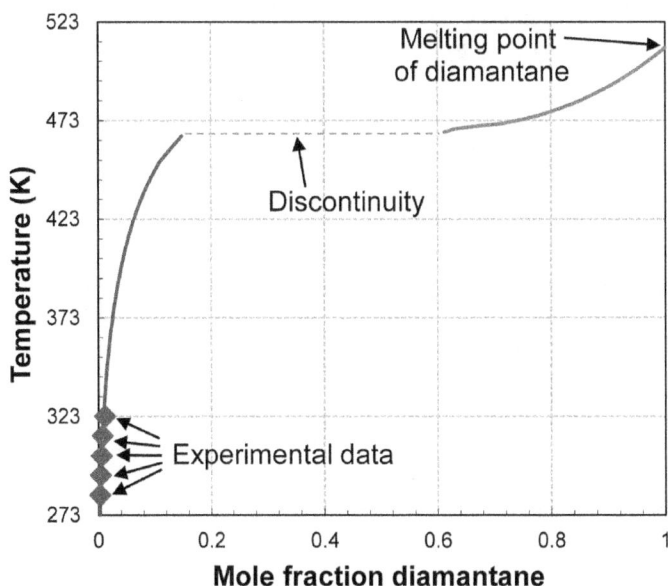

Figure 8.19: Calculated SLE phase diagram of diamantane/ethyl acetate over the entire composition range.

Example 8.8(I) – Regression of ternary system solubility data

Consider the ternary system involving raw material A, drug intermediate B, and solvent S discussed in Example 3.11. To determine the phase diagram, the melting points and heats of fusion of pure A and pure B were measured by DSC. Three sets of experiments were performed to determine the solubility of the two solutes in solvent S, by setting the concentration of solvent at three relatively constant levels (around 82 wt%, 85 wt%, and 92 wt% of S). The ratio of $B/(A+B)$ in each set was varied from 0 to 1 to cover the composition space. Each mixture was subjected to heating, and its dissolution temperature (at which all solids disappeared) was recorded. Basically, the experimental data represent scattered points on the solubility surfaces of A and B, plotted as black squares in Figure 8.20. The numbers next to the squares indicate the measured dissolution temperatures.

Using the dissolution temperature data and the measured values of melting point and heat of fusion, the parameters of the NRTL activity coefficient model were obtained, as summarized in Table 8.11. A reasonable fit between calculated and experimental values were achieved, as shown in the parity plot (Figure 8.21). The squares, circles, and triangles represent the data sets at 82 wt%, 85 wt%, and 92 wt% of S, respectively. The model was then used to calculate the isothermal cuts at different temperatures, as indicated in Figure 8.20.

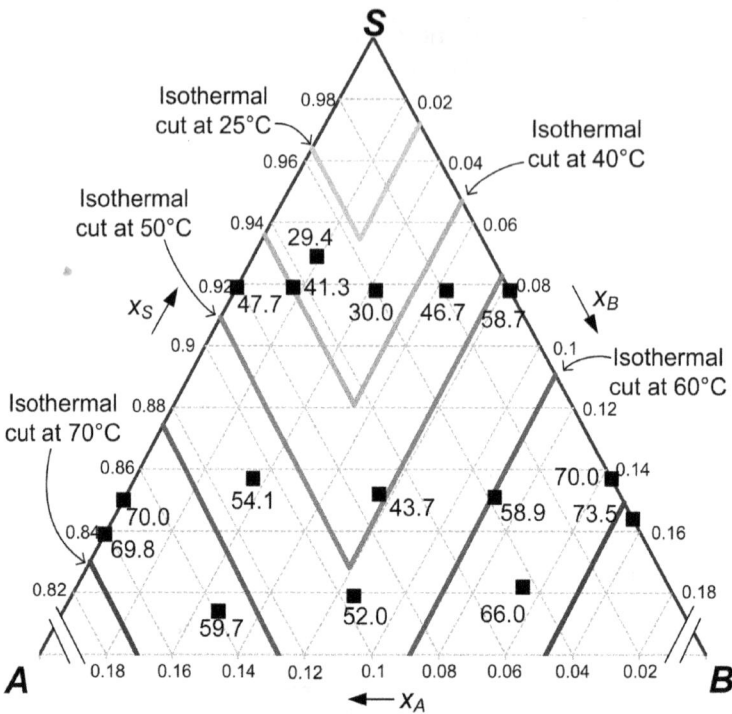

Figure 8.20: Dissolution temperature data for a drug intermediate system (data courtesy of Sunovion Pharmaceuticals, Process Engineering [286]).

Table 8.11: NRTL parameter values obtained from regression of dissolution temperature data.

Component i	Component j	a_{ij}	b_{ij}	α_{ij}
A	B	-2.555×10^6	0	0.3
A	S	2.885×10^6	0	0.3
B	A	-2.781×10^6	0	0.3
B	S	2.480×10^6	0	0.3
S	A	1.454×10^6	0	0.3
S	B	1.073×10^6	0	0.3

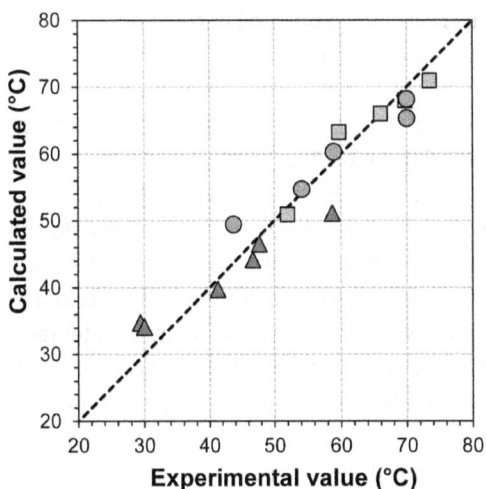

Figure 8.21: Parity plot between experimental and calculated values of the dissolution temperature.

8.4.5 Boundary verification

With the SLE phase diagram in hand, it is possible to synthesize the process and approximately determine the suitable operating conditions. Furthermore, the most important region that needs to be determined with better accuracy can be identified. For example, if the objective is to separate A and B from a solution as illustrated in Figure 8.22, the most important region is the neighborhood of the compartment boundary, near points 4, 5, and 6. This is where further verification is

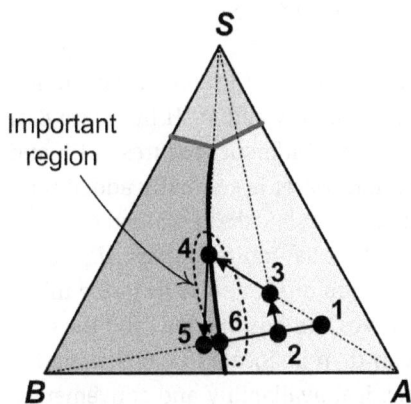

Figure 8.22: Important region for boundary verification based on process synthesis.

needed, since no data of ternary mixtures have been collected at this point. It is not necessary to take data on the S-rich side, because the process points are not expected to fall into that area.

Additional experiments can be performed to verify the location of the boundary near the important region. As illustrated in Figure 8.23, cuts at constant concentration of S can be taken, giving pseudo-binary diagrams featuring the solubility curves of A and B. Using the step-warming method, the intersection between the two curves can be identified by taking several data points on both curves. The identity of solids on opposite sides of the suspected intersection point can be checked to confirm the location. Alternatively, the double saturation points at various temperatures can be directly determined using the analytical method. In any case, the additional experiments provide a better estimate of the boundary location. These new data points can then be included in the next round of model parameter regression, so that a more accurate SLE model is obtained. If necessary, the design of the process and/or the crystallizer conditions are adjusted in the subsequent iteration, based on the updated image of the phase diagram.

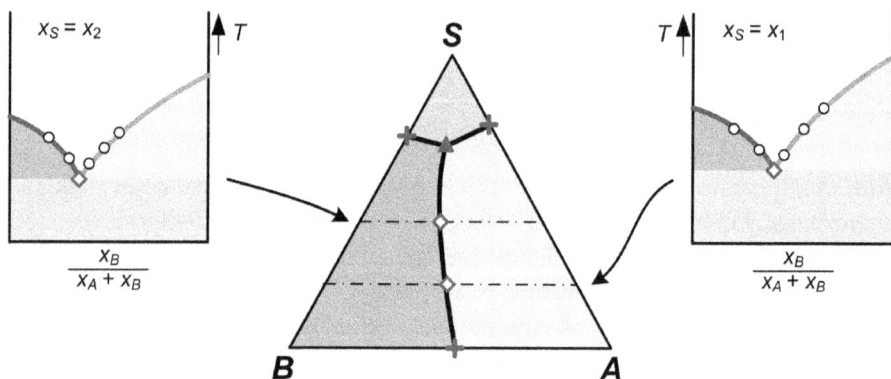

Figure 8.23: Experimental determination of SLE compartment boundary.

A similar strategy can be applied for electrolyte systems. For example, the isothermal phase diagram of the acid–base system involving HX and MOH in Figure 8.24 can be determined using the solvent addition method. Different mixtures of HX and MOH are prepared (e.g., point 1 in Figure 8.24), and water is gradually added until all solids dissolve. The amount of water needed for complete dissolution at constant temperature is determined, so that a point along the saturation curves AB, BC, or CD is obtained. By starting with a different proportion of HX and MOH in the initial mixture, different points along the saturation curves can be obtained, and the double saturation points such as points B and C can then be located by observing the intersection of those curves. Depending on chemical availability and convenience, mixtures of HX and MX or MX and MOH can also be used as a starting point.

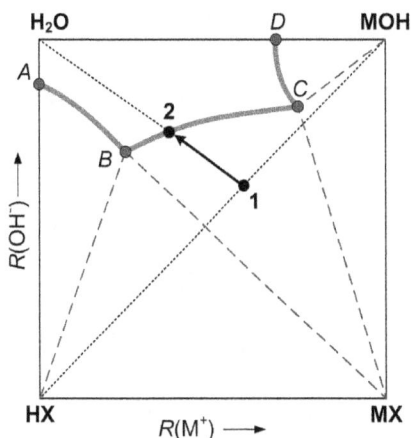

Figure 8.24: Synthetic method for obtaining SLE data of an electrolyte system.

Example 8.9 – Boundary verification in an adduct-forming system

Kwok et al. [118] measured the ternary phase diagram of phenol/BPA/water system. To verify the boundary between the adduct and BPA compartments, several cuts at constant water composition were taken, and the dissolution temperature of mixtures along the cuts were determined using the step-warming method. Figure 8.25 displays the solubility data taken along different cuts. Since the phenol-BPA adduct exhibits peritectic behavior, both the adduct and BPA solubility curves decrease in the direction of decreasing BPA content. Nonetheless, it is still possible to detect the point at which the gradient of the solubility curve changes, which corresponds to a point on the compartment boundary. For further verification, samples of solids from opposite sides of the apparent boundary were taken and analyzed by HPLC.

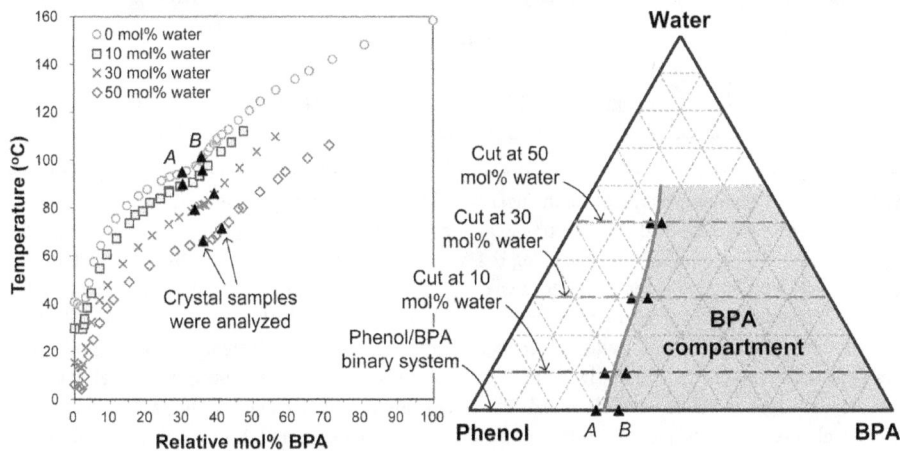

Figure 8.25: SLE experimental data for Phenol/BPA/water system.

An example of the HPLC results is shown in Figure 8.26. It was found that the solids from the left side of the boundary (point *A*) contained phenol and BPA (indicating that it was an adduct) while solids from the right side of the boundary (point *B*) contained only BPA. From this finding, it can be concluded that the change of gradient, indeed, corresponds to the boundary between adduct and BPA compartments.

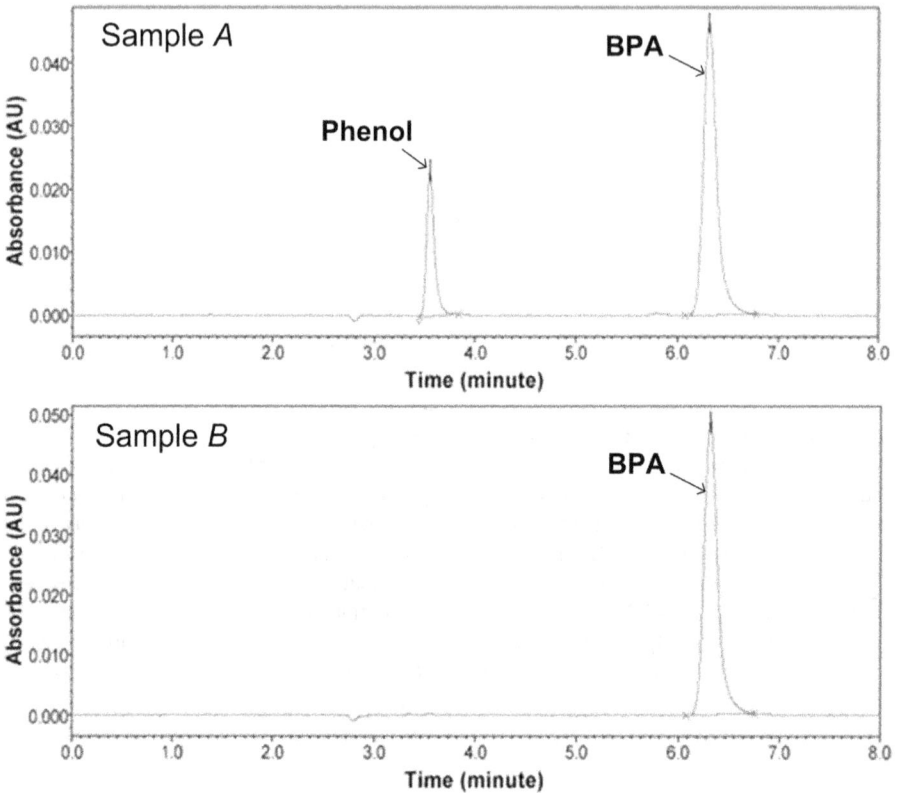

Figure 8.26: HPLC analysis results of dissolved crystal samples taken near the boundary [118].

Example 8.10 – Boundary verification in succinic acid/ammonium hydroxide system
As discussed in Example 3.10, the SLE phase diagram for succinic acid (SA)/sulfuric acid/ammonium hydroxide system is important for designing the succinic acid crystallization process. This SLE phase diagram was experimentally measured by Kwok [85]. Figure 8.27 illustrates the determination of solubility curves and verification of region boundaries for the cut at $SO_4^{2-}/OH^- = 0$. To obtain the solubility curves, different mixtures of SA and diammonium sulfate (DAS) were prepared. Two of such mixtures are depicted as points 1S and 2S, in Figure 8.27. Water was then added to each mixture at constant temperature until all solids were dissolved. The compositions at which solid mixtures 1S and 2S were completely dissolved are plotted on the phase diagram as points 1 and 2, respectively. Data points obtained from other solid mixtures are represented by open diamonds on the figure. Based on the trends of these data points, three solubility curves, corresponding to SA, MAS, and DAS, respectively, can be readily located. The intersection between two solubility curves defines the location of the double saturation point, which is then used to identify the different crystallization regions.

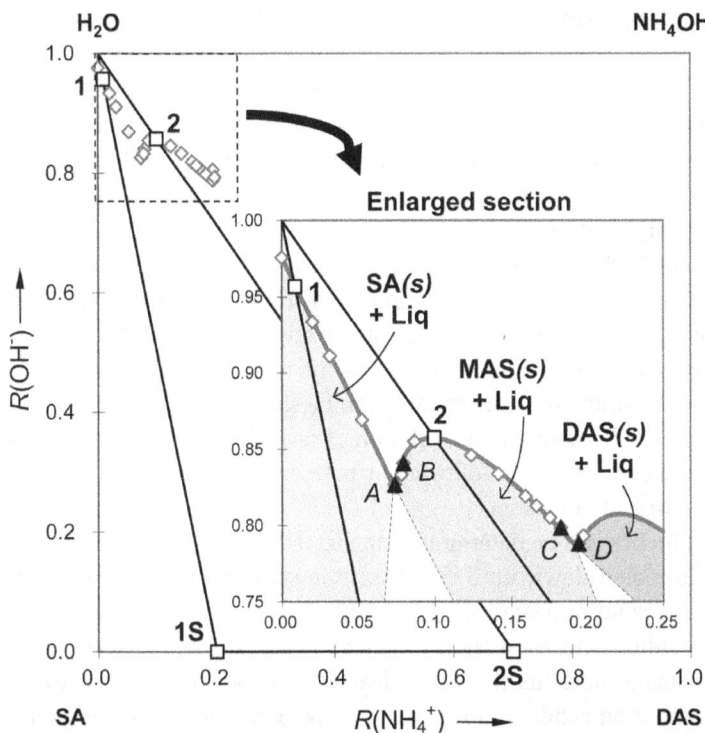

Figure 8.27: Determination of SLE phase diagram of succinic acid/NH$_4$OH system.

To verify the location of the double saturation points, four crystal samples were taken from mixtures close to the perceived locations, namely samples A and B near the SA/MAS double saturation point and samples C and D near the MAS/DAS double saturation point. These samples were obtained by slightly cooling down the mixture after complete dissolution, so that a small amount of crystals was formed. These crystal samples were analyzed by ion chromatography to determine the molar ratio of ammonium to succinate ions, so that the identity of the solids could be confirmed. A molar ratio of 0, 1, and 2 would correspond to SA, MAS, and DAS, respectively. The results, summarized in Table 8.12, indicate that the SA/MAS double saturation point is indeed located between A and B, and the MAS/DAS double saturation point is located between C and D.

Table 8.12: Results of ion chromatography analysis of crystal samples (data from [85]).

Sample	Succinate ion concentration		Ammonium ion concentration		NH$_4^+$/succinate molar ratio	Conclusion
	mg/L	mmol/L	mg/L	mmol/L		
A	34.65	0.30	0.17	0.01	0.03	SA
B	32.31	0.28	5.40	0.30	1.07	MAS
C	51.72	0.45	7.31	0.41	0.91	MAS
D	29.17	0.25	8.76	0.49	1.93	DAS

8.5 Experimental Determination of Crystallization Kinetics

Nucleation and growth kinetics, as well as the dependence of nucleation and growth rates on supersaturation, is needed to solve the population balance equations. While theoretical models are available, empirical power-law models are the most commonly used to represent growth kinetics. Experiment is required since the parameters in such empirical model cannot be estimated *a priori*. There are two basic experimental methods to determine nucleation and growth rates. The first method is to separately determine nucleation and growth rates by isolating crystal growth from nucleation process. For example, a single crystal subjected to a flow of saturated solution can be observed under a microscope to measure its growth rate [287]. Alternatively, nucleation and growth rates can be simultaneously determined in the same experiment, by observing the evolution of several output variables such as PSD or actual liquid concentration, followed by mathematical treatment to separate the effect of growth and nucleation.

The nucleation kinetics can be determined via metastable zone width measurement. The solution is cooled slowly until crystallization starts, which is determined by optical detection of solid appearance. Nývlt [288] suggested that at a constant supersaturation generation rate, for a short time when spontaneous nucleation occurs (that is, at the metastable limit), the nucleation rate can be assumed to be equal to the supersaturation generation rate. Since supersaturation is generated by cooling, it follows that

$$B = - \frac{dc^*}{dt} = \frac{dc^*}{dT} \left(-\frac{dT}{dt} \right) \tag{8.78}$$

where dc^*/dT is the slope of solubility curve and $-dT/dt$ is the cooling rate. Following eq. (6.1), the supersaturation, which is constant during this short period of time, is:

$$\Delta c = \frac{dc^*}{dT} \cdot \Delta T_{max} \tag{8.79}$$

where ΔT_{max} is the metastable zone width (the maximum supercooling before crystallization occurs), which depends on the cooling rate. In general, ΔT_{max} increases with the cooling rate. Substituting eqs. (8.78) and (8.79) into the power law expression for B (eq. (6.4)) and taking natural logarithm:

$$\ln \left(-\frac{dT}{dt} \right) = \ln k_b + (b-1)\ln \left(\frac{dc^*}{dT} \right) + b \ln \Delta T_{max} \tag{8.80}$$

Therefore, by plotting $\ln \Delta T_{max}$ against $\ln(-dT/dt)$, the parameters of the nucleation kinetic equation can be estimated from the slope and intercept. Nagy et al. [289] extended this method to simultaneously determine the nucleation and growth kinetic parameters, taking into account concentration and PSD information measured online. With advances in process analytical technology, *in situ* measurement of particle size and count in the slurry using probes such as focused beam reflectance measurement (FBRM), attenuated total reflection Fourier transform infrared spectroscopy (ATR-FTIR), and Raman spectroscopy is becoming standard practice [290]. It has been shown that the data from such measurement can be used quantitatively in a calibration-free manner to estimate crystal growth rate [291].

Alternatively, growth rate kinetics can be obtained in batch setup by performing seeded crystallization using seed crystals of a known size at a supersaturation level, within the metastable zone. Assuming that there is no nucleation and the growth rate is constant, the growth kinetics can be determined by measuring the change in crystal size or mass via analysis of the desupersaturation curve [292]. Expressing the supersaturation as a function of time, the parameters of the power-law expression for growth rate (eq. (6.6)) can be determined as

$$g = \frac{2k_a \Delta c_0}{3\rho_s k_v L_s A_s} + \frac{\Delta \dot{c}_0 \Delta \ddot{c}_0}{\Delta \dot{c}_0^{\,2}} \tag{8.81}$$

$$k_g = \frac{-\Delta \dot{c}_0}{A_s \Delta c_0^{\,g}} \tag{8.82}$$

where $\Delta \dot{c}$ and $\Delta \ddot{c}$ are the first and second derivatives, respectively, of the supersaturation with respect to time, and index zero refers to the initial value. ρ_s is the crystal density, k_a and k_v are shape factors, L_s is the average size of the seeds, and A_s is the surface area of the seeds.

One example of a continuous laboratory setup for nucleation and growth rate measurement is the MSMPR crystallizer shown in Figure 8.28. A feed solution is introduced to the crystallizer at a constant rate, and the perfectly mixed slurry is

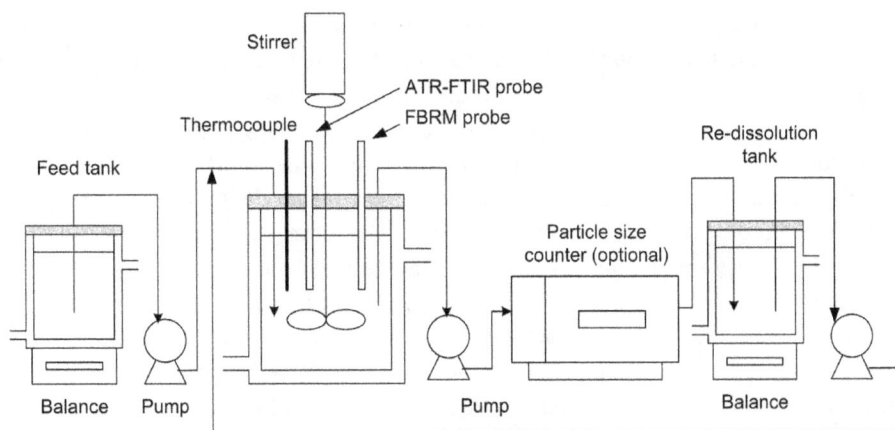

Figure 8.28: Continuous setup for measuring crystallization kinetics.

pumped out at the same rate. The slurry is reused after passing it through a re-dissolution tank to dissolve the crystals. The PSD is measured after a steady-state has been achieved (normally about 3 times the residence time), and the nucleation and growth rates are evaluated from the population density plot [293].

The expression for population density as a function of size (eq. (7.7)) can be written as

$$n(L) = n_0 \exp\left(-\frac{L}{G\tau}\right) \tag{8.83}$$

where $n_0=B/G$ is the population density at $L=0$, and $\tau=Q_{out}/V$ is the residence time. Under ideal conditions, the relationship between $\ln n$ and L is linear, with a slope of $-1/G\tau$ and an intercept of $\ln n_0$. In practice, the distribution often deviates from the theoretical straight line for very small particle sizes due to difficulties in accurate detection, but an effective rate of nucleation is obtained by extrapolating the straight portion to $L = 0$. By repeating the experiment at different flow rates, different supersaturation levels in the crystallizer can be obtained, so that the dependence of nucleation and growth rates on supersaturation can be evaluated.

8.6 Summary

This chapter covers how SLE phase diagrams and crystallization kinetics can be obtained via an integrative approach combining modeling and experimental activities. Using equations stemming from the thermodynamic criterion for stability together with an activity coefficient model allows the calculation of the SLE behavior and phase diagrams. However, experiments are needed to determine the model parameters and to validate model predictions. In line with the overall objective, the experimental effort focuses on key features such as compartment boundary. When experiment is difficult or impossible, estimation methods can be used to predict melting point, heat of fusion, and solubility, but at the end, there is really no substitute for good data in crystallization conceptual design. By starting from minimum information and iterating among the activities with increasing accuracy, crystallization process development can be done with minimum time, effort, and resources.

Various software tools are available to support the modeling activities in crystallization process development. Process simulators often include the capability of physical property prediction, using various methods. Specially developed software packages for estimation of physical properties, including heat of fusion, melting point, and solubility parameter, are also available in the market. For example, ICAS (Integrated Computer Aided System) [264] includes tools for property prediction based on various methods as well as other computer-aided tools for process design and synthesis, while Cranium [294] uses Joback group contribution method to estimate solubility parameter as well as various other properties. COSMOtherm [295] calculates activity coefficients and solubility using the COSMO-RS method. A soft-ware package called SLEEK (Solid–Liquid Equilibrium Engineering Kit) [296] has also been developed to assist in modeling of SLE phase behavior, creation and visu-alization of SLE phase diagrams, as well as regression of experimental SLE data to thermodynamically consistent models.

Exercises

8.1. The melting point and heat of fusion data for *ortho*- and *para*-nitrochlorobenzenes are given below.

Component	T_m (K)	ΔH_f (J/kmol)
o-NCB	306.15	1.92×10^7
p-NCB	356.65	1.41×10^7

Assuming ideal liquid phase solution and no solid compounds are present,
a. Calculate the eutectic temperature and composition.
   Hint: Use the solubility equation, with $y = 1$, to calculate the saturation curves of o-NCB and p-NCB. Note that these two curves intersect at the eutectic point.
b. Plot the binary SLE phase diagram (temperature *vs.* composition) for this system.

8.2. The physical properties of the xylenes system, which is known to exhibit near-ideal simple eutectic behavior, are as follows:

Component	T_m (K)	ΔH_f (J/kmol)
p-Xylene	286.41	1.711×10^7
m-Xylene	225.3	1.157×10^7
o-Xylene	247.98	1.36×10^7

a. Calculate the composition and temperature of the binary and ternary eutectics.
b. Plot the two-dimensional polythermal projection of the ternary SLE phase dia-gram involving these three components.
   Hint: Locate the eutectic points on a composition triangle, and then plot the bi-nary eutectic troughs connecting them. Temperatures should be ignored in this projection.
c. Plot an isothermal cut at −40 °C (233.15 K) of the ternary SLE phase diagram, and then identify the single and double saturation regions on the diagram.

Hint: Since this temperature is above the melting point of *m*-xylene, there is no saturation region of *m*-xylene in the isothermal cut.

8.3. Consider a conjugate salt system involving ions A^+, B^+, X^- and Y^-, which only forms simple salts at 25 °C. The values of K_{sp} at that temperature are as follows:

Component	K_{sp}
AX	2.46×10^{-6}
BX	1.24×10^{-5}
AY	8.25×10^{-6}
BY	7.33×10^{-7}

a. Assuming ideal behavior, plot the Jänecke projection of the phase diagram at 25 °C, showing the double and triple saturation points as well as the saturation troughs.
b. How would the Jänecke projection change if the Debye–Hückel limiting law (eq. (8.50)) is used to calculate the activity coefficients? The value of Debye–Hückel constant (A) for aqueous solution at 25 °C can be taken as 1.172 $mol^{-1/2}kg^{1/2}$.

8.4. Use Marrero and Gani's method to predict the melting point of *o*-xylene, *m*-xylene, and *p*-xylene. The following values for relevant first- and second-order group contribution toward the melting point are listed in the original publication [259].

First-order groups	aCH	0.586
	aC-CH$_3$	1.0068
Second-order groups	AROMRINGS^{1}s^2	−0.6388
	AROMRINGS^{1}s^3	−0.6218
	AROMRINGS^{1}s^4	0.984

There is no third-order group that needs to be considered. For melting point, the left-hand side of eq. (8.64) is defined as $f(X)=\exp(T_m/T_{m,0})$, with $T_{m,0}=147.450$ K.

Hint: Count the number of occurrences of each group in the molecules, and then use eq. (8.64) to calculate $f(X)$. Refer to Example 8.1 and the original publication for the meaning of the group names.

8.5. The solubility of aspirin in acetone, based on the experimental data reported by Maia and Giulietti [205], is shown as follows:

Temperature (K)	281.9	290.6	297.9	304.4	310.6	315.3	319.8	323.3	326.3
Mole fraction aspirin	0.061	0.075	0.088	0.101	0.114	0.127	0.139	0.151	0.162

a. Show that for a binary system, the NRTL activity coefficients are given by

$$\ln \gamma_1 = x_2^2 \left[\tau_{21} \left(\frac{G_{21}}{x_1 + x_2 G_{21}} \right)^2 + \frac{\tau_{12} G_{12}}{(x_1 G_{12} + x_2)^2} \right] \tag{8.84}$$

$$\ln \gamma_2 = x_1^2 \left[\tau_{12} \left(\frac{G_{12}}{x_1 G_{12} + x_2} \right)^2 + \frac{\tau_{21} G_{21}}{(x_1 + x_2 G_{21})^2} \right] \tag{8.85}$$

Hint: Apply eq. (8.34) to the expression for excess Gibbs free energy given by eq. (8.44).

b. Using the deviation between experimental and calculated mole fraction as the objective function to be minimized, that is:

$$\min F = \sum_k \left(x_i^{calc} - x_i^{expt} \right)_k^2 \tag{8.86}$$

determine a suitable set of NRTL parameters that can fit the given data. For simplicity, set $b_{ij}=0$ and $\alpha_{ij}=\alpha_{ji}=0.3$. The melting point and heat of fusion of aspirin is 408.15 K and 2.98×10^7 J/kmol, respectively.

Hint: Simultaneously solve the solubility equation, eq. (8.29), and the expression for NRTL activity coefficient, eq. (8.84), to obtain the calculated mole fraction at each temperature.

8.6. To develop the SLE phase diagram of a binary system consisting of A and B, several samples with different compositions are analyzed using DSC. The DSC curves are shown in Figure 8.29 along with the onset temperature and enthalpy data.

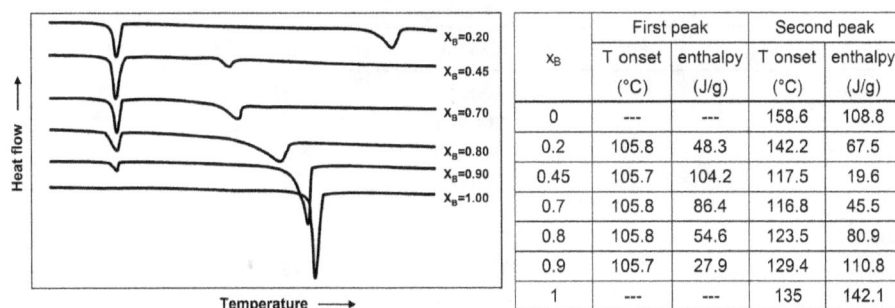

x_B	First peak		Second peak	
	T onset (°C)	enthalpy (J/g)	T onset (°C)	enthalpy (J/g)
0	---	---	158.6	108.8
0.2	105.8	48.3	142.2	67.5
0.45	105.7	104.2	117.5	19.6
0.7	105.8	86.4	116.8	45.5
0.8	105.8	54.6	123.5	80.9
0.9	105.7	27.9	129.4	110.8
1	---	---	135	142.1

Figure 8.29: DSC data for a binary system (problem 8.6).

a. Which peaks are pure component peaks and which ones are eutectic peaks?
b. Based on the information, determine the composition of the eutectic point.
c. Compare the results with calculation results, assuming ideal behavior.

8.7. Karnaukhov [297] investigated the ternary SLE phase behavior of $NaNO_3/NH_4NO_3/H_2O$ system at 25 °C. Different mixtures of the salts and water were prepared and kept at constant temperature until equilibrium was reached, and then both the liquid and solid phase compositions were measured. Since the solid phase was not washed, some liquid phase was included in each solid phase sample. A part of the data is shown below.

Data point	Liquid phase composition (wt%)			Solid phase composition (wt%)		
	$NaNO_3$	NH_4NO_3	H_2O	$NaNO_3$	NH_4NO_3	H_2O
1	47.29	0.00	52.71	87.91	0.00	12.09
2	42.49	11.47	46.04	85.58	2.87	11.55
3	35.77	22.83	41.40	73.95	9.20	16.85
4	32.87	30.13	37.00	76.06	9.04	14.90
5	29.65	36.45	33.90	69.40	14.40	16.20
6	25.72	42.98	31.30	72.83	16.60	10.57

Based on the data, does $NaNO_3$ form a solid solution with NH_4NO_3?

8.8. A continuous crystallization experiment is performed in an MSMPR bench-scale crystallizer with a slurry volume of 300 mL. The feed solution is introduced to the crystallizer at a constant rate of 300 mL/h, and the slurry is pumped out at the same rate. After a steady-state has been achieved, the solid content in the slurry is measured to be 100.5 g. The crystals can be assumed to be spherical in shape, with a density of 1,426 kg/m^3. The particle size distribution is measured by microscopic observation. The results are represented as the number of particles having a size within certain ranges, as summarized as follows:

Size range (µm)	Representative size (m)	Count
0–100	0.00005	950
100–200	0.00015	735
200–300	0.00025	492
300–400	0.00035	311
400–500	0.00045	270
500–600	0.00055	206
600–700	0.00065	153
700–800	0.00075	103
800–900	0.00085	72
900–1,000	0.00095	49
1,000–1,100	0.00105	31
1,100–1,200	0.00115	22
1,200–1,300	0.00125	11

(continued)

Size range (µm)	Representative size (m)	Count
1,300–1,400	0.00135	7
1,400–1,500	0.00145	4

From the data, prepare a population density plot and determine the nucleation rate B [no/m^3.s] and growth rate G [m/s].

Hint: Population density $n(L)$ [no/m^4] is the number of particles in the range represented by size L per unit volume or slurry divided by the interval size ΔL. To determine this, first calculate the total volume of particles in the sample as well as the magma density (mass of solids per unit volume of slurry).

9 Concluding Remarks

9.1 A Hierarchical, Multiscale, Integrative Approach

The flowsheet and operating conditions of a chemical plant are determined in conceptual design. It takes place at the early stage of the plant design and is followed by detailed engineering design, procurement, and construction. The amount of effort for conceptual design is small compared to the total effort but the ultimate economic performance of a plant is largely determined at this stage. Thus, it is imperative to select the best conceptual design alternative for crystallization processes. To achieve this goal systematically and efficiently, process synthesis, modeling, and experiments are used in an integrative manner. In addition, the designer is encouraged to tackle crystallization as a multiscale problem by considering issues that might seem to be outside of the conceptual design. For example, company strategy at the enterprise scale is taken into account because the economic performance of a plant has to be aligned with the company's overall financial and strategic goals. Molecular design is considered because the molecular structure of the product compound and its interactions with other molecules (e.g., the formation of an adduct) would affect the SLE phase behavior and plant design. In general, multiscale issues are tackled in a hierarchical manner. Issues classified by their length and time scales are resolved successively from coarse to fine scales.

The integrative approach has three major components as summarized in Table 9.1. Topics discussed in this book that are related to each component are shown in Figure 9.1. The first component involves the representation of the solid–liquid equilibrium phase behavior in the form of phase diagrams for molecular or electrolyte and reactive or nonreactive systems using the proper coordinates. Phase diagrams with up to four components can be readily visualized, sometimes using polythermal projections and isothermal cuts. These standard types of phase diagrams are sufficient for most systems that generally contain a limited number of major components that might undergo crystallization. For process designers interested in systems with more components than what can be represented using the well-established phase diagrams, a general method has been developed for cutting and projecting a high-dimensional phase diagram into a series of low dimensional projections and cuts. With practice, the high-dimensional phase diagram can be mentally visualized using the low-dimensional ones.

The second component is the crystallization strategy that produces process alternatives by purposefully moving around on the phase diagram. Table 9.2 summarizes the features to look for and the required information for several different process objectives. For example, if the purpose is to crystallize a desired product, the process point has to travel to the corresponding single saturation crystallization region. If it is necessary to crystallize another product thereafter, the double saturation variety that

https://doi.org/10.1515/9781501519901-009

Table 9.1: Components of the integrative approach to conceptual design of crystallization processes.

Component	Explanation
SLE phase behavior representation	Using phase diagram visualization techniques along with thermodynamic models to identify crystallization regions in composition space where one or more species can crystallize out
Crystallization strategy	Moving around in composition space, crossing eutectics if necessary, and crystallizing the desired product(s) by combining basic operations that include temperature, pressure, or composition swing
Multiscale plant design	Accounting for kinetic effects and crystallization downstream processing

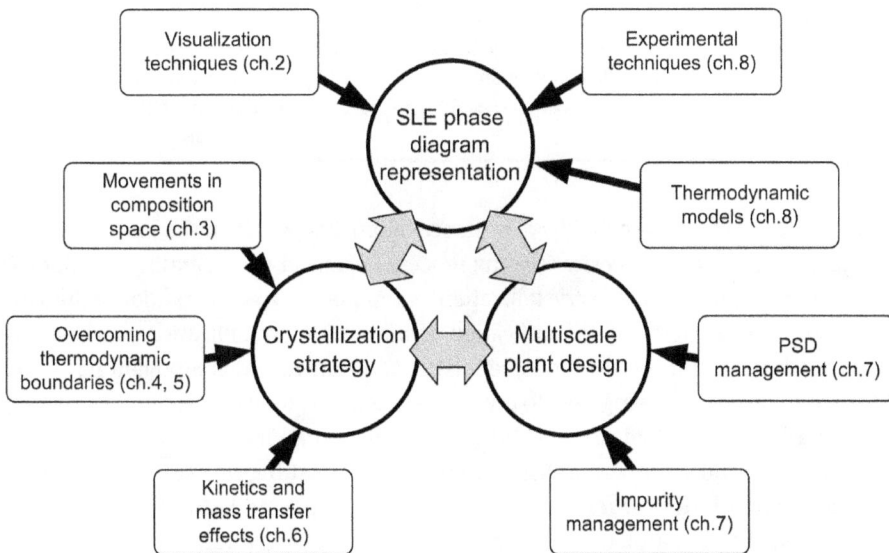

Figure 9.1: Integrative approach to conceptual design of crystallization processes.

needs to be crossed to reach the crystallization region of the other product has to be identified. The length of the crystallization path between the chosen feed point location (i.e., feed composition) and the boundary of the crystallization region has to be as long as possible to maximize the product yield. Similarly, the end point of the crystallization path (i.e. the exit mother liquor composition) has to be placed sufficiently far away from the boundary of the crystallization region to avoid the co-precipitation of the neighboring component.

After deciding on the features to look for in the phase diagram, moves are made by operations such as cooling, heating, component addition and removal. Then, the

Table 9.2: Features and required information for achieving different objectives.

Objective	Feature to look for	Required information
Crystallize the desired product	Crystallization region of the desired product	Identification of the movement to crystallize out the desired product
Crystallize another product in the next step	Location of the double saturation variety between the two products	Feasibility of moving into the crystallization region of the other product by crossing the double saturation variety, upon temperature or composition swing
Achieve maximum yield of a product	Distance between the feed composition and the boundary of the crystallization region	How to locate the feed point to maximize the distance between the feed point and the boundary along the crystallization path
Achieve the desired purity of a product	Distance between the exit mother liquor composition and the boundary of the crystallization region	How to locate the end point to minimize the distance between the end point and the boundary along the crystallization path without compromising purity

process alternative represented by a sequence of process points on the phase diagram can be translated into the corresponding process flowsheet and operating conditions. By including all the available crystallization techniques on a single platform, the synthesis strategy promotes the identification of innovative crystallization processes. It may be counter-intuitive, but for some mixtures, an unwanted component has to be crystallized out before being able to crystallize out the desired component. Of course, all the traditional crystallization techniques can be represented by a combination of the moves, as shown in Table 9.3. Drowning-out or salting-out are specific cases of component addition where an anti-solvent or a salt is added to the solution to initiate crystallization of the desired product. Reactive crystallization is yet another special case of component addition where a reaction is formed after combining two or more streams.

The third component of the integrative approach completes the design for the whole plant by moving up and down the length scale. At the smaller length scale, it takes into account nucleation and growth kinetics, reaction kinetics, and transport. In general, nucleation and growth kinetics tend to be scale-dependent and are hard to predict. Thus, the related crystallization parameters tend to be measured in industry. Similarly, reaction kinetics are often determined experimentally. A lot is known about mixing in a crystallizer, albeit empirically. Also, while computer simulation can reliably predict the multiphase fluid flow field in a crystallizer, it is still a challenge to include particle breakage and agglomeration in modelling the dynamics of a crystallizer. At the plant scale, the crystallizer is designed in conjunction with the

Table 9.3: Traditional classification of crystallization processes versus movements in composition space.

Traditional classification	Movements in composition space
Cooling crystallization	Temperature swing to reach the saturation variety of the desired product
Evaporative crystallization	Solvent removal to reach the saturation variety of the desired product
Drowning-out or salting-out crystallization	Addition of a mass separating agent to reach the saturation variety of the desired product
Reactive crystallization	Combination of reactant streams to reach the saturation variety of the desired product
Adductive crystallization	Facilitate product recovery by adding another component such that an adduct is formed
Extractive crystallization	Bypassing the crystallization boundary with the creation of an additional degree of freedom of movements by adding another component
Fractional crystallization	Moving from one compartment to another by combination of heating, cooling, solvent addition, and solvent removal

downstream processing system consisting of filters, washers, and dryers. In fact, the third component represents the transition between process synthesis and process development in which pilot plant testing and technology transfer are emphasized. The reader is encouraged to consult a large collection of books and articles on how to build a real plant from the conceptual design developed using this hierarchical, multiscale, integrative approach [30, 31, 298, 299, 300, 301].

9.2 Outlook

This book focuses on the conceptual design of primarily well-established crystallization processes to produce organics, inorganics, and pharmaceuticals that are in solid form. The design methods and the systems thinking in this approach can be applied in other areas that may increase in significance, two of which are discussed in the paragraphs that follow.

One area is sustainability. In a circular economy, crystallization is expected to play an increasingly significant role in the recovery and reuse of materials with finite reserves. A case in point is the recovery of metal salts from spent lithium-ion batteries in electric vehicles (EV) [302]. Given the growth in the number of EVs worldwide, it is essential that the metals in these batteries be recycled. The process for recovering metals from lithium nickel manganese cobalt oxide (NMC for short), which is a preferred cathode material for automotive batteries, is illustrated in Figure 9.2 [303]. Lithium is first selectively dissolved by adding oxalic acid solution, leaving behind Ni, Co, and Mn oxalates in the residual solids. Since some manganese is also dissolved along with Li, KOH is added to the liquid to selectively precipitate $Mn(OH)_2$ before crystallizing out Li_2CO_3 by adding K_2CO_3. Ni and Co can similarly be recovered by selectively dissolving them in ammonium solvent and then separately precipitating them as $Ni_2C_2O_4 \cdot 2H_2O$ and $Co(OH)_2$, respectively. Insights on the SLE phase behavior of relevant systems, especially the saturation boundaries and compartments for recovering various components, can be used to identify the maximum yield in dissolution and precipitation processes, as well as to come up with process optimization alternatives. For example, the identification of the Li_2CO_3 crystallization region in the SLE phase diagram involving Li^+, K^+, CO_3^{2-}, $C_2O_4^{2-}$, OH^-, and H_2O helps determine the optimum crystallization condition for recovering lithium in high yield and high purity [304].

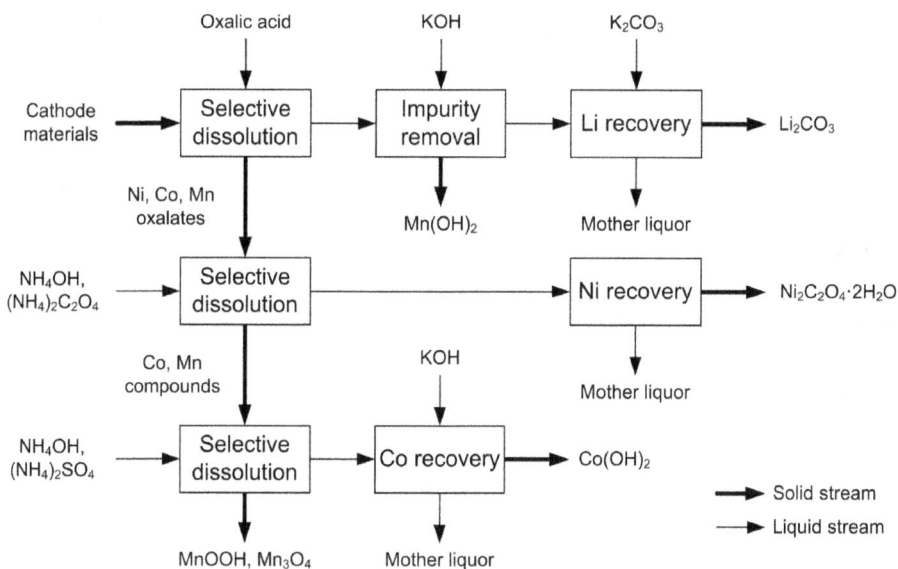

Figure 9.2: Process flowsheet for metals separation and recovery from cathode materials (adapted from [303]).

Recycling of rare earth elements (REE) (the 15 elements in the lanthanide series plus Scandium and Yttrium), which are another example of materials with finite reserves, from REE-containing industrial residues and end-of-life products such as fluorescent lamps and permanent magnets is also an important issue in sustainability. The separation techniques employed for recovering REE from natural sources, such as flotation, leaching, and solvent extraction, are also applicable to the recovery and purification of REE from end-of-life products [305]. Although commercial separation of individual REEs is mostly done by solvent extraction [306], selective dissolution and precipitation often plays a role in separating impurities from REEs as well as recovering the REEs from purified solutions. For example, iron can be removed from an industrial residue with high iron concentration and low rare earth content by selectively precipitating the REEs as phosphate salts [307]. Insights on the relevant SLE phase behavior would be useful in maximizing the REE recovery and minimizing the precipitation of iron. Kinetically-controlled reactive crystallization processes have also been suggested as a possible alternative to solvent extraction. For example, the mixture of neodymium (Nd) and samarium (Sm) chlorides can be separated by adding boric acid and controlling the reaction time to selectively crystallize Nd borate–chloride salt over its Sm counterpart [308].

Another interesting potential application of crystallization-based separation and purification techniques in sustainability is the recovery of valuable chemicals from biosources. For example, hyaluronic acid (HA), a popular skin-hydrating cosmetic ingredient, can be isolated from eggshell, which is produced in large amounts by the egg processing industries [309, 310]. After mechanically separating the eggshell membrane from a cracked eggshell, HA can be obtained via enzymatic digestion or chemical hydrolysis of the eggshell membrane. This yields a crude extract that also contains other biomaterials such as proteins and nucleic acids. It has been reported that adding salt and alcohol to an HA-containing extract led to the precipitation of a mixture of HA and protein [311]. After redissolving this precipitate in a mixture of water and alcohol, protein can be selectively precipitated by adjusting the salt and alcohol content, leaving behind an HA-enriched supernatant. Therefore, the application of anti-solvent and protein crystallization techniques with the understanding of the relevant SLE phase behavior, as discussed in Chapter 5, would be useful to come up with purification process alternatives to isolate high-purity HA from the crude extract.

In the past two decades, chemical engineering as a profession has gradually expanded from process design to product design [312, 313]. As part of this development, another possible growth area for crystallization is design of chemical products, many of which are solids. An example is the lithium nickel manganese cobalt oxide cathode material, $LiNi_xMn_yCo_{1-x-y}O_2$ ($0 < x+y < 1$), which is made by sintering of $LiCO_3$ and $NMC(OH)_2$ [314]. The compound salt, $NMC(OH)_2$, is in turn produced by co-precipitation of a mixture of $NiSO_4$, $MnSO_4$, and $CoSO_4$ in stoichiometric ratio [315]. Common NMCs include NMC111, NMC532, NMC622, and NMC911, where the three-digit number indicates the molar ratio between nickel, manganese, and cobalt. Another example involves metal selenides which are attractive anode materials for Li-ion batteries. Using solvent-directed self-assembly, NiSe-based anode with diverse morphologies, including nanosheet, porous flower, hexagonal plate, and multishell microsphere, can be obtained after selenization. These anode materials possess distinct electrochemical performance [316]. Yet another example is the formation of inorganic and metallic nanoparticles. Indium tin oxide nanoparticle and its inclusion in the polyvinyl butyral interlayer of solar control windows has been a commercial product for a while [317, 318]. The synthesis of copper nanoparticles and its use in a conductive inkjet ink for printed electronics have been an active R&D area [319, 320]. The formation of these solid products has been primarily empirical in nature.

The body of knowledge in this book will hopefully contribute to finding solutions to these and many other challenges in the conceptual design of crystallization, and more broadly solid formation, processes.

Appendix: Solutions to Selected Exercise Problems

Problem 2.4

In the absence of other information, the connections from binary eutectic points $E1$, $E2$, and $E3$ to the ternary eutectic point $E4$ can be assumed as straight lines.

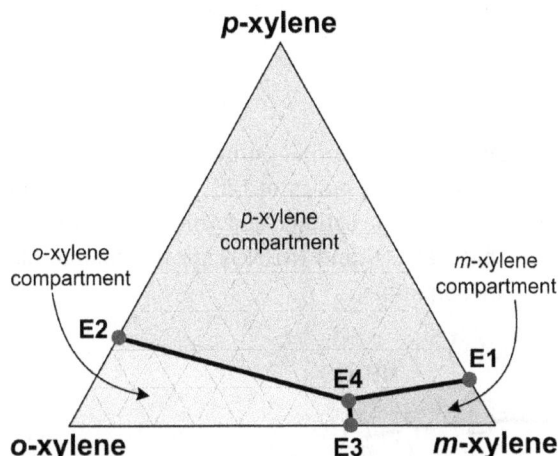

Problem 2.5

From eq. (2.14), $N_{P, \, min} = \left(\frac{5-1}{2}\right)^2 + 0 = 4$. Therefore, three other projections are needed.

The two-dimensional projection with B and C as the reference components can be obtained by simply ignoring x_B and x_C, and then normalizing the remaining mole fractions to obtain

$$X_A = \frac{x_A}{x_A + x_D + x_E}, \quad X_D = \frac{x_D}{x_A + x_D + x_E}, \quad \text{and} \quad X_E = \frac{x_E}{x_A + x_D + x_E}.$$

For example, the normalized mole fractions of the quinary eutectic point $ABCDE$ are $X_A = 0.054/(0.054 + 0.210 + 0.711) = 0.055$, $X_D = 0.210/(0.054 + 0.210 + 0.711) = 0.215$, and $X_E = 0.711/(0.054 + 0.210 + 0.711) = 0.730$. The normalized mole fractions of all the other vertices can be obtained in the same way and plotted on a triangle. The edges connecting those vertices can then be drawn according to the given adjacency information.

Problem 2.6

The system involves six components ($C = 6$) and 2 reactions ($R = 2$), so the dimension of the phase diagram is $C - R - 1 = 3$. With D and E as reference components,

https://doi.org/10.1515/9781501519901-010

$$v_A^T = [\,-1 \quad 0\,], \; v_B^T = [\,0 \quad -1\,], \; v_C^T = [\,-1 \quad -1\,], \; v_{TOT}^T = [\,-1 \quad -1\,]$$

$$V_{\text{ref}} = \begin{bmatrix} v_{D,1} & v_{D,2} \\ v_{E,1} & v_{E,2} \end{bmatrix} = \begin{bmatrix} 1 & 0 \\ 0 & 1 \end{bmatrix}$$

Therefore, substituting into eq. (2.16), the canonical coordinates are

$$X_1 = \frac{x_A + x_D}{1 + x_D + x_E}, \; X_2 = \frac{x_B + x_E}{1 + x_D + x_E} \; \text{and} \; X_3 = \frac{x_C + x_D + x_E}{1 + x_D + x_E}$$

Problem 2.8

To plot the Jänecke projection, only the double and triple saturation points need to be identified. From the SLE data, the number of moles of Na^+, NH_4^+, NO_3^-, and HCO_3^- for each point can be calculated so that the cationic and anionic coordinates can be obtained. The plot indicates that $NaHCO_3$ and NH_4NO_3 are the compatible salts, while $NaNO_3$ and NH_4HCO_3 are incompatible.

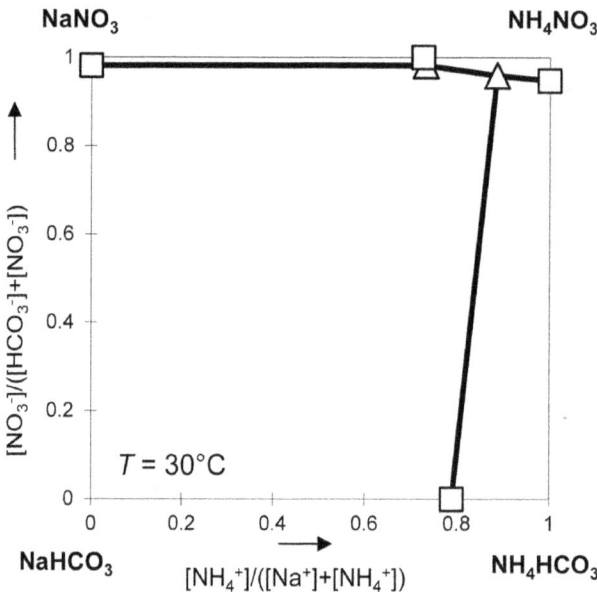

Problem 3.6

Cooling would cause crystallization of component A because the point is located in the compartment of A. Since crystallization essentially removes A from the liquid, the liquid composition has to move in the direction away from vertex A. Maximum recovery of pure solid A is achieved at point 2.

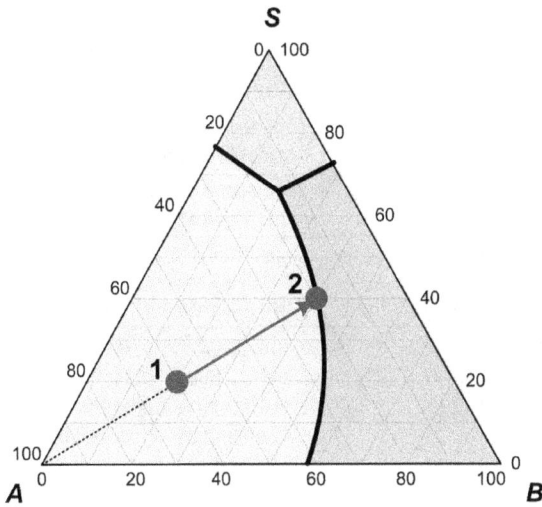

By lever rule, the ratio of solid to liquid is (length of 1–2)/(length of A–1) = 1. Hence, 1 kg of the initial mixture (containing 60%, or 0.6 kg of A) would produce 0.5 kg of solids, which is pure A. Therefore, the maximum recovery of the pure solid is 0.5/0.6 = 83.3%.

Problem 3.7

The overall mixture would have the composition 25% A, 25% B, and 50% C, as indicated by point F. Since it falls inside the double saturation region of A and B, both A and B solids crystallize out.

Pure product B can be obtained by adding more solvent to the mixture so that point F moves into the single saturation region of B. Alternatively, increasing the temperature would move the saturation curves away from C (since the solubility increases with temperature) so that point F would be located inside the single saturation region of B at the higher temperature.

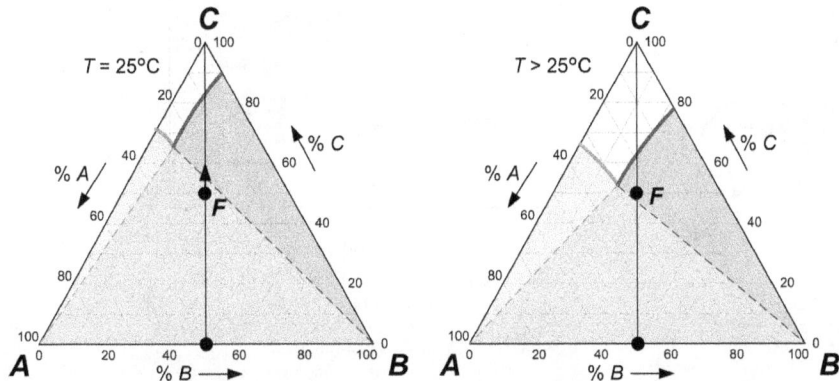

Problem 3.8

The composition can be converted to mass fraction by taking a basis for the total ionic concentration, for example $[H^+] + [Na^+] = [OH^-] + 2[Glu^{2-}] = 1$ kmol. The ionic concentration can then be expressed in terms of H_2O, H_2Glu, and NaOH by noting that $[NaOH] = [Na^+]$ and $[H_2Glu] = [Glu^{2-}]$. The water concentration is then calculated from the remaining $[H^+]$ or $[OH^-]$ (either way gives the same result).

Point	Concentration (kmol/kmol)				Concentration (kg/kmol)			Mass fraction		
	H^+	Na^+	OH^-	Glu^{2-}	H_2O	H_2Glu	NaOH	H_2O	H_2Glu	NaOH
1	1.0000	0.0000	0.8824	0.0588	15.882	8.653	0.000	0.6473	0.3527	0.0000
2	0.9685	0.0315	0.8819	0.0591	15.307	8.687	1.260	0.6061	0.3440	0.0499
3	0.9371	0.0629	0.8952	0.0524	14.981	7.709	2.516	0.5944	0.3058	0.0998
4	0.8421	0.1579	0.7368	0.1316	10.421	19.355	6.316	0.2887	0.5363	0.1750
5	0.9048	0.0952	0.8810	0.0595	14.143	8.756	3.810	0.5295	0.3278	0.1426

Stream 1 is obtained by only adding water to 100 kg/h H_2Glu → $F_1 = 100/0.3527 = 283.5$ kg/h.

The product, MSG, can be considered as a combination of NaOH + H_2Glu (1:1 molar ratio) and by overall balance, the NaOH portion must be equal to NaOH feed → the flow rate of 50 wt% NaOH solution added is $F_N = (100*40/141.7)/0.5 = 54.4$ kg/h.

Balance over NaOH addition: $F_2 x_{H2Glu,2} = F_3 x_{H2Glu,3}$ and $F_2 + F_N = F_3$ → $F_2 = 436.1$ kg/h. Therefore, $F_5 = F_2 - F_1 = 436.1 - 283.5 = 152.5$ kg/h.

Problem 4.1

The flowsheet can be obtained starting from the evaporation of S1 from stream F to crystallize A at 80 °C, followed by the addition of S1 to reach the saturation region of B at 40 °C. Stream 4, which is obtained after crystallizing B, is simply recycled to close the loop. Since the feed of the evaporator shifts from point F to point 5, the location of point 1 also shifts but everything else stays the same.

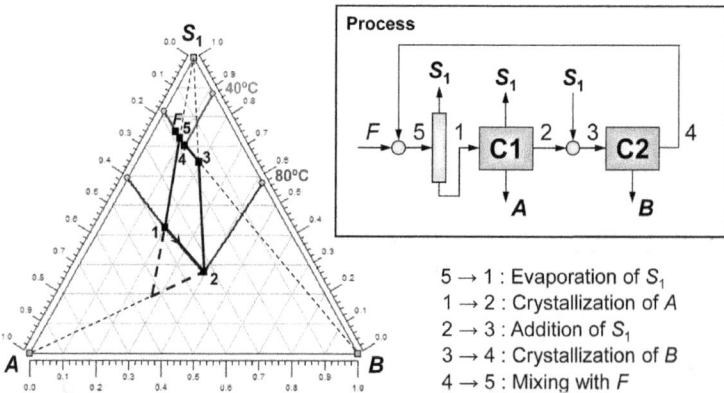

5 → 1 : Evaporation of S_1
1 → 2 : Crystallization of A
2 → 3 : Addition of S_1
3 → 4 : Crystallization of B
4 → 5 : Mixing with F

Problem 4.2

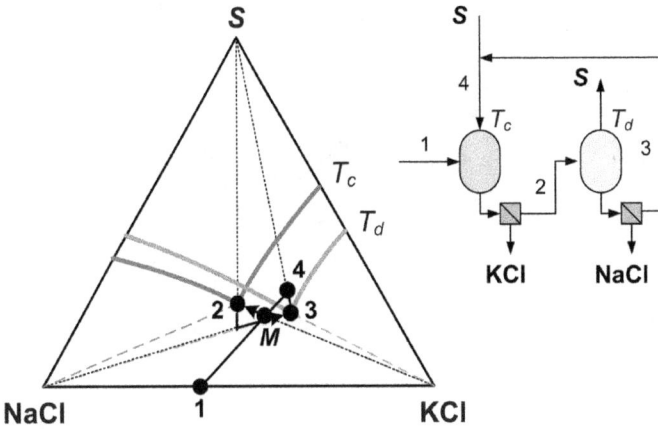

The composition of NaCl and KCl in the feed (the location of point 1) is likely to determine which process alternative is more economical.

Problem 4.3

$F_F = 1,000$ kg/h ($F_{A,F} = 200$ kg/h and $F_{B,F} = 800$ kg/h)

By overall balance, $F_7 = 800$ kg/h, $F_8 = 200$ kg/h, and $F_5 = F_6$.

Mass balance of A around dissolver: $F_{A,F} = F_1 x_{A,1} \rightarrow F_1 x_{A,1} = 200$

Mass balance of A around mixing point: $F_1 x_{A,1} + F_4 x_{A,4} = F_2 x_{A,2}$ (a)

Mass balance of A around crystallizer C1: $F_2 x_{A,2} = F_3 x_{A,3}$ (b)

Combining (a) and (b): $F_1 x_{A,1} + F_4 x_{A,4} = F_3 x_{A,3}$ (c)

Reading from the phase diagram, $x_{A,3} = 0.3$, $x_{A,4} = 0.35 \rightarrow 200 + 0.35 F_4 = 0.3 F_3$ (d)

Mass balance of B around crystallizer C2: $F_3 x_{B,3} = F_4 x_{B,4}$

Reading from the phase diagram, $x_{B,3} = 0.2$, $x_{B,4} = 0.35 \rightarrow 0.2 F_3 = 0.35 F_4$ (e)

Combining (d) and (e): $200 = 0.1 F_3 \rightarrow F_3 = 2,000$ kg/h, $F_4 = 400/0.35 = 1,143$ kg/h.

Overall balance around crystallizer C2: $F_3 = F_4 + F_6 + F_8 \rightarrow F_6 = 2,000 - 1,143 - 200 = 657$ kg/h.

Therefore, $F_1 = 1,000 + 657 = 1,657$ kg/h $\rightarrow x_{A,1} = 200/1,657 = 0.12$, $x_{B,1} = 800/1,657 = 0.48$.

$x_{S,1} = 1 - 0.12 - 0.48 = 0.6$, which defines the location of stream 1 on the phase diagram.

Since $T_c < T_h$, any saturation curve passing through point 1 should correspond to a temperature above T_h. In other words, the saturation temperature of stream 1 is above T_h. As stream 1 is an unsaturated solution, the dissolver temperature should be above the saturation temperature, which implies that it must be above T_h as well.

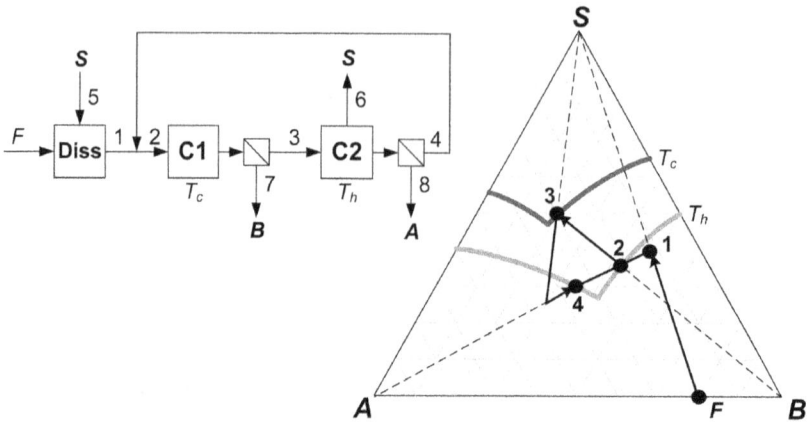

Problem 4.4

The process is similar to the one shown in Figure 4.12.

The feed can be considered as a mixture of x mol of HA and y mol of Na_2B in water (to make things simpler, 1 kg of water is used as the basis). Therefore, $[H^+] = [A^-] = x$, $[Na^+] = 2y$, and $[B^{2-}] = y$.

From the diagram, the composition of point F is given by

$R(Na^+) = [Na^+]/([Na^+] + [H^+]) = 0.25 \rightarrow x = 6y$

$R(A^-) = [A^-]/([A^-] + 2[B^{2-}]) = 0.75 \rightarrow x = 6y$ (same conclusion)

By overall balance, the same amount of HA and Na_2B are obtained as products. Therefore, the molar ratio of HA/Na_2B products = 6.

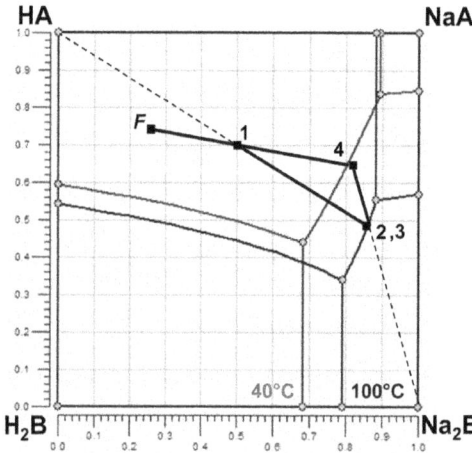

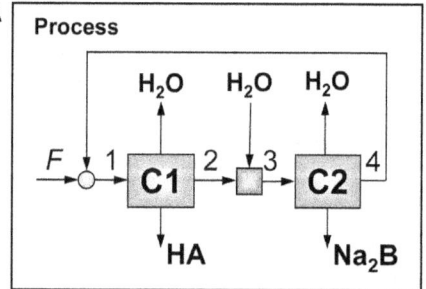

Process

$F \rightarrow 1$: Mixing with recycle
$1 \rightarrow 2$: Crystallization of HA
$2 \rightarrow 3$: Addition of water
$3 \rightarrow 4$: Crystallization of Na_2B

Problem 4.5

The total concentration of ions in the brine = 10 wt% → 1,000 kg of brine = 100 kg salts + 900 kg water.

From the graph in Figure 4.14, evaporation of 8 kg water/kg salts leads to about 64% recovery of NaCl.

Therefore, by mass balance, the mother liquor would contain $36.4 (1 - 0.64) =$ 13.1 kg Na^+, $56.2 (1 - 0.64) = 20.2$ kg Cl^-, 3.33 kg K^+, 4.00 kg SO_4^{2-}, and $900 - 800 = 100$ kg water.

Mother liquor composition: 9.32 wt% Na^+, 2.37 wt% K^+, 14.39 wt% Cl^-, 2.84 wt% SO_4^{2-}, 71.08 wt% water.

Problem 4.7

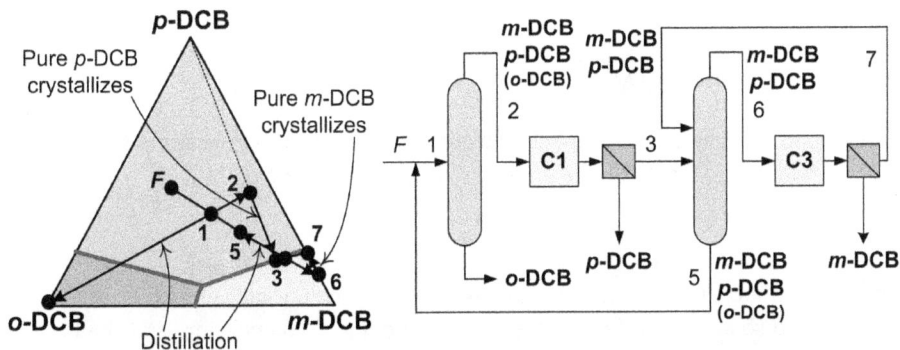

Note that the locations of the process points shown above are not meant to be exact. The key point is that points 6 and 7 lie inside the compartment of m-DCB. Also, points 3, 5, and the m-DCB vertex have to be collinear, because by mass balance around the second half of the process (second distillation column and crystallizer C3), the inlet flow rate (stream 3) must be equal to the total outlet flow rate (stream 5 plus pure m-DCB).

Problem 4.8

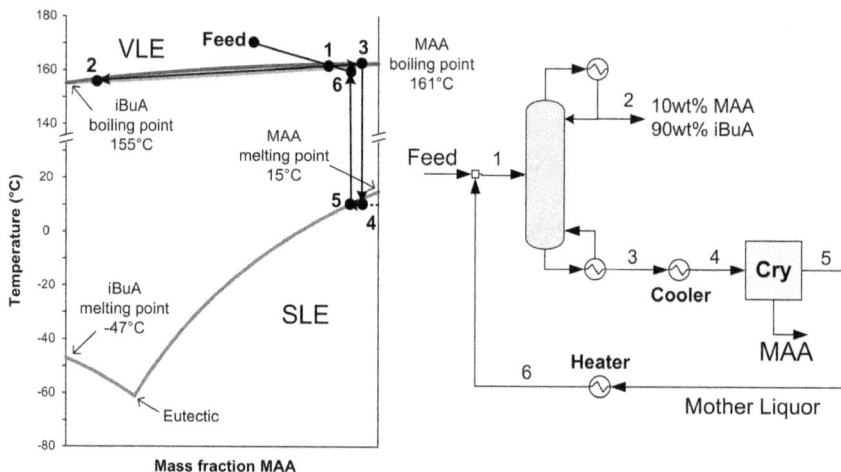

There is no information on the eutectic composition but it does not matter because the crystallizer does not operate anywhere near the eutectic point.

Problem 5.1

The compositions are plotted on the phase diagram. The crystallizer feed is located inside the two solids region (A(s) + C(s) + Liq). The liquid composition is at the double saturation point and the solid composition (from the tie-line) is 60% A and 40% C.

Pure C solids can be obtained by adding more B such that the crystallizer feed composition falls inside the saturation region of pure C.

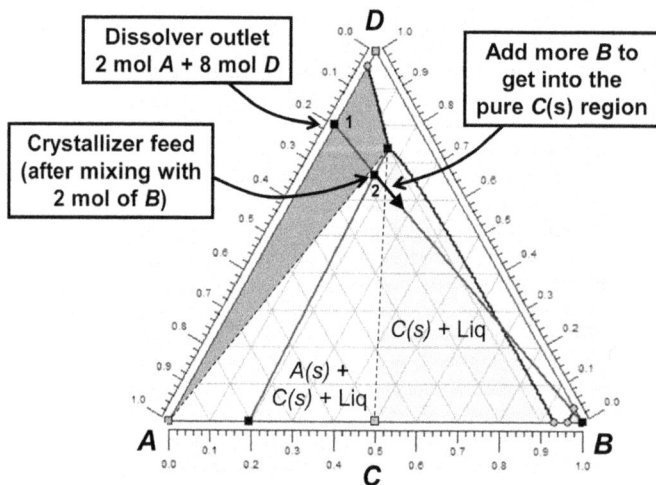

Problem 5.2

To crystallize B first, mix the feed with a recycle stream to produce a mixture located in compartment B. Due to the location of compartment boundaries, such a recycle stream has to be obtained by removing solvent from the mother liquor of the AS crystallizer.

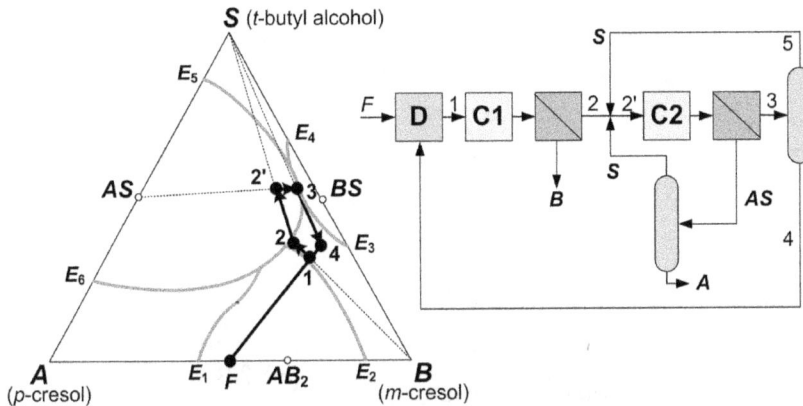

Problem 5.3

To obtain pure (+)-MA, the feed from SMB needs to be located inside the compartment of (+)-MA. As discussed, the boundary of this compartment pretty much aligns with a straight line connecting the water vertex to the eutectic between (+)-MA and the racemate. Therefore, the ratio of (+) to (−)-MA should be at least 69:31.

A solution with total MA concentration of 50 wt% and a ratio of (+) to (−)-MA of 9:1 is given by point F. Tracing the process path as (+)-MA crystallizes, the maximum recovery is obtained when the path intersects the compartment boundary at point L.

Composition of the crystallizer feed (point F): $x_{(+),F} = 0.45$, $x_{(-),F} = 0.05$, $x_{W,F} = 0.5$.

Along the process path, (−)-MA to water ratio remains constant $\rightarrow x_{(-),L}/x_{W,L} = 0.05/0.5 = 0.1$.

Since point L is on the compartment boundary, $x_{(+),L}/x_{(-),L} = 0.69/0.31 = 2.23$.

As $x_{(+),L} + x_{(-),L} + x_{W,L} = 1$, $x_{(-),L} = 1/(2.23 + 1 + 1/0.1) = 0.0756$, and $x_{(+),L} = 0.168$.

By mass balance, $F_F x_{(-),F} = F_L x_{(-),L} \rightarrow F_L/F_F = 0.05/0.0756 = 0.661$ and $F_S/F_F = 1 - 0.661 = 0.339$.

Maximum recovery of (+)-MA $= F_S/F_F x_{(+),F} = 0.339/0.45 = 75.3\%$.

Looking at the isothermal cuts, the temperature of point L should be between 25 and 30 °C (approximately 26 °C).

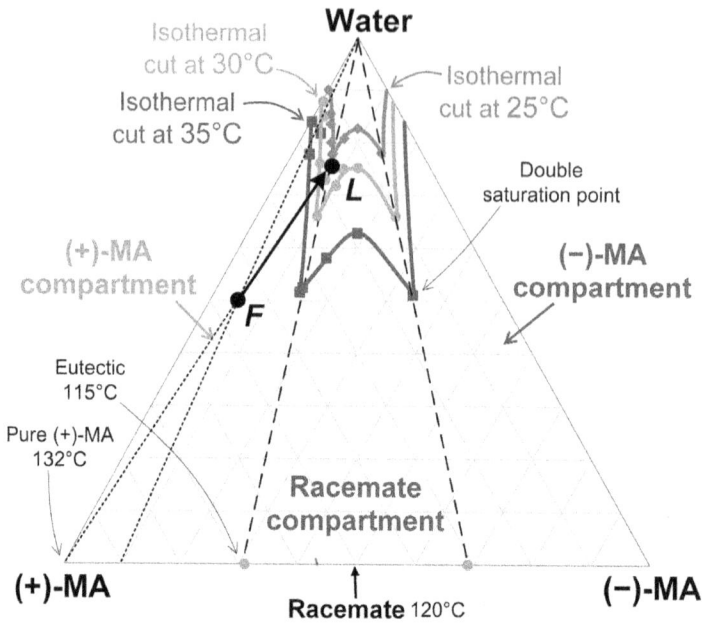

Problem 5.4

Evaporating water from point 1 gives point 2 inside the DL-Met compartment at T_c so that DL-Met is crystallized in the first crystallizer. Evaporating water from the mother liquor (point 3) gives point 4 located in the L-Met compartment at T_h, allowing crystallization of pure L-Met from the second crystallizer.

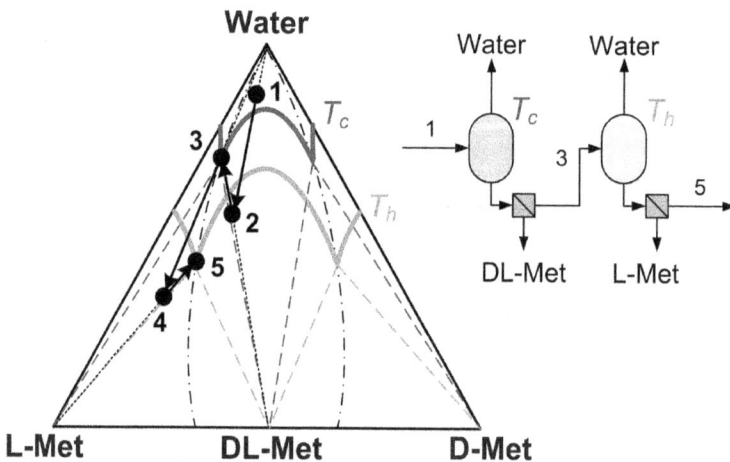

Problem 5.5

Take a basis of 100 kg (solvent-free) feed $\rightarrow$ 83.3 kg (−)Sch B and 16.7 kg (±)γ-Sch.
Let the solvent-free flow rate of the liquid and solid products be F_L and F_S, respectively.

From (−)Sch B balance, $83.3 = 0.764 F_L + 0.916 F_S$
From (±)γ-Sch balance, $16.7 = 0.236 F_L + 0.084 F_S$
Simultaneously solving the two equations $\rightarrow F_L = 54.6$ kg and $F_S = 45.4$ kg
Therefore, (-)Sch B crystallization yield = $(0.916)(45.4)/(0.833)(100) = 0.499$.

Problem 5.7

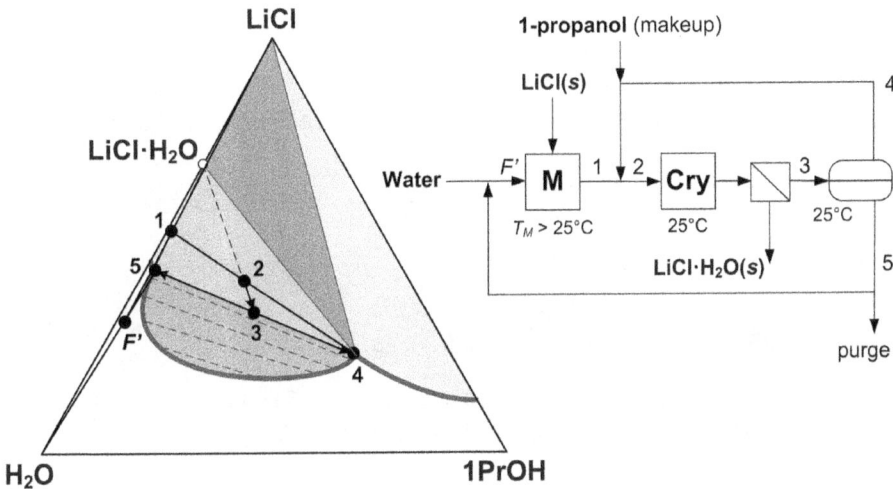

Problem 5.8

Let the flow rates of the purge, aqueous recycle, product, and fresh water streams be F_{5P}, F_{5R}, F_{prod}, and F_W, respectively.
From overall NaCl balance, $100 = F_{prod} + F_{5P}x_{NaCl,5} \rightarrow F_{prod} = 100 - F_{5P}x_{NaCl,5}$
From overall water balance, $F_W = F_{5P}x_{W,5}$
NaCl balance around block M: $100 = F_{5R}x_{NaCl,5} + F_1 x_{NaCl,1}$ (a)
Water (W) balance around block M: $F_{5R}x_{W,5} + F_W = F_1 x_{W,1}$ (b)
2-Propanol (D) balance around block M: $F_{5R}x_{D,5} = F_1 x_{D,1} \rightarrow F_1 = F_{5R}x_{D,5}/x_{D,1}$ (c)
Substituting (c) into (a) and rearranging: $100/F_{5R} - x_{D,5}\cdot x_{NaCl,1}/x_{D,1} = -x_{NaCl,5}$ (d)
Substituting (c) into (b) and rearranging: $F_W/F_{5R} - x_{D,5}\cdot x_{W,1}/x_{D,1} = -x_{W,5}$ (e)
From solubility relationship: $x_{NaCl,1} = 0.27 - 0.5x_{D,1} \rightarrow x_{NaCl,1}/x_{D,1} = 0.27/x_{D,1} - 0.5$ (f)
Since $x_{NaCl,1} + x_{W,1} + x_{D,1} = 1 \rightarrow x_{W,1}/x_{D,1} = 0.73/x_{D,1} - 0.5$ (g)

Substituting (f) into (d) and (g) into (e) gives two linear equations in $1/F_{5R}$ and $1/x_{D,1}$, which can be readily solved to give F_{5R}. If $F_{5P} = 20$ kg/h → $F_{5R} = 1,822$ kg/h and $x_{D,1} = 0.128$.

As the purge stream increases, the recycle stream decreases, which would lead to a smaller crystallizer and lower capital cost. At the same time, the product flow rate decreases and the makeup solvent flow rate increases, leading to a lower profit margin.

Problem 6.1

150 g of 40 wt% aspirin solution contains 60 g aspirin and 90 g ethanol.

If 35 g aspirin is crystallized, the remaining amount is 60 − 35 = 25 g → final concentration = 25/90 = 0.278 g aspirin/g ethanol.

Solubility at 25°C: x = 0.061 or 0.255 g aspirin/g ethanol → supersaturation =0.278 − 0.255 = 0.023 g aspirin/g ethanol.

From the solubility expression, the temperature at which the solubility is 0.278 g aspirin/g ethanol (x = 0.066) is 300.39 K (27.2 °C) → degree of supercooling = 27.2 − 25 = 2.2 °C.

Problem 6.2

Equation (6.8) can be rearranged to give

$$\frac{c_{imp,L}}{1 - G/G_0} = \frac{1}{\alpha K_{imp}} + \frac{1}{\alpha} c_{imp,L}$$

Plotting $c_{imp,L}/(1-G/G_0)$ against $c_{imp,L}$ gives a slope of 0.2795 and an intercept of 0.0855. Therefore, α = 1/slope = 3.578 and K_{imp} = slope/intercept = 3.267.

Problem 6.3

The key is to place all points within the metastable zone, with points 2 and 4 below the extension of the solubility curves.

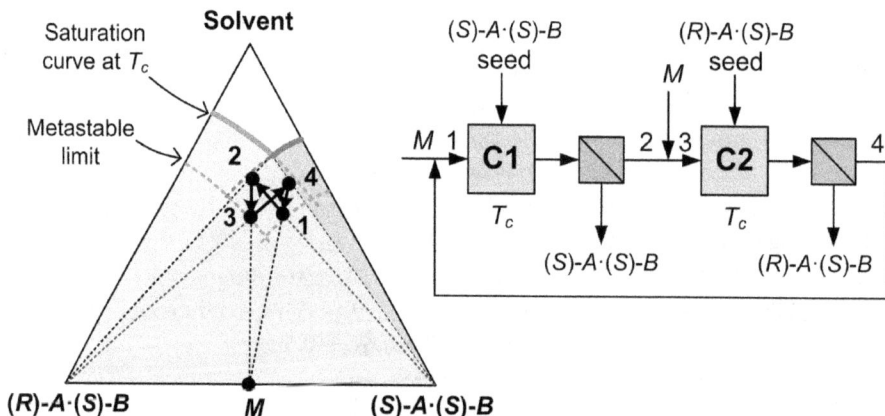

Problem 6.4

Since M is a racemic mixture, it contains the same amount of D-Thr and L-Thr. Total feed to the process $= 2 \times 100 = 200$ kg/h $\rightarrow$ by overall mass balance (at steady state), both D-Thr and L-Thr product streams are $200 \times 0.5 = 100$ kg/h.

From overall mass balance around C1, $F_1 = 100 + F_2$ (a)

From L-Thr balance around C1, $F_1 x_{L\text{-Thr},1} = 100 + F_2 x_{L\text{-Thr},2}$ (b)

Overall mass balance over the mixer prior to C1: $100 + F_4 = F_1$ (c)

L-Thr mass balance over the mixer prior to C1: $100 \times 0.5 + F_4 x_{L\text{-Thr},4} = F_1 x_{L\text{-Thr},1}$ (d)

From (a) and (c), $F_2 = F_4$, and from (b) and (d), $50 + F_2 x_{L\text{-Thr},2} = F_4 x_{L\text{-Thr},4}$ (e)

Therefore, $F_2 = F_4 = 50/(x_{L\text{-Thr},4} - x_{L\text{-Thr},2})$.

From Figure 6.23, $x_{L\text{-Thr},2} = 0.093$ and $x_{L\text{-Thr},4} = 0.1 \rightarrow F_2 = 7{,}143$ kg/h and $F_1 = 7{,}243$ kg/h.

From (b), $F_1 x_{L\text{-Thr},1} = 764$ kg/h $\rightarrow$ per-pass crystallization yield of L-Thr in C1 $= 100/764 = 13\%$.

Problem 6.5

Step 1: Starting from the racemic mixture in solution (point 1), perform asymmetric transformation for about 1 h (3,600 s) at temperature T_c. According to curves iii(b) in Figure 6.8, the crystals would be essentially pure S and correspondingly, the liquid composition would move to point 2. At the end of this period, quickly separate the crystals from the mother liquor.

Step 2: Heat up the mother liquor to T_h to make it unsaturated and let the racemization reaction proceed until the composition reaches point 3 ($R/S = 50/50$).

Step 3: Remove some solvent (for example, using a rotary evaporator) to bring the solution composition back to point 1. Upon cooling down to T_c, this mixture can be mixed with fresh feed for processing in the next batch.

Since there is no purge and essentially all R is transformed to S, the theoretical recovery is 100%.

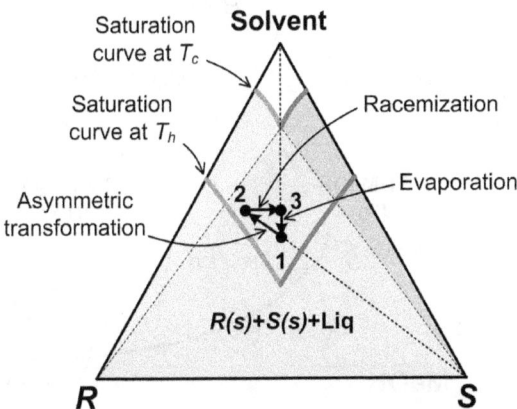

Problem 6.6

Path I: Cooling below the solubility curve of β form (but above the solubility curve of α) and introducing seeds of β upon further cooling → β crystallizes out and the process path approaches the solubility curve of β → only β form is obtained as final product.

Path II: Cooling below the metastable limit of α form → α crystallizes spontaneously and the process path approaches the solubility curve of α → as cooling continues toward the final temperature, α begins to transform into the more stable β form (cooling rate is slower than transformation rate) → the final product is mostly β form.

To obtain α form as the final product, cooling can be done faster and the solids should be isolated quickly before they have a chance to transform into β.

Problem 6.7

Form B is more stable because it has a higher melting point. Since there is no change in relative stability at any temperature, the system exhibits monotropic behavior.

Problem 6.8

A sketch of the phase diagram is shown below. It may not be exactly correct since the crossover temperature is not known from the information; it could be above or below the incongruent melting point of BH.

Starting from a solution of BPT in methanol, a possible process to obtain crystals of form A would include increasing the temperature to higher than 328 K (until A becomes the most stable form), removing some methanol (by evaporation), and adding water up to $V_{MeOH} = 0.8$.

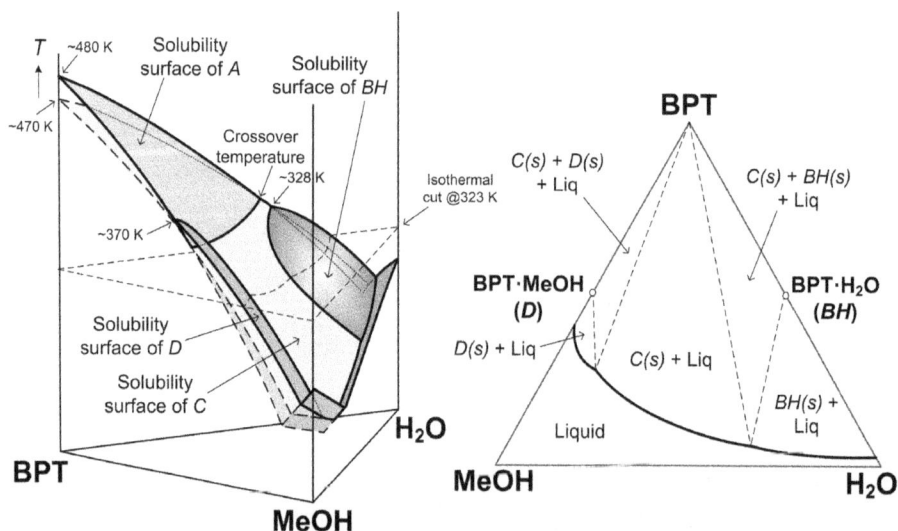

Problem 7.1

Crystals in the shape of a rectangular prism with sides having a ratio of 1:1:2 and longest side $= L \rightarrow$ volume $= LD^2 = L^3/4$. Therefore, $k_v = 0.25$.

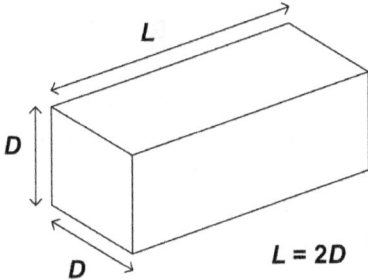

Since the total number of particles per kg of powder is equal to the zeroth moment, M_0, the population density [no/m.kg] can be written as

$$n(L) = M_0 f(L)$$

The total volume of particles per kg of powder is the reciprocal of solid true density,

$$k_v \int_0^\infty n(L)L^3 dL = \frac{1}{\rho_S}$$

Thus, $M_0 = \dfrac{1}{\rho_S k_v \int_0^\infty f(L)L^3 dL}$.

By numerical integration, $\int_0^\infty f(L)L^3 dL = 4.168 \cdot 10^{-9} \text{m}^3 \rightarrow M_0 = 7.39 \cdot 10^5 \text{no/kg}$.

Problem 7.2

From the definition of $W(L)$,

$$W(L) = \rho_S k_v n(L)L^3 = \frac{\rho_S k_v B}{G} L^3 \exp\left(-\frac{Q_{out}L}{GV}\right)$$

At the maximum point, the differential with respect to L is zero,

$$\frac{dW(L)}{dL} = -\frac{\rho_S k_v B}{G}\frac{Q_{out}}{GV}L^3 \exp\left(-\frac{Q_{out}L}{GV}\right) + \frac{3\rho_S k_v B}{G}L^2 \exp\left(-\frac{Q_{out}L}{GV}\right) = 0$$

Solving for L gives $L_d = \dfrac{3GV}{Q_{out}}$.

Magma density can be obtained by integrating the expression for $W(L)$,

$$M_T = \int W(L)dL = \frac{\rho_S k_v B}{G}\int_0^\infty L^3 \exp\left(-\frac{Q_{out}L}{GV}\right) = \frac{6\rho_S k_v B}{G}\left(\frac{GV}{Q_{out}}\right)^4.$$

It can also be expressed in terms of L_d as $M_T = \dfrac{2\rho_s k_v B}{27G} L_d^4$.

Problem 7.3

Let the mass of the solid product at maximum theoretical yield (equilibrium is achieved) = m_P.

Based on mass balance, $(5,000)(0.3) = (5,000 - m_P)(0.05) + m_P \rightarrow m_P = 1,315.8$ kg.

If there is no nucleation and crystal growth is uniform, eq. (7.17) can be used to calculate the required seed mass based on the seed and product crystal size: $m_S = m_P (50/1,000)^3 = 0.164$ kg.

Problem 7.4

From the given solubility expression, the AA concentration in the feed is $c^* = 0.00575$ $(70) - 0.1775 = 0.225$ kg AA/kg H_2O, and in the mother liquor is $c^* = 0.00575(50) -$ $0.1775 = 0.11$ kg AA/kg H_2O.

Overall mass balance: $F_F = F_S + F_L$ (F = feed, S = crystals, L = mother liquor).

AA mass balance: $F_F*0.225/(1 + 0.225) = F_S + F_L*0.11/(1 + 0.11) \rightarrow$ since $F_S = 4,600$ kg/h, $F_F = 49,000$ kg/h.

Magma density is the mass of crystals per unit volume of slurry. The volumetric flow rate of the slurry is $4,600/1,360 + (49,000 - 4,600)/1,300 = 37.5$ m³/h $\rightarrow M_T = 4,600/37.5 = 122.5$ kg/m³.

Since $M_T = \dfrac{2\rho_s k_v B}{27G} L_d^4$ and $B = 5.36 \times 10^{31} G^{3.5} M_T^{0.4} \rightarrow M_T^{0.6} = \frac{2}{27}\rho_s k_v L_d^4 (5.36 \times 10^{31}) G^{2.5}$

Substituting appropriate values, $G = 1.77 \times 10^{-7}$ m/s.

The required crystallizer volume is $V = Q_{out} L_d/3 G = 2.95$ m³.

V is proportional to L_d/G, but $L_d^4 G^{2.5}$ has to stay constant to give the same $M_T \rightarrow V \propto L_d^{2.6}$.

Therefore, if V increases, L_d would also increase.

Problem 7.5

The vessel diameter, liquid height, and impeller diameter can be obtained using geometric similarity relationships (eq. (7.20)) and the agitation speed is calculated using eq. (7.23). Note that in the expressions for N_{Re} and N_P, the units of N need to be rps (rotation per second).

The power per unit volume is lower in the larger crystallizer.

	Pilot scale	Commercial scale
Volume, V (m³)	0.1	10
Vessel diameter, T (m)	0.5	2.321
Liquid height, Z (m)	0.509	2.364
Impeller diameter, D (m)	0.25	1.16
Agitation speed, N (rpm)	100	21.5
Reynolds number, N_{Re}	59,896	278,012
Power number, N_P	4	4
Required power, P	0.0208	0.448
Power per unit volume, P/V (kW/m³)	0.208	0.0448

Problem 7.7

The slurry contains 30% solids → dry cake flow rate = 500*0.3 = 150 kg/h.

The cake contains 0.15 kg water per kg dry cake → water in cake = 150*0.15 = 22.5 kg/h.

Therefore, filtrate flow rate = 500 − (150 + 22.5) = 327.5 kg/h; density = 1,050 kg/m³ → q_F = 327.5/1,050 = 0.312 m³/h and c = 150/0.312 = 480.9 kg/m³.

From plot of t_f/V_f [s/m³] against V_f [m³], slope = 3.85 · 10⁹ s/m⁶ and intercept = 7,566 s/m³.

Therefore, α = 5.6 · 10¹⁰ m/kg and R_m = 2.65 · 10⁹/m. Substituting to eq. (7.33) → L_F = 1.477 m.

If d_p decreases, α would increase → L_F would increase.

Problem 7.8

By fitting the data to the exponential function, f_{max} = 0.962 and b = 1.89.

Taking a basis of 1 kg dry solid in the slurry (including 5 mg OxA), the wet cake after filtration contains 0.15 kg liquid (including 0.045 kg AsA and 30 mg OxA).

If the wash ratio is W, the washing liquid requirement is 0.15 W kg, which includes 1.5 W mg OxA.

With a fraction removal of f, the OxA content in the residual liquid after washing is 200(1 − f) + 10 f = 200 − 190 f ppm, so the residual liquid (0.15 kg) includes 0.045 kg AsA and 30 − 28.5 f mg OxA.

After deliquoring, the residual liquid (0.1 kg) includes 0.03 kg AsA and 20 − 19 f mg OxA.

Drying removes all the water, while all AsA and OxA from the residual liquid becomes a part of the dried solids → amount of solids = 1 + 0.03 +(20 − 19 f)/10⁶ kg, including 5 + 20 − 19 f = 25 − 19 f mg OxA.

To achieve the target of 10 ppm, the value of f should be at least 0.774, corresponding to W = 0.863.

Therefore, the required washing liquid amount is 0.15*0.863 = 0.129 kg/kg dry solid.

Problem 8.1

The ideal solubility equations for o- and p-NCB can be written as:

$$x_o = \exp\left[\frac{\Delta H_{f,o}}{R}\left(\frac{1}{T_{m,o}} - \frac{1}{T}\right)\right] \text{ and } x_p = \exp\left[\frac{\Delta H_{f,p}}{R}\left(\frac{1}{T_{m,p}} - \frac{1}{T}\right)\right]$$

At the eutectic point, $x_o + x_p = 1 \rightarrow$ substituting the melting point and heat of fusion values and solving the equations simultaneously gives $x_p = 0.337$ ($x_o = 0.663$) and $T = 290.31$ K.

To plot the phase diagram, use the solubility equation to calculate the mole fraction of o-NCB and p-NCB versus temperature. A few points between the eutectic temperatures, the respective pure component melting points would suffice.

Problem 8.2

The eutectic points can be calculated in the same manner as in Problem 8.1. For the ternary eutectic, apply $x_{PX} + x_{MX} + x_{OX} = 1$.

Eutectic	PX/MX binary	PX/OX binary	MX/OX binary	PX/MX/OX ternary
Temperature (K)	220.80	238.32	211.84	209.59
Mole fraction PX	0.118	0.235	–	0.072
Mole fraction MX	0.882	–	0.675	0.629
Mole fraction OX	–	0.765	0.325	0.299

Problem 8.3

The basic equations to use are the K_{SP} equations:

$$\left(\gamma_A^* + m_{A+}\right)\left(\gamma_X^* - m_{X-}\right) = K_{SP,AX}$$

$$\left(\gamma_B^* + m_{B+}\right)\left(\gamma_X^* - m_{X-}\right) = K_{SP,BX}$$

$$\left(\gamma_A^* + m_{A+}\right)\left(\gamma_Y^* - m_{Y-}\right) = K_{SP,AY}$$

$$\left(\gamma_B^* + m_{B+}\right)\left(\gamma_Y^* - m_{Y-}\right) = K_{SP,BY}$$

With the ideal assumption, all y^* values are set to 1.

First, calculate the four double saturation points involving salts with a common ion. For example, for AX/BX double saturation point, $m_{A+}m_{X-} = K_{SP,AX}$ and $m_{B+}m_{X-} = K_{SP,BX}$. Therefore, $R(B^+) = \frac{m_{B+}}{m_{A+} + m_{B+}} = \frac{K_{SP,BX}}{K_{SP,AX} + K_{SP,BX}} = \frac{1.24 \times 10^{-5}}{2.46 \times 10^{-6} + 1.24 \times 10^{-5}} = 0.834$.

Because this mixture only has X^-, $R(Y^-) = \frac{m_{Y-}}{m_{X-} + m_{Y-}} = 0$.

The results for the four double saturation points are summarized as follows:

Double saturation point	AX/BX	AY/BY	AX/AY	BX/BY
$R(B^+)$	0.834	0.082	0	1
$R(Y^-)$	0	1	0.770	0.056

From the location of these four points (AX/AY is close to AY/BY and AX/BX is close to BX/BY), it can be deduced that AX/BY is the compatible pair. Therefore, the two triple saturation points are $AX/AY/BY$ and $AX/BX/BY$. The two triple saturation points involve compatible pairs and the salt that shares a common ion with both of them. Since the K_{SP} equations for the three salts must be simultaneously satisfied at the triple saturation points, the resulting coordinates are:

Triple saturation point	AX/AY/BY	AX/BX/BY
$R(B^+)$	0.082	0.834
$R(Y^-)$	0.770	0.056

The last step is to calculate the double saturation troughs connecting the points. Except the trough connecting the two triple saturation points, all others are simply straight lines.

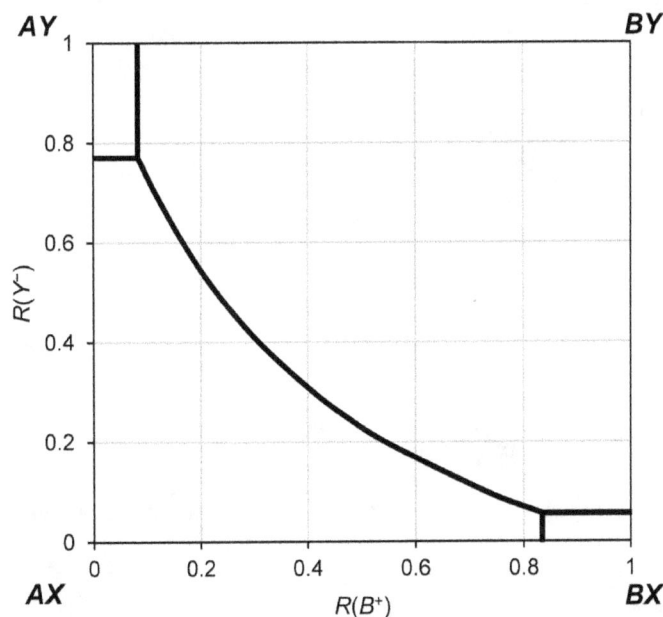

The trough connecting $AX/AY/BY$ and $AX/BX/BY$ (which is not a straight line) is calculated by recognizing that AX and BY are simultaneously saturated along this trough. Thus, $m_{A+} m_{X-} = K_{SP,AX}$ and $m_{B+} m_{Y-} = K_{SP,BY}$.

It follows that $R(B^+) = \dfrac{m_{B+}}{m_{A+} + m_{B+}} = \dfrac{K_{SP,BY}/m_{Y-}}{K_{SP,AX}/m_{X-} + K_{SP,BY}/m_{Y-}} = \dfrac{K_{SP,BY}}{K_{SP,AX}m_{Y-}/m_{X-} + K_{SP,BY}}.$

$$\frac{1}{R(Y^-)} = \frac{m_{X-} + m_{Y-}}{m_{Y-}} = \frac{1}{m_{Y-}/m_{X-}} + 1 \rightarrow \frac{m_{Y-}}{m_{X-}} = \frac{1}{1/R(Y^-) - 1}$$

Therefore, $R(B^+) = \dfrac{K_{SP,BY}}{K_{SP,AX}/[1/R(Y^-) - 1] + K_{SP,BY}}.$

The trough can then be plotted by varying $R(Y^-)$ from 0.056 to 0.770 (between the two triple saturation points) and calculating the corresponding values of $R(B^+)$.

Taking into account the non-ideality, the equation for AX/BX double saturation point becomes

$$R(B^+) = \frac{m_{B+}}{m_{A+} + m_{B+}} = \frac{K_{SP,BX}/\gamma_{B+}^*}{K_{SP,AX}/\gamma_{A+}^* + K_{SP,BX}/\gamma_{B+}^*}$$

Since $z_{A+} = z_{B+} = 1$, Debye–Hückel limiting law gives $\gamma_{A+}^* = \gamma_{B+}^*$. Therefore, the locations of the double saturation points on the Jänecke projection are the same as the ideal case. Similarly, the triple saturation points as well as the double saturation troughs remain at the same locations.

However, the solvent content at each point on the phase diagram depends on the activity coefficient. For example, for AX/BX double saturation point, $m_{B+} m_{X-} = K_{SP,BX}/\gamma_{B+}^* \gamma_{X-}^*$.

Because of neutrality, $m_{A+} + m_{B+} = m_{X-}$, so $\dfrac{m_{B+}}{m_{X-}} = \dfrac{K_{SP,BX}/\gamma_{B+}^*}{K_{SP,AX}/\gamma_{A+}^* + K_{SP,BX}/\gamma_{B+}^*}.$

Combining the two equations, it can be obtained that

$$m_{B+} = \sqrt{\frac{\left(K_{SP,BX}/\gamma_{B+}^*\right)\left(K_{SP,BX}/\gamma_{B+}^* + \gamma_{X-}^*\right)}{K_{SP,AX}/\gamma_{A+}^* + K_{SP,BX}/\gamma_{B+}^*}},$$

which simplifies to $m_{B+} = \sqrt{\dfrac{\left(K_{SP,BX}\right)\left(K_{SP,BX}/\gamma_{X-}^*\right)}{K_{SP,AX} + K_{SP,BX}}}$ when $\gamma_{A+}^* = \gamma_{B+}^*$.

Therefore, the value of the solvent coordinate depends on the values of the activity coefficients.

Problem 8.6

The enthalpy of the eutectic peak can be plotted against composition to locate the eutectic point. Based on the intersection of the two lines, the eutectic composition is $x_B = 0.55$.

From the data:

$T_{m,A} = 158.6\ ^\circ C = 431.8\ K$, $\Delta H_{m,A} = (108.8/1{,}000)(180.15) = 19.6\ kJ/mol$

$T_{m,B} = 135\ ^\circ C = 408.2\ K$, $\Delta H_{m,A} = (142.1/1{,}000)(180.15) = 25.6\ kJ/mol$

At the eutectic point, $x_{A,sat} + x_{B,sat} = 1$

Substituting the solubility equation (eq. (8.29)) for both $x_{A,sat}$ and $x_{B,sat}$ with $\gamma = 1$ and solving for T gives $T = 377.6\ K$ and $x_{B,sat} = 0.543$, which is close to the experimental result.

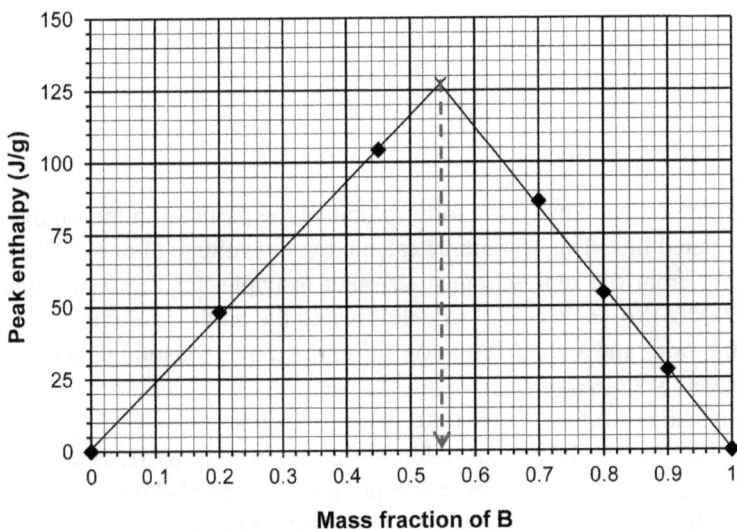

Problem 8.7

The liquid and wet solid phases can be plotted on the composition triangle, and each tie-line can be extended to the $NaNO_3$–NH_4NO_3 edge to determine the water-free composition of the solid phase. Since the water-free solid phases are all close to the $NaNO_3$ vertex, it can be concluded that they are pure $NaNO_3$ and not a solid solution.

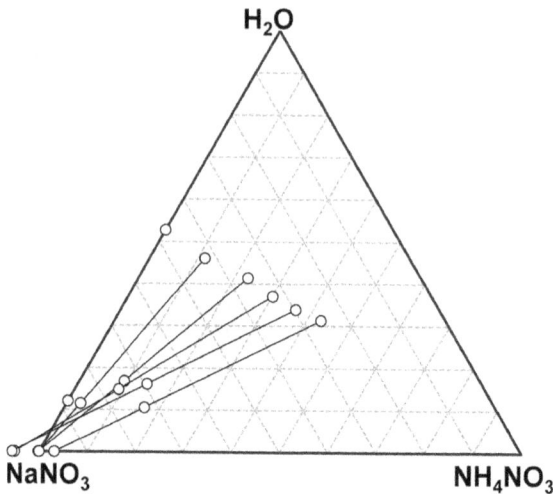

Problem 8.8

To calculate the population density, the total volume of the particles in the sample needs to be calculated first. This is done by calculating the volume of one particle in each size range from the representative size. For example, for the 0–100 μm range, $V_p = \pi(0.00005)^3/6 = 6.545 \cdot 10^{-14}$ m³. Multiplying the particle volume by the count in each size range and summing up over the entire range, the total volume of particles in the sample (3,416 particles) = $1.961 \cdot 10^{-7}$ m³.

The magma density is $(0.1005 \text{ kg})/(0.0003 \text{ m}^3) = 335$ kg/m³.

Therefore, the volume of solid per unit volume of slurry = $335/1,426 = 0.235$ m³/m³.

The number of particles in each size range per unit volume of slurry can then be obtained by multiplying the count by a factor of $0.235/1.961 \cdot 10^{-7}$.

Since $\Delta L = 0.0001$ m (100 μm), $n(L) = \text{count} * (0.235/1.961 \cdot 10^{-7})/0.0001$.

From the plot of $n(L)$ versus L, the slope = $-1/(G\tau) = -3,783$ and the intercept = $n_0 = 1.71 \cdot 10^{13}$.

The residence time $\tau = V/Q = 300$ mL/(300 mL/h) = 1 h = 3,600 s → $G = 1/(3,783)(3,600) = 7.34 \cdot 10^{-8}$ m/s and $B = n_0 G = 1.257 \cdot 10^6$ no/m³.s.

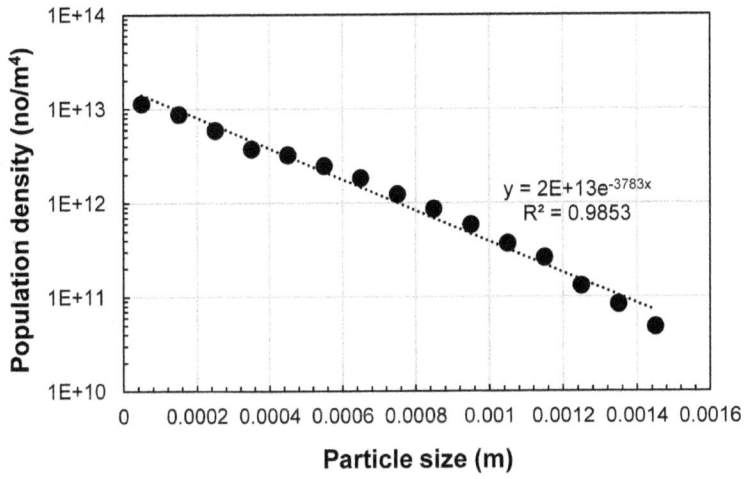

References

[1]   Geertman RM. Sodium Chloride: Crystallization. In Wilson ID, ed. Encyclopedia of
        Separation Science. Cambridge, MA, USA, Academic Press, 2000, 4127–4134.
[2]   Variankaval N, Cote AS, Doherty, MF. From Form to Function: Crystallization of Active
        Pharmaceutical Ingredients. AIChE J 2008, 54, 1682–1688.
[3]   Mumford RW. Potassium Chloride from the Brine of Searles Lake. Ind Eng Chem 1938, 30,
        872–878.
[4]   Rajagopal S, Ng KM, Douglas JM. Design of Solids Processes: Production of Potash. Ind Eng
        Chem Res 1988, 27, 2071–2078.
[5]   Wilkomirsky I. Production of Lithium Carbonate from Brines. US Patent 5,993,759, 1999.
[6]   Liang L, Du K, Peng Z, Cao Y, Duan J, Jian J. Co-precipitation Synthesis of $Ni_{0.6}Co_{0.2}Mn_{0.2}$
        $(OH)_2$ Precursor and Characterization of $LiNi_{0.6}Co_{0.2}Mn_{0.2}O_2$ Cathode Material for Secondary
        Lithium Batterles. Electrochim Acta 2014, 130, 82–89.
[7]   Haines Jr HW. p-Xylene from Petroleum. Ind Eng Chem Res 1955, 47, 1096–1103.
[8]   Moyers CGJ. Industrial Crystallization for Ultrapure Products. Chem Eng Prog 1986. 82(5),
        42–46.
[9]   Dermer OC. Bisphenol A. In McKetta JJ, ed. Encyclopedia of Chemical Processing and
        Design, vol. 4. Boca Raton, FL, USA, CRC Press, 1977, 406–430.
[10]   Bernis AG, Dindorf JA, Horwood B, Samans C. Phthalic Acids and Other
        Benzenepolycarboxylic Acids. In Kirk Othmer Encyclopedia of Chemical Technology.
        3rd ed., vol. 17. New York, NY, USA, John Wiley & Sons, 1982, 732–770.
[11]   Oppenheim JP, Dickerson GL. Adipic Acid. In Kirk-Othmer Encyclopedia of Chemical
        Technology. 5th ed., vol. 1. Hoboken, NJ, USA, Wiley, 2003, 553–582.
[12]   Dahl JE, Liu SG, Carlson RMK. Isolation and Structure of Higher Diamondoids,
        Nanometer-sized Diamond Molecules. Science 2003, 299, 96–99.
[13]   Guccione E. The Basics of Azo Dye Syntheses. Chem Eng 1963, 70(17), 138–140.
[14]   Kawakita T. Amino Acids, Glutamate. In Flickinger MC, Drew SW, eds. Encyclopedia of
        Bioprocess Technology: Fermentation, Biocatalysis, and Bioseparation, New York, NY, USA,
        Wiley, 1999, 77–88.
[15]   Kuellmer V. Ascorbic Acid. In Kirk-Othmer Encyclopedia of Chemical Technology. 4th
        ed., vol. 25. New York, NY, USA, Wiley, 2000, 17–47.
[16]   Thomas MR. Salicylic Acid and Related Compounds. In Kirk-Othmer Encyclopedia of
        Chemical Technology. 4th ed., vol. 21. New York, NY, USA, Wiley, 1997, 601–626.
[17]   Tung HH, Waterson S, Reynolds S, Paul E. Resolution of Ibuprofen via Stereospecific
        Crystallization. AIChE Symp Ser 1991, 87, 64–71.
[18]   Ng KM. MOPSD: A Framework Linking Business Decision-Making to Product and Process
        Design. Comput Chem Eng 2005, 29, 51–56.
[19]   Ng KM. A Multiscale-Multifaceted Approach to Process Synthesis and Development. In Gani
        R, Jørgensen SB, eds. Computer Aided Chemical Engineering, vol. 9. Amsterdam,
        Netherlands, Elsevier, 2001, 41–54.
[20]   Panthani MG, Korgel BA. Nanocrystals for Electronics. Annu Rev Chem Biomol Eng 2012,
        3, 287–311.
[21]   Jarvis M, Khrisnan V, Mitragotri S. Nanocrystals: A Perspective on Translational Research
        and Clinical Studies. Bioeng Transl Med 2019, 4, 5–16.
[22]   Wilson E. Giant Crystals Invade National Lab. Chem Eng News 2000, 78(8), 8.
[23]   Lerou JJ, Ng KM. Chemical Reaction Engineering: A Multiscale Approach to a Multiobjective
        Task. Chem Eng Sci 1996, 51, 1595–1614.

https://doi.org/10.1515/9781501519901-011

[24]   Seider WD, Lewin DR, Seader JD, Widagdo S, Gani R, Ng KM. Product and Process Design Principles: Synthesis, Analysis and Evaluation. 4th ed. New York, NY, USA, Wiley, 2017.

[25]   Ng KM, Wibowo C. Beyond Process Design: The Emergence of a Process Development Focus. Korean J Chem Eng 2003, 20, 791–798.

[26]   Yu LX. Pharmaceutical Quality by Design: Product and Process Development, Understanding, and Control. Pharm Res 2008, 25, 781–791.

[27]   Douglas JM. Conceptual Design of Chemical Processes. New York, NY, USA, McGraw Hill, 1988.

[28]   Bermingham SK, Neumann AM, Kramer HJM, Verheijen PJT, van Rosmalen GM, Grievink J. A Design Procedure and Predictive Models for Solution Crystallization Processes. AIChE Symp Ser 2000, 323, 250–264.

[29]   Larsen PA, Patience DB, Rawlings JB. Industrial Crystallization Process Control. IEEE Contr Syst Mag 2006, 26(4), 70–80.

[30]   Mersmann A. Crystallization Technology Handbook. 2nd ed., New York, NY, USA, Marcel Dekker, 2001.

[31]   Lewis AE, Seckler M, Kramer H, van Rosmalen G. Industrial Crystallization: Fundamentals and Applications. Cambridge, UK, Cambridge University Press, 2015.

[32]   Browne CA. A Handbook of Sugar Analysis: A Practical and Descriptive Treatise for Use in Research, Technical and Control Laboratories. 2nd ed. New York, NY, USA, John Wiley & Sons, 1912.

[33]   Hall DL, Sterner SM, Bodnar RJ. Freezing Point Depression of NaCl-KCl-H_2O Solutions. Econ Geol 1988, 83, 197–202.

[34]   Washburn EW, West CJ, Hull C. International Critical Tables of Numerical Data, Physics, Chemistry and Technology: Volume IV. New York, NY, USA, McGraw-Hill, 1928.

[35]   Zhao G, Yan W. Solubilities of Betulin in Chloroform + Methanol Mixed Solvents at $T = (278.2, 288.2, 293.2, 298.2, 308.2$ and $313.2)$ K. Fluid Phase Equil 2008, 267, 79–82.

[36]   Glasstone S. Textbook of Physical Chemistry. 2nd ed. New York, NY, USA, D. Van Nostrand Company, Inc., 1946.

[37]   Purdon FF, Slater VW. Aqueous Solution and the Phase Diagram. London, UK, Edward Arnold & Co., 1946.

[38]   Ricci JE. The Phase Rule and Heterogeneous Equilibrium. New York, NY, USA, D. Van Nostrand Company Inc., 1951.

[39]   Moore WJ. Physical Chemistry. 4th ed. Englewood Cliffs, NJ, USA, Prentice-Hall, 1963.

[40]   Haase R, Schönert H. Solid-Liquid Equilibrium. New York, NY, USA, Pergamon, 1969.

[41]   Hilgeman FR, Mouroux FYN, Mok D, Holan MK. Phase Diagrams of Binary Solid Azole Systems. J Chem Eng Data 1989, 34, 220–222.

[42]   Moyers CGJ. Industrial Crystallization for Ultrapure Products. Chem Eng Prog 1986, 82(5), 42–46.

[43]   Wyrzykowska-Stankiewicz D, Palczewska-Tulińska M. Solid-Liquid Equilibrium in a Binary System with Incongruent Melting Complex Compound. Thermochim Acta 1991, 190, 209–216.

[44]   Jadhav VK, Chivate MR, Tavare NS. Separation of Phenol from Its Mixture with o-Cresol by Adductive Crystallization. J Chem Eng Data 1992, 37, 232–235.

[45]   Tare JP, Chivate MR. Separation of Close Boiling Isomers by Adductive and Extractive Crystallization. AIChE Symp Ser 1976, 72, 95–99.

[46]   Singh NB, Giri DP, Singh NP. Solid-Liquid Equilibria for p-Dichlorobenzene + p-Dibromobenzene and p-Dibromobenzene + Resorcinol. J Chem Eng Data 1999, 44, 605–607.

[47]   Scholastica Kennard SM, McCusker PA. Some Systems of Carbon Halides with Dioxane, Pyridine and Cyclohexane. J Am Chem Soc 1948, 70, 3375–3377.

[48]   Rai US, George S. Some Physicochemical Studies on Binary Organic Eutectics and 1:2 Molecular Complexes. Cryst Res Technol 1991, 26, 511–519.

[49]   Sangster J. Phase Diagrams and Thermodynamic Properties of Binary Organic Systems Based on 1,2-, 1,3-, 1,4-Diaminobenzene or Benzidine. J Phys Chem Ref Data 1994, 23, 295–338.

[50]   Lincoln AT. Textbook of Physical Chemistry. Boston, MA, USA, D. C. Heath & Co., 1918.

[51]   Sloan GJ, McGhie AR. Techniques of Melt Crystallization. New York, NY, USA, D. Van Nostrand Company, Inc., 1988.

[52]   Campbell AN, Prodan LA. An Apparatus for Refined Thermal Analysis Exemplified by a Study of the System p-Dichlorobenzene – p-Dibromobenzene – p-Chlorobromobenzene. J Am Chem Soc 1948, 70, 553–561.

[53]   Sediawan WB, Gupta S, McLaughlin E. Solid-Liquid Phase Diagrams of Binary Aromatic Hydrocarbon Mixtures from Calorimetric Studies. J Chem Eng Data 1989, 34, 223–226.

[54]   Oonk HAJ, Tjoa KH, Brants FE, Kroon J. The Carvoxime System III. Differential Scanning Calorimetry Experiments: Heats of Melting, Phase Diagrams and Melting Behavior of dl-Carvoxime. Thermochim Acta 1977, 19, 161–171.

[55]   Nojima H, Akehi S. Studies of Zone Melting. V. The Solid-Liquid Equilibrium and Zone Melting of the Bibenzyl-Diphenylacetylene System. Bull Chem Soc Jpn 1980, 53, 2067–2073.

[56]   Joncich MJ, Bailey DR. Zone Melting and Differential Thermal Analysis of Some Organic Compounds. Anal Chem 1960, 32, 1578–1581.

[57]   Nojima H. Studies of Zone Melting. IV. The Equilibrium Distribution Coefficient of the Bibenzyl-Stilbene System. Bull Chem Soc Jpn 1978, 51, 2513–2517.

[58]   Kitagorodsky AI. Molecular Crystals and Molecules. New York, NY, USA, Academic Press, 1973.

[59]   Kitagorodsky AI. Mixed Crystals. Berlin, Germany, Springer-Verlag, 1984.

[60]   Mastrangelo SVR, Dornte RW. The System Naphthalene-Thianaphthene. Anal Chem 1957, 29, 794–797.

[61]   Matsuoka M. Developments in Melt Crystallization. In Garside J, Davey RJ, Jones AG, eds. Advances in Industrial Crystallization. Oxford, UK, Butterworth-Heinemann, 1991, 229–244.

[62]   Jacques J, Collet A, Wilen SH. Enantiomers, Racemates, and Resolutions. New York, NY, USA, John Wiley & Sons, 1981.

[63]   Ewing WW, Brandner JD, Slichter CB, Griesinger WK. The Temperature-Composition Relations of the Binary System Magnesium Nitrate-Water. J Am Chem Soc 1933, 55, 4822–4824.

[64]   Grosse Daldrup JB, Held C, Ruether F, Schembecker G, Sadowski G. Measurement and Modeling Solubility of Aqueous Multisolute Amino-Acid Solutions. Ind Eng Chem Res 2010, 49, 1395–1401.

[65]   Wibowo C, Ng KM. Visualization of High Dimensional Systems via Geometric Modeling with Homogeneous Coordinates. Ind Eng Chem Res 2002, 41, 2213–2225.

[66]   Cruickshank AJB, Haertsch N, Hunter TG. Liquid-liquid Equilibrium of Four-Component Systems. Ind Eng Chem 1950, 42, 2154–2158.

[67]   Samant KD, Berry DA, Ng KM. Representation of High-Dimensional, Molecular Solid-Liquid Phase Diagrams. AIChE J 2000, 46, 2435–2455.

[68]   Berry DA, Ng KM. Synthesis of Reactive Crystallization Processes. AIChE J 1997, 43, 1737–1750.

[69]   Ung S, Doherty MF. Vapor-Liquid Phase Equilibrium in Systems with Multiple Chemical Reactions. Chem Eng Sci 1995, 50, 23–48.

[70]   Pérez Cisneros ES, Gani R, Michelsen ML. Reactive Separation Systems – I. Computation of Physical and Chemical Equilibrium. Chem Eng Sci 1997, 52, 527–543.

[71]   Wibowo C, Samant KD, Ng KM. High Dimensional Solid-Liquid Phase Diagrams Involving Compounds and Polymorphs. AIChE J 2002, 48, 2179–2192.

[72]   Samant KD, Ng KM. Representation of High-Dimensional Solid-Liquid Phase Diagrams for Ionic Systems. AIChE J 2001, 47, 861–879.

[73]   Berry DA, Ng KM. Separation of Quaternary Conjugate Salt Systems by Fractional Crystallization. AIChE J 1996, 42, 2162–2174.

[74]   Nicolaisen H, Rasmussen P, Sørensen JM. Correlation and Prediction of Mineral Solubilities in the Reciprocal Salt System $(Na^+, K^+)(Cl^-, SO_4^{2-})$-$H_2O$ at 0–100 °C. Chem Eng Sci 1993, 48, 3149–3158.

[75]   Kwok KS, Ng KM, Taboada ME, Cisternas LA. Thermodynamics of Salt Lake System: Representation, Experiments, and Visualization. AIChE J 2008, 54, 706–727.

[76]   Gmehling J. Solid-Liquid Equilibria in Binary Mixtures of Organic Compounds. Fluid Phase Equilib 1995, 113, 117–125.

[77]   Cesar MAB, Ng KM. Improving Product Recovery in Fractional Crystallization Processes: Retrofit of an Adipic Acid Plant. Ind Eng Chem Res 1999, 38, 823–832.

[78]   Kocakuşak S, Köroğlu HJ, Akçay K, Savaşçì ÖT, Tolun R. Production of Sodium Perborate Monohydrate by Fluidized-Bed Dehydration, Ind Eng Chem Res 1997, 36, 2862–2865.

[79]   Genck WJ. Production of Strong, Spherulitic Sodium Perborate - An Industrial Case Study. AIChE Annual Meeting, Reno, NV, USA, 2001.

[80]   Harjo B, Ng KM, Wibowo C. Development of Amino Acid Crystallization Processes: L-Glutamic Acid. Ind Eng Chem Res 2007, 46, 2814–2822.

[81]   Serajuddin ATM. Salt Formation to Improve Drug Solubility. Adv Drug Deliv Rev 2007, 59, 603–616.

[82]   Lam KW, Xu J, Ng KM, Wibowo C, Lin G, Luo KQ. Pharmaceutical Salt Formation Guided by Phase Diagrams. Ind Eng Chem Res 2010, 49, 12503–12512.

[83]   Fumagalli C. Succinic Acid and Succinic Anhydride. In Kirk-Othmer Encyclopedia of Chemical Technology. 4th ed., vol. 22. New York, NY, USA, John Wiley & Sons, 1997, 1074–1102.

[84]   Yedur S, Berglund KA, Dunuwila DD. Succinic Acid Production and Purification. US Patent 6,265,190, 2001.

[85]   Kwok KS. Experimental Determination of Solid-Liquid Equilibrium (SLE) Phase Diagrams of Different Organic and Inorganic Chemical Systems (M.Phil. Thesis). Hong Kong, The Hong Kong University of Science and Technology, 2004.

[86]   Söhnel O, Garside J. Precipitation: Basic Principles and Industrial Applications. Oxford, UK, Butterworth-Heinemann, 1992.

[87]   Hanson DN, Lynn S. Methods for Crystallizing Salts from Aqueous Solutions. US Patent 4,879,042, 1989.

[88]   Weingaertner DA, Lynn S, Hanson DN. Extractive Crystallization of Salts from Concentrated Aqueous Solution. Ind Eng Chem Res 1991, 30, 490–501.

[89]   Berry DA, Dye SR, Ng KM. Synthesis of Drowning-Out Crystallization-Based Separations. AIChE J 1997, 43, 91–103.

[90]   Fitch B. How to Design Fractional Crystallization Processes. Ind Eng Chem 1970, 62(12), 6–33.

[91]   Cisternas LA, Rudd DF. Process Designs for Fractional Crystallization from Solution. Ind Eng Chem Res 1993, 32, 1993–2005.

[92]   Cisternas LA, Vásquez CM, Swaney RE. On the Design of Crystallization-Based Separation Processes: Review and Extension. AIChE J 2006, 52, 1754–1769.

[93] Dye SR, Ng KM. Fractional Crystallization: Design Alternatives and Tradeoffs. AIChE J 1995, 41, 2427–2438.

[94] Findlay RA, Weedman JA. Separation and Purification by Crystallization. In Kobe KA, McKetta JJ, eds. Advances in Petroleum Chemistry and Refining, Vol. I. New York, NY, USA, Interscience, 1958, 119–209.

[95] Dale GH. Crystallization, Extractive and Adductive. In McKetta JJ, ed. Encyclopedia of Chemical Processing and Design, vol. 13. Boca Raton, FL, USA, CRC Press, 1981, 456–506.

[96] Dye SR, Ng KM. Bypassing Eutectics with Extractive Crystallization: Design Alternatives and Tradeoffs. AIChE J 1995, 41, 1456–1470.

[97] van der Ham F. Eutectic Freeze Crystallization (Ph.D. Dissertation). Delft, Netherlands, TU Delft, 1999.

[98] Himawan C, Kramer HJM, Witkamp GJ. Study on the Recovery of Purified $MgSO_4 \cdot 7H_2O$ Crystals from Industrial Solution by Eutectic Freezing. Sep Purif Technol 2006, 50, 240–248.

[99] Reddy ST, Lewis AE, Witkamp GJ, Kramer HJM, van Sponsen J. Recovery of $Na_2SO_4 \cdot 10H_2O$ from a Reverse Osmosis Retentate by Eutectic Freeze Crystallisation Technology. Chem Eng Res Des 2010, 88, 1153–1157.

[100] Wibowo C, Ng KM. Unified Approach for Synthesizing Crystallization-based Separation Processes. AIChE J 2000, 46, 1400–1421.

[101] Takano K, Gani R, Ishikawa T, Kolar P. Conceptual Design and Analysis Methodology for Crystallization Processes with Electrolyte Systems. Fluid Phase Equil 2002, 194–197, 783–803.

[102] Ng KM, Systematic Separation of a Multicomponent Mixture of Solids Based on Selective Crystallization and Dissolution. Sep Technol 1991, 1, 108–120.

[103] Harjo B, Wibowo C, Zhang EJN, Luo KQ, Ng KM. Development of Process Alternatives for Separation and Purification of Isoflavones. Ind Eng Chem Res 2007, 46, 181–189.

[104] Bender DR, Demarco AM, McCauley JA. Compound Separation by Cyclic, Selective Dissolution. Isolation of Diastereomeric, 1β-Methylcarbapenem Key Intermediates. Sep Sci Technol 1993, 28, 1169–1176.

[105] Rao NN, Singh JR, Misra R, Nandy T. Liquid-Liquid Extraction of Phenol from Simulated Sebacic Acid Wastewater. J Sci Ind Res India 2009, 68, 823–828.

[106] O'Young L, Kelkar VV. Multi-Disciplinary Process Development: From Lab to Plant (AIChE/ASME Short Course Materials). Pomona, CA, USA, ClearWaterBay Technology, Inc., 2009.

[107] Takaizumi K, Wakabayashi T. The Freezing Process in Methanol-, Ethanol-, and Propanol-Water Systems as Revealed by Differential Scanning Calorimetry. J Soln Chem 1997, 26, 927–939.

[108] Anderson R, Chapoy A, Tanchawanich J, Haghighi H, Lachwa-Langa J, Tohidi B. Binary Ethanol–Methane Clathrate Hydrate Formation in the System $CH_4–C_2H_5OH–H_2O$: Experimental Data and Thermodynamic Modelling. In Proceedings of the 6th International Conference on Gas Hydrates (ICGH 2008), Vancouver, Canada, 2008.

[109] Flick EW. Industrial Solvents Handbook. 5th ed. Westwood, NJ, USA, Noyes Data Corporation, 1998.

[110] Wissner A, Hauser W, Bitners F, Wambach R. Process for the Preparation of Dichlorobenzenes. US Patent 4,300,004, 1981.

[111] Doyle RA, Miller JT, Wilsak RA, Roberts SA, Zoia G. Para-xylene Production Process Integrating Pressure Swing Adsorption and Crystallization. US Patent 6,600,083, 2003.

[112] Bauer Jr W, Mason RM, Upmacis RK. Process for Purifying Unsaturated Carboxylic Aids Using Distillation and Melt Crystallization. US Patent 5,523,480, 1996.

[113] Vishweshwar P, McMahon JA, Bis JA, Zaworotko MJ. Pharmaceutical Co-Crystals. J Pharm Sci 2006, 95, 499–516.

[114] Hickey MB, Peterson ML, Scoppettuolo LA, et al. Performance Comparison of a Co-crystal of Carbamazepine with Marketed Product. Eur J Pharm Biopharm 2007, 67, 112–119.

[115] Ainouz A, Authelin J, Billot P, Lieberman H. Modeling and Prediction of Cocrystal Phase Diagrams. Int J Pharm 2009, 374, 82–89.

[116] Lee KC, Kim KJ. Effect of Supersaturation and Thermodynamics on Co-Crystal Formation. Chem Eng Technol 2011, 34, 619–623.

[117] Blagden N, de Matas M, Gavan PT, York P. Crystal Engineering of Active Pharmaceutical Ingredients to Improve Solubility and Dissolution Rates. Adv Drug Deliv Rev 2007, 59, 617–630.

[118] Kwok KS, Chan HC, Chan CK, Ng KM. Experimental Determination of Solid-Liquid Equilibrium Phase Diagrams for Crystallization-based Process Synthesis. Ind Eng Chem Res 2005, 44, 3788–3798.

[119] O'Young DL, Hsieh ST, Kelkar V. System and Method for Producing Bisphenol-A (BPA) Using Direct Crystallization of BPA in a Single Crystallization Stage. US Patent 7,163,582, 2007.

[120] Cross LC, Klyne W. Rules for the Nomenclature of Organic Chemistry: Section E: Stereochemistry (Recommendations 1974). Pure Appl Chem 1976, 45, 11–30.

[121] Stick R, Williams S. Carbohydrates: The Essential Molecules of Life. 2nd ed. Amsterdam, Netherlands, London, UK, Elsevier Science, 2009.

[122] van Eikeren P. Commercial Manufacture of Chiral Pharmaceuticals. In Ahuja S, ed. Chiral Separations: Applications and Technology, Washington, DC, USA, American Chemical Society, 1997.

[123] Dextropropoxyphene. Wikipedia. (Accessed May 31, 2020, at http://en.wikipedia.org/wiki/Dextropropoxyphene.)

[124] Levopropoxyphene. Wikipedia. (Accessed May 31, 2020, at http://en.wikipedia.org/wiki/Levopropoxyphene.)

[125] Molecule of the Week Archive - Thalidomide. American Chemical Society. (Accessed May 31, 2020, at http://www.acs.org/content/acs/en/molecule-of-the-week/archive/t/thalidomide.html.)

[126] Schroer JW, Wibowo C, Ng KM. Synthesis of Chiral Crystallization Processes. AIChE J 2001, 47, 369–388.

[127] Wang Y, Chen AM. Enantioenrichment by Crystallization. Org Process Res Dev 2008, 12, 282–290.

[128] Le Minh T, Lorenz H, Seidel-Morgenstern A. Enantioselective Crystallization Exploiting the Shift of Eutectic Compositions in Solid-Liquid Phase Diagrams. Chem Eng Technol 2012, 35, 1003–1008.

[129] Lorenz H, Kaemmerer H, Polenske D, Seidel-Morgenstern A. Process for Enantioseparation of Chiral Systems with Compound Formation Using Two Subsequent Crystallization Steps. US Patent 8,992,783 B2, 2015.

[130] Seidel-Morgenstern A, Lorenz H. Processes to Separate Enantiomers. Angew Chem Int Ed 2014, 53, 1218–1250.

[131] Collins AN, Sheldrake GN, Crosby J, eds. Chirality in Industry II: Developments in the Manufacture and Applications of Optically Active Compounds. West Sussex, UK, John Wiley & Sons, 1992.

[132] Tung HH, Waterson S, Reynolds SD. Formation and Resolution of Ibuprofen Lysinate. US Patent 4,994,604, 1991.

[133] Schlonner G, Lodewijk E, Withers G. Resolution of Ibuprofen. US Patent 5,621,140, 1994.

[134] Lam WH, Ng KM. Diastereomeric Salt Crystallization Synthesis for Chiral Resolution of Ibuprofen. AIChE J 2007, 53, 429–437.

[135] Fung KY, Ng KM, Wibowo C. Synthesis of Chromatography-Crystallization Hybrid Separation Processes. Ind Eng Chem Res 2005, 44, 910–921.

[136] Wibowo C, O'Young L. A Hybrid Route to Chirally Pure Products. Chem Eng Prog 2005, 101(11), 22–27.

[137] Lorenz H, Sheehan P, Seidel-Morgenstern A. Coupling of Simulated Moving Bed Chromatography and Fractional Crystallisation for Efficient Enantioseparation. J Chromatogr A 2001, 908, 201–214.

[138] Lorenz H, Seidel-Morgenstern A. Binary and Ternary Phase Diagrams of Two Enantiomers in Solvent Systems. Thermochim Acta 2002, 382, 129–142.

[139] Lin SW, Ng KM, Wibowo C. Synthesis of Crystallization Processes for Systems Involving Solid Solutions. Comput Chem Eng 2008, 32, 956–970.

[140] Luk KF, Ko KM, Ng KM. Separation and Purification of Schisandrin B from Fructus Schisandrae. Ind Eng Chem Res 2008, 47, 4193–4201.

[141] Lau YT, Ng KM, Lau DTW, Wibowo C. Quality Assurance of Chinese Herbal Medicines: Procedure for Single-herb Extraction. AIChE J 2013, 59, 4241–4254.

[142] Lau YT, Ng KM, Chen N, Lau DTW, Leung PC, Wibowo C. Quality Assurance of Chinese Herbal Medicines: Procedure for Multiple-herb Extraction. AIChE J 2014, 60, 4014–4026.

[143] Luk KF, Ko KM, Ng KM. Separation and Purification of (–)Schisandrin B from Schisandrin B Stereoisomers. Biochem Eng J 2008, 42, 55–60.

[144] Thompson BC, Frechet JMJ. Organic Photovoltaics — Polymer-Fullerene Composite Solar Cells. Angew Chem Int Ed 2008, 47, 58–77.

[145] Zhou X, Liu J, Jin Z, Gu Z, Wu Y, Sun Y. Solubility of Fullerene C_{60} and C_{70} in Toluene, o-Xylene and Carbon Disulfide at Various Temperatures. Fullerene Sci Technol 1997, 5, 285–290.

[146] Kwok KS, Chan YC, Ng KM, Wibowo C. Separation of Fullerenes C_{60} and C_{70} Using A Crystallization-based Process. AIChE J 2010, 56, 1801–1812.

[147] Ng KM. Separation of Fullerene C_{60} and C_{70} Using Crystallization. US Patent 7,875,086 B2, 2011.

[148] Odagiri T, Chan YC, Kwok KS, Ng KM. A Novel Evaporative Crystallization Column for the Purification of Fullerene C_{60}. AIChE J 2007, 53, 531–534.

[149] Mori S, Iitani K, Yamamoto M, et al. Process for Recovering L-Amino Acid from Fermentation Liquors. US Patent 5,017,480, 1991.

[150] Tam SK, Chan HC, Ng KM, Wibowo C. Design of Protein Crystallization Processes Guided by Phase Diagrams. Ind Eng Chem Res 2011, 50, 8163–8175.

[151] Kuehner DE, Engmann J, Fergg F, Wernick M, Blanch HW, Prausnitz JM. Lysozyme Net Charge and Ion Binding in Concentrated Aqueous Electrolyte Solutions. J Phys Chem B 1999, 103, 1368–1374.

[152] Hasegawa M, Yoshida K, Miyauchi S, Terazono M. Process for Crystallizing Egg White Lysozyme. US Patent 4,504,583, 1985.

[153] Lai SM, Yuen MY, Siu LKS, Ng KM, Wibowo C. Experimental Determination of Solid-Liquid-Liquid Equilibrium Phase Diagrams. AIChE J 2007, 53, 1608–1619.

[154] Taboada ME, Graber TA, Cisternas LA, Cheng YS, Ng KM. Process Design for Drowning-out Crystallization of Lithium Hydroxide Monohydrate. Chem Eng Res Des 2007, 85, 1325–1330.

[155] Kim S, Wei C, Kiang S. Crystallization Process Development of an Active Pharmaceutical Ingredient and Particle Engineering via the Use of Ultrasonics and Temperature Cycling. Org Process Res Dev 2003, 7, 997–1001.

[156] Rousseau RW. Crystallization. In Kirk-Othmer Encyclopedia of Chemical Technology. 4th ed., vol. 7. New York, NY, USA, Wiley Interscience, 1994, 683–730.

[157] Mullin JW. Crystallization. 4th ed. Boston, MA, USA, Butterworth-Heinemann, 2001.

[158] Marcilla A, Ruiz F, Garcia AN. Liquid-Liquid-Solid Equilibria of the Quaternary system Water-Ethanol-Acetone-Sodium Chloride at 25 °C. Fluid Phase Equilib 1995, 112, 273–289.

[159] Gomis V, Ruiz F, De Vera G, López E, Saquete MD. Liquid-Liquid-Solid Equilibria for the Ternary Systems Water-Sodium Chloride or Potassium Chloride-1-Propanol or 2-Propanol. Fluid Phase Equilib 1994, 98, 141–147.

[160] Davison RR, Smith WHJ, Hood DW. Phase Equilibria of Desalination Solvents: Water-NaCl-Amines. J Chem Eng Data 1966, 11, 304–309.

[161] Subramaniam B, Rajewski RA, Snavely K. Pharmaceutical Processing with Supercritical Carbon Dioxide. J Pharm Sci 1997, 86, 885–890.

[162] Teja AS, Eckert CA. Commentary on Supercritical Fluids: Research and Applications. Ind Eng Chem Res 2000, 39, 4442–4444.

[163] Harjo B, Ng KM, Wibowo C. Synthesis of Supercritical Crystallization Processes. Ind Eng Chem Res 2005, 44, 8248–8259.

[164] van Konynenburg PH, Scott RL. Critical Lines and Phase Equilibria in Binary van der Waals Mixtures. Philos Trans R Soc Lond A 1980, 298, 495–540.

[165] Lu BCY, Zhang D. Solid-Supercritical Fluid Phase Equilibria. Pure Appl Chem 1989, 61, 1065–1074.

[166] Dobbs JM, Johnston KP. Selectivities in Pure and Mixed Supercritical Fluid Solvents. Ind Eng Chem Res 1987, 26, 1476–1485.

[167] Dixon DJ, Johnston KP, Bodmeier RA. Polymeric Materials Formed by Precipitation with a Compressed Fluid Antisolvent. AIChE J 1993, 39, 127–139.

[168] Tom JW, Debenedetti PG. Particle Formation with Supercritical Fluids: A Review. J Aerosol Sci 1991, 22, 555–584.

[169] Weber A, Yelash LV, Kraska T. Effect of the Phase Behavior of the Solvent-Antisolvent Systems on the Gas-Antisolvent-Crystallization of Paracetamol. J Supercrit Fluids 2005, 33, 107–113.

[170] Lin CSM. The Relationship of Systematic Effect of Solubility on Particle Size Using PCA (M. Phil. Thesis). Hong Kong, The Hong Kong University of Science and Technology, 2004.

[171] Lin C, Ng KM, Wibowo C. Producing Nano-particles Using Precipitation with Compressed Antisolvent. Ind Eng Chem Res 2007, 46, 3580–3589.

[172] Ke J, Mao C, Zhong M, Han B, Yan H. Solubilities of Salicylic Acid in Supercritical Carbon Dioxide with Ethanol Cosolvent. J Supercrit Fluids 1996, 9, 82–87.

[173] Zaidul ISM, Norulaini NAN, Mohd Omar AK, Sato Y, Smith Jr RL. Separation of Palm Kernel Oil from Palm Kernel with Supercritical Carbon Dioxide Using Pressure Swing Technique. J Food Eng 2007, 81, 419–428.

[174] Mukhopadhyay M, Rao GVR. Thermodynamic Modeling for Supercritical Fluid Process Design. Ind Eng Chem Res 1993, 32, 922–930.

[175] Polenske D, Lorenz H. Solubility and Metastable Zone Width of the Methionine Enantiomers and Their Mixtures in Water. J Chem Eng Data 2009, 54, 2277–2280.

[176] Gomis V, Ruiz F, Asensi JC, Cayuela P. Liquid-Liquid-Solid Equilibria for the Ternary Systems Water-Lithium Chloride-1-Propanol or 2-Propanol at 25 °C. Fluid Phase Equilib 1996, 119, 191–195.

[177] Kashchiev D, van Rosmalen GM. Review: Nucleation in Solutions Revisited. Cryst Res Technol 2003, 38, 555–574.

[178] Green DW, Perry RH. Perry's Chemical Engineers' Handbook. 8th ed. New York, NY, USA, McGraw-Hill, 2008.

[179]   Nielsen AE. Kinetics of Precipitation. Oxford, UK, Pergamon Press, 1964.

[180]   Burton WK, Cabrera N, Frank FC. The Growth of Crystals and the Equilibrium Structure of Their Surfaces. Proc R Soc Lond A 1951, 243, 299–358.

[181]   Grootscholten PAM, Asselbergs CJ, de Jong EJ. Secondary Nucleation. In Jančić SJ, de Jong EJ, eds. Industrial Crystallization 81. Amsterdam, Netherlands, North-Holland, 1982, 189–197.

[182]   Abegg CF, Stevens JD, Larson MA. Crystal Size Distributions in Continuous Crystallizers When Growth Rate is Size Dependent. AIChE J 1968, 14, 118–122.

[183]   Kubota N, Mullin JW. A Kinetic Model for Crystal Growth from Aqueous Solution in the Presence of Impurity. J Cryst Growth 1995, 152, 203–208.

[184]   Mudryy R, Damavarapu R, Stepanov V, Halder R. Crystallization of High Bulk Density Nitroguanidine. Picatinny Arsenal, NJ, USA, US Army Armament Research, Development and Engineering Center, 2011.

[185]   Sheldon RA. Chirotechnology: Industrial Synthesis of Optically Active Compounds, New York, NY, USA, Marcel Dekker, 1993.

[186]   Vetter T, Burcham CL, Doherty MF. Separation of Conglomerate Forming Enantiomers Using a Novel Continuous Preferential Crystallization Process. AIChE J 2015, 61, 2810–2823.

[187]   Rougeot C, Hein JE. Application of Continuous Preferential Crystallization to Efficiently Access Enantiopure Chemicals. Org Process Res Dev 2015, 19, 1809–1819.

[188]   Alvarez Rodrigo A, Lorenz H, Seidel-Morgenstern A. Online Monitoring of Preferential Crystallization of Enantiomers. Chirality 2004, 16, 499–508.

[189]   Küenburg B, Czollner L, Fröhlich J, Jordis U. Development of a Pilot Scale Process for the Anti-Alzheimer Drug (–)-Galanthamine Using Large-Scale Phenolic Oxidative Coupling and Crystallisation-Induced Chiral Conversion. Org Process Res Dev 1999, 3, 425–431.

[190]   Czollner L, Frantsits W, Küenburg B, Hedenig U, Fröhlich J, Jordis U. New Kilogram-Synthesis of the Anti-Alzheimer Drug (–)-Galanthamine. Tetrahedron Lett 1998, 39, 2087–2088.

[191]   Schroer JW, Ng KM. Process Paths of Kinetically Controlled Crystallization: Enantiomers and Polymorphs. Ind Eng Chem Res 2003, 42, 2230–2244.

[192]   Kelkar VV, Ng KM. Design of Reactive Crystallization Systems Incorporating Kinetics and Mass-Transfer Effects. AIChE J 1999, 45, 69–81.

[193]   Kelkar VV. Development of Reactive Crystallization Systems (Short Course Materials). Pomona, CA, ClearWaterBay Technology, Inc., 2004.

[194]   Threlfall TL. Analysis of Organic Polymorphs. Review Analyst 1995, 120, 2435–2460.

[195]   Chemburkar SR, Bauer J, Deming K, et al. Dealing with the Impact of Ritonavir Polymorphs on the Late Stages of Bulk Drug Process Development. Org Process Res Dev 2000, 4, 413–417.

[196]   Lin SW, Ng KM, Wibowo C. Integrative Approach for Polymorphic Crystallization Process Synthesis. Ind Eng Chem Res 2007, 46, 518–529.

[197]   Gracin S, Rasmuson ÅC. Polymorphism and Crystallization of p-Aminobenzoic Acid. Cryst Growth Des 2004, 4, 1013–1023.

[198]   Threlfall T. Structural and Thermodynamic Explanations of Ostwald's Rule. Org Process Res Dev 2003, 7, 1017–1027.

[199]   Nývlt J. The Ostwald Rule of Stages. Cryst Res Tech 1995, 30, 443–449.

[200]   de Gennes PG, Prost J. The Physics of Liquid Crystals. Oxford, UK, Clarendon Press, 1993.

[201]   Li L, Salamończyk M, Shadpour S, Zhu C, Jákli A, Hegmann T. An Unusual Type of Polymorphism in a Liquid Crystal. Nat Commun 2018, 9:714.

[202]   Wibowo C, Novoa R, Hasson M. Development of a Process for Crystallizing the Desired Crystal Form of An API. AIChE Annual Meeting, Philadelphia, PA, USA, 2008.

[203] Starbuck C, Spartalis A, Wai L, et al. Process Optimization of a Complex Pharmaceutical Polymorphic System via In Situ Raman Spectroscopy. Cryst Growth Des 2002, 2, 515–522.

[204] Beckmann W. Seeding the Desired Polymorph: Background, Possibilities, Limitations, and Case Studies. Org Process Res Dev 2000, 4, 372–383.

[205] Maia GD, Giulietti M. Solubility of Acetylsalicylic Acid in Ethanol, Acetone, Propylene Glycol, and 2-Propanol. J Chem Eng Data 2008, 53, 256–258.

[206] Batra H, Penmasta R, Phares K, Staszewski J, Tuladhar SM, Walsh DA. Crystallization Process Development for a Stable Polymorph of Treprostinil Diethanolamine (UT-15C) by Seeding. Org Process Res Dev 2009, 13, 242–249.

[207] Kitamura M. Controlling Factors and Mechanism of Polymorphic Crystallization. Cryst Growth Des 2004, 4, 1153–1159.

[208] Wibowo C, Chang WC, Ng KM. Design of Integrated Crystallization Systems. AIChE J 2001, 47, 2474–2492.

[209] Randolph DA, Larson MA. Theory of Particulate Process. 2nd ed. San Diego, CA, USA, Academic Press, Inc., 1988.

[210] Berglund KA. Analysis and Measurement of Crystallization Utilizing the Population Balance. In Myerson AS, ed. Handbook of Industrial Crystallization. 2nd ed. Boston, MA, USA, Butterworth-Heinemann, 2001, 101–114.

[211] Hounslow MJ, Lewis AE, Sanders SJ, Bondy R. Generic Crystallizer Model: I. A Model Framework for a Well-Mixed Compartment. AIChE J 2005, 51, 2942–2955.

[212] Bamforth AW. Industrial Crystallization. London, UK, Leonard-Hill, 1965.

[213] Kubota N, Doki N, Yokota M, Sato A. Seeding Policy in Batch Cooling Crystallization. Powder Technol 2001, 121, 31–38.

[214] Yu ZQ, Chew JW, Chow PS, Tan RBH. Recent Advances in Crystallization Control: An Industrial Perspective. Chem Eng Res Des 2007, 85, 893–905.

[215] Kwok KS. Development of Crystallization Processes Aided by Phase Diagrams and Workflow (Ph.D. Thesis). Hong Kong, The Hong Kong University of Science and Technology, 2011.

[216] Lakerveld R. Development of a Task-based Design Approach for Solution Crystallization Processes (Doctoral Thesis). Delft, Netherlands, Delft University of Technology, 2010.

[217] Samant KD, O'Young L. Understanding Crystallization and Crystallizers. Chem Eng Prog 2006, 102(10), 28–37.

[218] Armstrong Engineering Associates, Inc. Armstrong Continuous Crystallizers for Organic Chemicals (brochure), West Chester, PA, USA, Armstrong Engineering Associates, Inc., 2005.

[219] Eek RA, Dijkstra S, van Rosmalen GM. Dynamic Modeling of Suspension Crystallizers, Using Experimental Data. AIChE J 1995, 41, 571–584.

[220] Koresawa E, Nakamura M, Wibowo C, O'Young L. Increasing Productivity of Ammonium Sulfate Crystallization Plant via Particle Size Distribution Modeling. In Proceedings of the 10th APCChE Congress. Kitakyushu, Japan, 2004.

[221] Green D. Crystallizer Mixing: Understanding and Modeling Crystallizer Mixing and Suspension Flow. In Myerson AS, ed. Handbook of Industrial Crystallization. 2nd ed. Boston, MA, USA, Butterworth-Heinemann, 2001, 181–199.

[222] Couper JR, Penney WR, Fair JM, Walas SM. Chemical Process Equipment: Selection and Design. Revised 2nd ed. Amsterdam, Netherlands, Elsevier, 2009.

[223] Genck WJ. Optimizing Crystallizer Scaleup. Chem Eng Prog 2003, 99(6), 36–44.

[224] Yi YJ, Myerson AS. Laboratory Scale Batch Crystallization and the Role of Vessel Size. Chem Eng Res Des 2006, 84, 721–728.

[225] Torbacke M, Rasmuson ÅC. Mesomixing in Semi-Batch Reaction Crystallization and Influence of Reactor Size. AIChE J 2004, 50, 3107–3119.

[226] Zhang GGZ, Grant DJW. Incorporation Mechanism of Guest Molecules in Crystals: Solid Solution or Inclusion? Int J Pharm 1999, 181, 61–70.

[227] Hall RN. Segregation of Impurities during the Growth of Germanium and Silicon Crystals. J Phys Chem 1953, 57, 836–839.

[228] Denbigh KT, White ET. Studies on Liquid Inclusions in Crystals. Chem Eng Sci 1966, 21, 739–754.

[229] Slaminko P, Myerson AS. The Effect of Crystal Size on Occlusion Formation During Crystallization from Solution. AIChE J 1981, 27, 1029–1031.

[230] Gu CH, Grant DJW. Relationship between Particle Size and Impurity Incorporation during Crystallization of (+)-Pseudoephedrine Hydrochloride, Acetaminophen, and Adipic Acid from Aqueous Solution. Pharm Res 2002, 19, 1068–1070.

[231] Cheng YS, Lam KW, Ng KM, Wibowo C. Workflow for Managing Impurities in An Integrated Crystallization Process. AIChE J 2010, 56, 633–649.

[232] Burton JA, Prim RC, Slichter WP. The Distribution of Solute in Crystals Grown from the Melt. Part I. Theoretical. J Chem Phys 1953, 21, 1987–1996.

[233] Sangwal K, Pałcyńska T. On the Supersaturation and Impurity Concentration Dependence of Segregation Coefficient in Crystals Grown from Solutions. J Cryst Growth 2000, 212, 522–531.

[234] Chang WC. Synthesis of Integrated Chemical Systems (Ph.D. Dissertation). Amherst, MA, USA, University of Massachusetts, 1998.

[235] Sulzer Chemtech. Fractional Crystallization (brochure). Winterthur, Switzerland, Sulzer Chemtech, 2006.

[236] Wynn NP. Separate Organics by Melt Crystallization. Chem Eng Prog 1992, 88(3), 52–60.

[237] Chang WC, Ng KM. Synthesis of Processing System Around a Crystallizer. AIChE J 1998, 44, 2240–2251.

[238] Wakeman RJ, Tarleton ES. Filtration: Equipment Selection, Modeling and Process Simulation. Oxford, UK, Elsevier, 1999.

[239] Hill PJ, Ng KM. Simulation of Solids Processes Accounting for Particle-Size Distribution. AIChE J 1997, 43, 715–726.

[240] Zeitsch K. Centrifugal Filtration. In Svarovsky L, ed. Solid-Liquid Separation. 3rd ed. London, UK, Butterworths, 1990, 476–532.

[241] Wakeman RJ. The Performance of Filtration Post-treatment Processes: 1. The Prediction and Calculation of Cake Dewatering Characteristics. Filtr Separat 1979, 16, 655–660.

[242] Eloneva S, Said A, Fogelholm CJ, Zevenhoven R. Preliminary Assessment of a Method Utilizing Carbon Dioxide and Steelmaking Slags to Produce Precipitated Calcium Carbonate. Appl Energy 2012, 90, 329–334.

[243] Wyrsta MD, Komon ZJA. Systems and Methods for Alkaline Earth Production. US Patent 9,738,950 B2, 2017.

[244] GEA Group AG. Melt Crystallization: The Efficient Purification Alternative (brochure). 's-Hertogenbosch, Netherlands, GEA Group AG, 2012.

[245] UOP LLC. Badger/Niro Para-Xylene Crystallization Process. Des Plaines, IL, USA, UOP LLC, 2001.

[246] Otawara K, Matsuoka T. Axial Dispersion in a Kureha Crystal Purifier (KCP). J Cryst Growth 2002, 237–239, 2246–2250.

[247] Arkenbout GF. Melt Crystallization Technology. Lancaster, PA, USA, Technomic Publishing Company, Inc., 1995.

[248] McCabe WL, Thiele EW. Graphical Design of Fractionating Columns. Ind Eng Chem 1925, 17, 605–611.

[249] Levenspiel O. Modeling in Chemical Engineering. Chem Eng Sci 2002, 57, 4691–4696.

[250] Tester JW, Modell M. Thermodynamics and Its Applications. 3rd ed. Upper Saddle River, NJ, USA, Prentice Hall, 1996.

[251] Yang H, Thati J, Rasmusson ÅC. Thermodynamics of Molecular Solids in Organic Solvents. J Chem Thermodyn 2012, 48, 150–159.

[252] Linke WF, Seidell A. Solubilities of Inorganic and Metal Organic Compounds. Washington, DC, USA, American Chemical Society, 1965.

[253] Silcock HL. Solubilities of Inorganic and Organic Compounds. Oxford, UK, Pergamon, 1979.

[254] Zemaitis Jr JF, Clark DM, Rafal M, Scrifner NC. Handbook of Aqueous Electrolyte Thermodynamics. New York, NY, USA, AIChE, 1986.

[255] Karapet'yants MK, Karapet'yants ML. Thermodynamic Constants of Inorganic and Organic Compounds. Ann Arbor, MI, USA, Humphrey Science Publishers, 1970.

[256] Tanveer S, Hao Y, Chen CC. Introduction to Solid-Fluid Equilibrium Modeling. Chem Eng Prog 2014, 110(9), 37–47.

[257] Gross J, Sadowski G. Perturbed-Chain SAFT: An Equation of State Based on a Perturbation Theory for Chain Molecules. Ind Eng Chem Res 2001, 40, 1244–1260.

[258] Joback KG, Reid RC. Estimation of Pure-Component Properties from Group Contributions. Chem Eng Commun 1987, 57, 233–243.

[259] Marrero J, Gani R. Group-contribution Based Estimation of Pure Component Properties. Fluid Phase Equilib 2001, 183–184, 183–208.

[260] Fredenslund A, Jones RL, Prausnitz JM. Group-contribution Estimation of Activity Coefficients in Nonideal Liquid Mixtures. AIChE J 1975, 21, 1086–1099.

[261] Chen CC, Song Y. Solubility Modeling with a Nonrandom Two-Liquid Segment Activity Coefficient Model. Ind Eng Chem Res 2004, 43, 8354–8362.

[262] Florey K. Aspirin. In Florey K, ed. Analytical Profiles of Drug Substances, vol. 8. London, UK, Academic Press, Inc., 1979, 1–46.

[263] Kirklin DR. Enthalpy of Combustion of Acetylsalicylic Acid. J Chem Thermodyn 2000, 32, 701–709.

[264] KT Consortium - Software. Technical University of Denmark. (Accessed May 31, 2020, at http://www.kt.dtu.dk/english/research/kt-consortium/software.)

[265] Klamt A, Eckert F. COSMO-RS: A Quantum Chemistry Based Alternative to Group Contribution Methods for the Prediction of Activity Coefficients in Multi-component Mixtures. Fluid Phase Equilib 2000, 172, 43–72.

[266] Matsuoka M, Ozawa R. Determination of S-L Phase Equilibria of Binary Organic Systems by DSC. J Cryst Growth 1989, 96, 596–604.

[267] Mettler Toledo. Two-component Phase Diagram and the Determination of the Eutectic Composition of Dimethyl Terephthalate (DMT) and Benzoic Acid. USER COM 1999, 2/99, 18–19.

[268] Ozawa R, Matsuoka M. Determination of Solid-Liquid Phase Equilibrium of Organic Ternary Eutectic Mixtures by Differential Scanning Calorimeter - The o-, m- and p-Nitroaniline System. J Cryst Growth 1989, 98, 411–419.

[269] Jakob A, Joh R, Rose C, Gmehling J. Solid-Liquid Equilibria in Binary Mixtures of Organic Compounds. Fluid Phase Equilib 1995, 113, 117–126.

[270] Wittig R, Constantinescu D, Gmehling J. Binary Solid-Liquid Equilibria of Organic Systems Containing -Caprolactone. J Chem Eng Data 2001, 46, 1490–1493.

[271] Zhang L, Gui Q, Lu X, Wang Y, Shi J. Measurement of Solid-Liquid Equilibria by a Flow-Cloud-Point Method. J Chem Eng Data 1998, 43, 32–37.

[272] Crystal 16. Technobis Crystallization Systems. (Accessed May 31, 2020, at http://www.crystallizationsystems.com/crystal16.)

[273] Parallel Chemistry. HEL Group. (Accessed May 31, 2020, at http://helgroup.com/products/parallel-chemistry/.)

[274] Marchand P, Lefèbvre L, Querniard F, et al. Diastereomeric Resolution Rationalized by Phase Diagrams under the Actual Conditions of the Experimental Process. Tetrahedron Asymmetry 2004, 15, 2455–2465.

[275] Chan YC. Solubility Measurement Apparatus for Rapid Determination of Solid-Liquid Equilibrium Behavior (M.Phil. Thesis). Hong Kong, The Hong Kong University of Science and Technology, 2008.

[276] Apelblat A, Manzurola E. Solubility of Oxalic, Malonic, Succinic, Adipic, Maleic, Malic, Citric, and Tartaric Acids in Water from 278.15 K to 338.15 K. J Chem Thermodyn 1987, 19, 317–320.

[277] Gaivoronskii AN, Granzhan VA. Solubility of Adipic Acid in Organic Solvents and Water. J Appl Chem 2005, 78, 404–408.

[278] Lin HM, Tien HY, Hone YT, Lee MJ. Solubility of Selected Dibasic Carboxylic Acids in Water, in Ionic Liquid of [Bmim][BF4], and in Aqueous [Bmim][BF4] Solutions. Fluid Phase Equilib 2007, 253, 130–136.

[279] Fuyuhiro A, Yamanari K, Shimura Y. Solubility Isotherms of Reciprocal Salt-Pairs of Optically Active Cobalt(III) Complexes. Bull Chem Soc Jpn 1979, 52, 90–93.

[280] Wyatt DK, Grady LT. Determination of Solubility. In Rossiter BW, Baetzold RG, eds. Physical Methods of Chemistry, Vol. VI. Determination of Thermodynamics Properties. 2nd ed. New York, NY, USA, John Wiley & Sons, 1996.

[281] Schreinemakers FAH. Graphische Ableitungen aus den Lösungsisothermen eines Doppelsalzes und seiner Componenten. Z Phys Chem 1893, 11, 75–109.

[282] Schott H. A Mathematical Extrapolation for the Method of Wet Residues. J Chem Eng Data 1961, 6, 324–325.

[283] Samant KD, O'Young L, Kwok M, Ng KM. Workflow and Regression Methods for Determining Solid-Liquid Phase Diagrams. In FOCAPD 2004 Proceedings, Princeton, NJ, USA, 2004, 385–389.

[284] Dahl JE, Moldowan JM, Peters KE, et al. Diamondoid Hydrocarbons as Indicators of Natural Oil Cracking. Nature 1999, 399, 54–57.

[285] Chan YC, Choy KKH, Chan AHC, et al. Solubility of Diamantane, Trimantane, Tetramantane, and Their Derivatives in Organic Solvents. J Chem Eng Data 2008, 53, 1767–1771.

[286] Wibowo C. Developing Crystallization Processes. Chem Eng Prog 2011, 109(3), 21–31.

[287] Omar W, Al-Sayed S, Sultan A, Ulrich J. Growth Rate of Single Acetaminophen Crystals in Supersaturated Aqueous Solution under Different Operating Conditions. Cryst Res Technol 2008, 43, 22–27.

[288] Nývlt J. Kinetics of Nucleation in Solution. J Cryst Growth 1968, 3, 377–383.

[289] Nagy ZK, Fujiwara M, Woo XY, Braatz RD. Determination of the Kinetic Parameters for the Crystallization of Paracetamol from Water Using Metastable Zone Width Experiments. Ind Eng Chem Res 2008, 47, 1245–1252.

[290] Barrett P, Smith B, Worlitschek J, Bracken V, O'Sullivan B, O'Grady D. A Review of the Use of Process Analytical Technology for the Understanding and Optimization of Production Batch Crystallization Processes. Org Process Res Dev 2005, 9, 348–355.

[291] Cornel J, Mazzotti M. Estimating Crystal Growth Rates Using In Situ ATR-FTIR and Raman Spectroscopy in a Calibration-free Manner. Ind Eng Chem Res 2009, 48, 10740–10745.

[292] Garside J, Gibilaro LG, Tavare NS. Evaluation of Crystal Growth Kinetics from a Desupersaturation Curve Using Initial Derivatives. Chem Eng Sci 1982, 37, 1625–1628.

[293] de Jong EJ. Development of Crystallizers. Int Chem Eng 1984, 24, 419–431.

[294] CRANIUM - Property Estimation Software. Molecular Knowledge Systems, Inc. (Accessed May 31, 2020, at http://www.molknow.com/Cranium/cranium.htm.)

[295] COSMOtherm: Predicting Solutions Since 1999. COSMOlogic GmbH & Co. KG. (Accessed May 31, 2020, at http://www.cosmologic.de/products/cosmotherm.html.)

[296] SLEEK. ClearWaterBay Technology, Inc. (Accessed May 31, 2020, at http://www.cwbtech.com/SLEEK.html.)

[297] Karnaukhov AS. Investigation of the Ternary Systems $NaNO_3$-NH_4NO_3-H_2O, KNO_3-NH_4NO_3-H_2O, and $RbNO_3$-NH_4NO_3-H_2O via Methods of Physicochemical Analysis at 25°. J Gen Chem USSR 1956, 26, 1169–1176.

[298] Myerson AS. Handbook of Industrial Crystallization. 2nd ed. Boston, MA, USA, Butterworth-Heinemann, 2002.

[299] Jones AG. Crystallization Process Systems. Boston, MA, USA, Butterworth-Heinemann, 2002.

[300] Tung HH, Paul EL, Midler M, McCauley JA. Crystallization of Organic Compounds: An Industrial Perspective. Hoboken, NJ, USA, Wiley, 2009.

[301] Beckmann W. Crystallization: Basic Concepts and Industrial Applications. Weinheim, Germany, Wiley-VCH, 2013.

[302] Harper G, Sommerville R, Kendrick E, et al. Recycling Lithium-ion Batteries from Electric Vehicles. Nature 2019, 575, 75–86.

[303] Li Q, Fung KY, Ng KM. Separation of Ni, Co, and Mn from Spent $LiNi_{0.5}Mn_{0.3}Co_{0.2}O_2$ Cathode Materials by Ammonia Dissolution. ACS Sustain Chem Eng 2019, 7, 12718–12725.

[304] Li Q, Fung KY, Xu L, Wibowo C, Ng KM. Process Synthesis: Selective Recovery of Lithium from Lithium-Ion Battery Cathode Materials. Ind Eng Chem Res 2019, 58, 3118–3130.

[305] Binnemans K, Jones PT, Blanpain B, et al. Recycling of Rare Earths: A Critical Review, J Clean Prod 2013, 51, 1–22.

[306] Gupta CK, Krishnamurthy N. Extractive Metallurgy of Rare Earths. Boca Raton, FL, USA, CRC Press, 2004.

[307] Beltrami D, Deblonde GJP, Bélair S, Weigel V. Recovery of Yttrium and Lanthanides from Sulfate Solutions with High Concentration of Iron and Low Rare Earth Content. Hydrometallurgy 2015, 157, 356–362.

[308] Yin X, Wang Y, Bai X, et al. Rare Earth Separations by Selective Borate Crystallization. Nat Commun 2017, 8:14438.

[309] Long FD, Adams RG, DeVore DP. Preparation of Hyaluronic Acid from Eggshell Membrane. US Patent 6,946,551, 2005.

[310] Coculová A, Krajcovic H. Isolation of the Hyaluronic Acid from the Eggshell Membranes. Int J Creat Res Studies 2018, 2(5), 67–74.

[311] Murado MA, Montemayor MI, Cabo ML, Vázquez JA, González MP. Optimization of Extraction and Purification Process of Hyaluronic Acid from Fish Eyeball. Food Bioprod Proc 2012, 90, 491–498.

[312] Gani R, Ng KM. Product Design – Molecules, Devices, Functional Products, and Formulated Products. Comput Chem Eng 2015, 81, 70–79.

[313] Ng KM, Gani R. Chemical Product Design: Advances in and Proposed Directions for Research and Teaching. Comput Chem Eng 2019, 126, 147–156.

[314] Gao TP, Wong KW, Fung KY, Zhang W, Ng KM. A Novel Three-step Calcination Strategy for Synthesizing High-quality $LiNi_{0.5}Mn_{0.3}Co_{0.2}O_2$ Cathode Materials: the Key Role of Suppressing Li_2O Formation. Electrochim Acta 2018, 288, 153–164.

[315] Dai Q, Kelly JC, Dunn J, Benavides PT. Update of Bill-of-materials and Cathode Materials Production for Lithium-ion Batteries in the GREET® Model. Argonne National Laboratory, 2018 (Accessed May 31, 2020, at https://greet.es.anl.gov/publication-update_bom_cm.)

[316] Gao TP, Wong KW, Ng KM. Impacts of Morphology and N-doped Carbon Encapsulation on Electrochemical Properties of NiSe for Lithium Storage. Energy Stor Mater 2020, 25, 210–216.

[317] Luo S, Feng J, Ng KM. Effect of Fatty Acid on the Formation of ITO Nanocrystals via One-pot Pyrolysis Reaction. Cryst Eng Comm 2015, 17, 1168–1172.

[318] Ge S, Wang N, Fung KY, Wong KW, Chan CT, Ng KM. Product Design: Nanoparticle-Loaded Polyvinyl Butyral Interlayer for Solar Control. AIChE J 2018, 64, 3614–3624.

[319] Tam SK, Fung KY, Ng KM. Copper Pastes Using Bimodal Particles for Flexible Printed Electronics. J Mater Sci 2016, 51, 1914–1922.

[320] Tam SK, Fung KY, Ng KM. Product Design: Metal Nanoparticle-Based Conductive Inkjet Ink. AIChE J 2016, 62, 2740–2753.

Index

activity coefficient models
- NRTL 261, 281
- regular solution 261, 268
adjacency matrix 33
amino acid crystallization 151
asymmetric transformation 184

battery recycle 303

chiral resolution 131, 181
common ion effect 10
compatible salts 45, 102
conglomerate 133
congruent melting point 14
conjugate salt system 39, 102
crossover temperature 191
crystallization design approach
- hierarchical 6, 300
- integrative 4, 300
- multiscale 3, 300
crystallization downstream processing
     system 238
crystallization kinetics determination 292
crystallization kinetics
- growth 176, 208
- nucleation 176, 208
crystallization process objectives
- complete dissolution 79, 82
- crystallization of desirable product 73
- maximum purity 66, 69
- maximum recovery 62, 65, 68
- purity improvement 70
- yield improvement 72, 232
crystallization techniques
- adductive crystallization 90, 126, 303
- antisolvent crystallization 90, 156
- cooling crystallization 90, 303
- drowning-out crystallization 90, 303
- eutectic freeze crystallization 90
- evaporative crystallization 90, 303
- extractive crystallization 90, 303
- fractional crystallization 90, 144, 303
- preferential crystallization 181
- reactive crystallization 90, 186–187,
     218, 303

- selective dissolution and crystallization 97
- solvent switching 111
crystallization workflow 5, 213
crystallizer types
- draft tube baffle 209, 221
- falling film 234
- forced circulation 221
- MSMPR 208, 221
- Oslo 223
- scraped surface 223
- static melt 233
- surface-cooled 223

deliquoring 240
diastereomers 132
displacement washing 239
downstream processing models 241

enantiomer 132
enantiotropic behavior 191
equation of state 260
experimental SLE methods
- analytical method 279
- Differential scanning calorimetry 271
- Differential thermal analysis 271
- discontinuous isoperibolic thermal
     analysis 277
- Schreinemakers' wet residue method 279
- solvent addition method 277
- step-warming method 276

freezing point depression 9

Gibbs phase rule 11
group contribution methods
- Marrero and Gani method 264, 267
- NRTL-SAC model 266
- UNIFAC method 265
growth dispersion 211

hybrid separation process
- crystallization–adsorption 119
- crystallization–chromatography 138
- crystallization–distillation 114–116, 118

https://doi.org/10.1515/9781501519901-012

impurity inclusion models 229
inclusion impurities 228
incompatible salts 45, 104
incongruent melting point 14

lever rule 58
liquid crystal crystallization 197
liquor recirculation 220

magma recirculation 220
mass separating agent 57, 156
melt crystallization 233
metastable limit 176, 192
metastable zone 176, 292
method of moments 210
monotropic behavior 191
multicomponent solid mixture separation 100

oiling out 161
Ostwald ripening 211
Ostwald's rule of stages 196

particle size distribution 187, 207
pharmaceutical salt form 76
phase behavior
– compound formation 13, 76, 126, 133, 137
– peritectic system 39, 289
– simple eutectic system 12, 18, 23, 133
– solid solution 13, 15, 141
phase diagram features
– compartment 20, 24
– critical point 12
– double saturation point 22
– eutectic point 13, 20
– eutectic trough 20
– liquidus curve 13
– lower critical end point 163
– saturation curve 13
– saturation surface 20
– solidus curve 13
– solubility curve 13
– solubility surface 20
– triple point 12, 163
– upper critical end point 163
phase diagram visualization, 3D to 2D
– Cruickshank projection 31
– diagonal cut 42, 45
– isobaric cut 164

– isoplethal cut 22
– isothermal cut 21, 26, 42, 91, 126, 141, 150, 163
– Jänecke projection 26, 45, 47, 49, 97, 135
– polythermal projection 20, 24, 89, 126
phase diagram visualization, multicomponent
– canonical coordinates 29
– central projection 29
– orthogonal projection 29
– parallel projection 29
– projection ray 29
– projective space 29
– reaction-invariant projection 35
– saturation variety 28
– stoichiometric lines 35
– vertex 33
polymorphic crystallization 190
population balance equation 207
population density 207
precipitation with compressed antisolvent 166
process path 179, 193
protein crystallization 154, 305
purification by melt crystallization 244

quantum chemistry methods
– COSMO-RS model 270

racemization 136
rapid expansion of supercritical solution 167
rare earth recycle 304
reslurry washing 239
resolving agent 134
Roozeboom types 16

saturation variety matrix 33
scale up 225
seeding 213
selection of crystallization solvent 109
solubility equation 259
solubility product equation 259
stereoisomer 132
stream combination 57
stream splitting 57
supercooling 176
supercritical fluid crystallization 162
supersaturation 176
surface impurities 228

thermodynamic boundary 63, 89, 287
tie-line 13
transformed mole fraction 36
$T–x$ diagram 12

unstable zone 177

wash ratio 239

www.ingramcontent.com/pod-product-compliance
Lightning Source LLC
Chambersburg PA
CBHW080716220326
41598CB00033B/5440